Industrial Transformation

Advances in IoT, Robotics, and Cyber Physical Systems for Industrial Transformation

Series Editors: S. Balamurugan and Dinesh Goyal

Smart Home, Smart City, and Wearable Technologies are the most exponentially growing applications of Internet of Things (IoT). Wearable Technology is considered to be highly ubiquitous and the most phenomenal IoT application. Smart World and Wearable Technology exhibit a high potential to transform our lifestyle. Ranging from healthcare tracking applications to smart watches and smart bands for personal safety, IoT has become one of the most indispensable part of our lives. In a study by Business Insider's premium research service, by the end of 2022, the wearable IoT market is expected to grow and reach 162.9 million units. Some of the top-notch applications of IoT include smart parking, smart wearable technologies, smart clothing, smart safety, smart farming, smart industry, robotics, and so on, thereby building the next-generation smart world. An important priority for today's communication world seems to rest on the development of a smart world, where intelligent "things" are connected to serve people better. With the exponential growth of IoT and its applications, the implementation of a smart world becomes much more feasible. The sensor market plays a vital role in applying IoT to build a smart world. Applications such as monitoring parking space availability to optimally park vehicles (Smart Parking), detecting the frequency and intensity of traffic and optimally selecting routes (Traffic Congestion), prediction of trash levels in garbage collection containers (Smart Waste Management), managing the intensity of street lights based on weather condition and sunlight (Smart Lighting), detecting and controlling excess hazardous gases coming from industries and vehicles (Smart Air Pollution Controlling), detecting the mix of hazardous chemicals from factories that appear in drinking water (Smart Water Pollution Controlling), efficient monitoring and management of power consumption (Smart Grid), artificial intelligence-driven retail marketing (Smart Shopping), detecting health abnormalities (Smart Health) and Home Automation mechanisms fall under the category of applying IoT and Robotics in building a smart world. The purpose of this book series is to portray certain practical applications of IoT in building a smart world. With widespread applications in both smart cities and smart homes, this book series grows increasingly important.

Industrial Internet of Things
Technologies and Research Directions
Edited by Anand Sharma, Sunil Kumar Jangir, Manish Kumar, Dilip Kumar Choubey, Tarun Shrivastava, and S. Balamurugan

Industrial Transformation
Implementation and Essential Components and Processes of Digital Systems
Edited by Om Prakash Jena, Sudhansu Shekhar Patra, Mrutyunjaya Panda, Zdzislaw Polkowski, and S. Balamurugan

For more information on this series, please visit: https://www.routledge.com/Advances-in-IoT-Robotics-and-Cyber-Physical-Systems-for-Industrial-Transformation/book-series/CRCAIRCPSIT

Industrial Transformation

Implementation and Essential Components and Processes of Digital Systems

Edited by

Om Prakash Jena, Sudhansu Shekhar Patra,
Mrutyunjaya Panda, Zdzislaw Polkowski,
and S. Balamurugan

CRC Press
Taylor & Francis Group
Boca Raton London New York

CRC Press is an imprint of the
Taylor & Francis Group, an **informa** business

First edition published 2022
by CRC Press
6000 Broken Sound Parkway NW, Suite 300, Boca Raton, FL 33487-2742
and by CRC Press
4 Park Square, Milton Park, Abingdon, Oxon, OX14 4RN

CRC Press is an imprint of Taylor & Francis Group, LLC

Library of Congress Cataloging-in-Publication Data

Names: Jena, Om Prakash, editor. | Patra, Sudhansu Shekhar, editor. | Panda, Mrutyunjaya, editor. | Polkowski, Zdzislaw, editor. | Balamurugan, S. (Shanmugam), 1985- editor.
Title: Industrial transformation : implementation and essential components and processes of digital systems / edited by Om Prakash Jena, Sudhansu Shekhar Patra, Mrutyunjaya Panda, Zdzislaw Polkowski, and S. Balamurugan. Other titles: Industrial transformation (CRC Press)
Description: First edition. | Boca Raton, FL : CRC Press, 2022. | Series: Advances in IoT, robotics, and cyber physical systems for industrial transformation | Includes bibliographical references and index. |
Identifiers: LCCN 2021050681 (print) | LCCN 2021050682 (ebook) | ISBN 9781032133980 (hbk) | ISBN 9781032133997 (pbk) | ISBN 9781003229018 (ebk)
Subjects: LCSH: Industry 4.0. | Automation.
Classification: LCC T59.6 .I24 2022 (print) | LCC T59.6 (ebook) | DDC 658.4/0380285–dc23/eng/20211221
LC record available at https://lccn.loc.gov/2021050681
LC ebook record available at https://lccn.loc.gov/2021050682

ISBN: 978-1-032-13398-0 (hbk)
ISBN: 978-1-032-13399-7 (pbk)
ISBN: 978-1-003-22901-8 (ebk)

DOI: 10.1201/9781003229018

Typeset in Times
by KnowledgeWorks Global Ltd.

Contents

Preface

The concept of industrial transformation has been strongly aligned with the computerization of manufacturing and the concept of requisite skills to be built by the new workforce to reap the benefits of high-paying jobs. A number of developments, such as the Internet of Things (IoT), Machine Learning (ML), Big Data Analytics, and Blockchain Technology, are today considered essential elements of the transformation of industry development and implementation. At the forefront of this trend are technical developments in the ability to gather, move, and interpret large quantities of data very efficiently. Industrial transformation leverages these new technologies to build smart factories that rapidly adapt and respond to changes in consumer demands for high-quality goods.

Artificial intelligence (AI) in conventional manufacturing sectors aids in design and development. The role of AI is in the management of resource flow and decision-making in Industrial Transformation operations and supply chain management. AI helps maintain sustainability and efficiency in industrial transformation with IoT, ML, data analysis and Blockchain technology. These reforms signal the trend and transition to seamlessly interconnected and automated entire development, management and regulatory frameworks. This transformation has been acknowledged by companies, who have turned their attention to how industrial transformation will influence their business; many others are changing things and investing for a future where their business is escalating through smart machines with intelligence techniques. Industrial transformation is the so-called fourth Industrial Revolution in manufacturing, logistics and the supply chain, the chemical industry, resource utilization, transportation, utilities, gas and oil, mining, metals, and other sectors, including the mineral industries, healthcare, medical products, sustainable development, wastage management, energy optimization, and smart cities.

This book focuses on industrial development, design, and implementation using AI, ML, and IoT. It aims to incorporate complex processes and functions as a central component of digital systems, including certain management, marketing, procurement, development, refinance, client relationship management, human resources, logistics, systems engineering, and many other activities. This book will explore market change trends, technologies (AI, ML, and Efficiency Technology) that are accelerating industrial transformation, and the effect of the transition to the Fourth Industrial Revolution on industrial infrastructure, biodiversity, and productivity-enhancement.

Chapter 1 provides various types of industrial applications such as identification of the approximate temperature of a boiler flame, the design of a Fuzzy-based controller, which automatically tunes its variable parameters to give the optimum performance even under process disturbances, a combination of automated Image Processing (IP) algorithms, AI techniques, and human visual inspection to guarantee the product quality and a non-contact, investigative method which uses deep learning algorithms to detect the flaws during the fabrication of steel sheets, which takes place inside a furnace at very high temperature.

Chapter 2 shows how ML is easing the industrial automation process and reviews the state-of-the-art of ML applications in a variety of Additive Manufacturing (AM) applications. ML has the potential to drastically minimize material waste, production time, and reduce the number of components required for assembly.

Chapter 3 describes the impact of Total Productive Maintenance (TPM) in the production performance of the steel industry. Through TPM, there is an increase in productivity and quality improvement. The findings show that maintenance is not an extra cost for the company; rather it is a way to increase productivity.

Chapter 4 discusses the plans for resource monitoring and saving the limited and finite resources in the smart cities and the use of AI in the field of smart cities to create systems and devices that can learn from the past experiences.

Chapter 5 describes the Load Frequency Control (LFC) scheme to control the frequency oscillation in single-area wind electric power systems supplemented by Fuel Cell (FC) and Battery Energy Storage System (BESS). The Ant Colony Optimization (ACO) technique is utilized to acquire optimal controller gain parameters of the proportional–integral–derivative (PID) controller.

Chapter 6 gives a custom model in deep learning, ML, which explores the network-on-chip (NoC)-based AI chips' contribution to a variety of perspectives on the accuracy, quality, speed up, prediction error, etc., towards high-speed communication and intricate, high-volume computation by multiprocessor systems-on-chip (MPSoCs). The AI technology dramatically improves the speed and accuracy of predicting various performance metrics.

Chapter 7 provides the various scenarios of hydrogen leakage in the air. The boundary layer approach to estimate the thickness of the hydrogen concentration layer adjacent to a ceiling wall is introduced. Both planner and axisymmetric cases of flow are considered. The vertical and horizontal hydrogen-free turbulent plumes are discussed. The momentum-dominated and buoyancy-dominated regimes of the hydrogen plume are highlighted.

Chapter 8 discusses the assignment problem (AP) in which the objective is to allocate a number of tasks to an equal number of machines/people/facilities with optimal decision parameters. It proposes a Genetic Algorithm (GA)-based approach to solve a multi-objective interval assignment problem (MOIAP) using a risk attitude parameter (RAP).

Chapter 9 focuses on the family of evolutionary algorithms and their real-life applications. Genetic algorithms, genetic programming, differential evolution, evolution strategies, evolutionary programming, Non-dominated sorting genetic algorithm, Teaching-Learning-Based Optimization Algorithm, Non-dominated Sorting Teaching-Learning Based Optimization Algorithm, Jaya Algorithm, Multi-objective Jaya Algorithm, and Rao Algorithms are discussed.

Chapter 10 provides a model which can be easily embedded with any existing casting simulation tool, and will be very useful for predicting the occurrence of porosity as well as improvement in the quality of castings. Castings typically face quality issues as defects form where porosity is present, especially in ASTM A413 (LM6) and ASTM A356 (LM25) alloys.

Chapter 11 presents Type D fuzzy representation for a few Cellular Automata (CA) rules in view of its selection towards MapReduce design in Industry 4.0 scenarios to accelerate load-reduced Big Data processing, and thus it ensures a cost-effective and easy physical implementation along with self-organizing capability and support towards design for uncertainty-based networks.

Chapter 12 discusses a patch-based framework for fine-grained classification for localizing and aggregating distinct points between similar regions based on correlations between the discriminative natures of highly localized regions. A triplet of similar feature points with logical constraints have been proposed to improve the accuracy of region localization and automatically find discriminative geometrically-constrained triplets for classification.

Chapter 13 proposes a system based on analysing visual data provided by the front camera of a vehicle. Convolutional Neural Networks (CNN) are used to process the provided data and predict traffic patterns. To overcome the limitation of CNN, the use of a lightweight model is proposed, which uses compression techniques to reduce the size and speed up the processing time on the embedded device. The single-shot multi-box detection (SSD), an object detection framework, is used for task detection.

Chapter 14 examines the relationship between CO2 emissions, financial development, fossil fuel energy consumption, and renewable energy consumption in Asian countries. Using the statistical method of pooled ordinary least square, random effect model, and fixed effect model, results demonstrate that the contribution of renewable energy consumption to total energy consumption in most Asian economies has significantly declined except from a few advanced economies, for instance, China, Japan, and South Korea.

Chapter 15 focuses on how recurrent neural networks can help farmers to seek advice for their location on a given weather forecast. The main purpose of this work is to put forward the concept of using the data created over the years by various agricultural and meteorological divisions to aid computers in learning the pattern and sequences and to generate the advisory independently of human input.

Chapter 16 shows PID controllers, fuzzy logic controllers, and buck-boost controllers that are capable of boosting the charging speed of electrical vehicles. The electric charging station will essentially be capable of supplying the charging of numerous electric vehicles with different capacities at a significantly reduced time scale. The optimal performance of the circuit is displayed with the aid of simulation.

Chapter 17 provides a combination of orthogonal arrays as well as experimental designs, which is used to provide optimal welding process parameters that give the best response to join composites in the aluminium metal matrix. The most significant process variables are welding speed (WS), tool rotational speed (TRS), and axial force (AF), which influence welding performance. The maximum obtained tensile strength and hardness of a developed weld joint are 307.48MPa and 123.736 HV respectively.

Chapter 18 proposes SSA ("Squirrel Search Algorithm") for resolving the solar-wind-hydro-thermal power generation's scheduling, considering pumped-storage-hydraulic (PSH) units as well as the insecurity of energy sources. Owing to the introduction of sources of renewable energy, it has become vital for inspecting their impact on optimal power generation scheduling. Solar-wind-hydro-thermal-systems are considered integrating PSH units in this work.

Contributors

A. Kalaivani
Saveetha School of Engineering, SIMATS
Chennai, India

Abdessalem Ben Abdelali
University of Monastir
Monastir, Tunisia

Mouna Afif
University of Monastir
Monastir, Tunisia

Mehtab Alam
Jamia Hamdard
New Delhi, India

B. Anand
Hindusthan College of Engineering and Technology
Coimbatore, Tamil Nadu, India

Riadh Ayachi
University of Monastir
Monastir, Tunisia

Santosh Kumar Behera
Indian Institute of Information Technology Pune
Pune, Maharashtra, India

N.P.G. Bhavani
Saveetha School of Engineering, SIMATS
Chennai, India

Biswajit R. Bhowmik
National Institute of Technology Karnataka
Karnataka, India

D. Boopathi
Paavai Engineering College
Tamil Nadu, India

Radhamadhab Dalai
Birla Institute of Technology Mesra
Ranchi, India

Jayesh M. Dhodiya
S.V. National Institute of Technology
Surat, India

Mohamed F. El-Amin
Effat University
Jeddah, Saudi Arabia

Vaibhav Gangwar
Greater Noida Institute of Technology
Greater Noida, UP, India

Anjali Jain
Amity University
Noida, Uttar Pradesh, India

Chitralekha Jena
KIIT University
Odisha, India

Om Prakash Jena
Ravenshaw University
Cuttack, Orissa, India

Bibin Jose
Vellore Institute of Technology
Vellore, India

Ajay Kumar
Inderprastha Engineering College
Ghaziabad, UP, India

K. Gokul Kumar
Vellore Institute of Technology
Vellore, India

K. Jagatheesan
Paavai Engineering College
Tamil Nadu, India

N. Kanimozhi
A.V.C. College of Engineering
Mayiladuthurai, Tamilnadu, India

Ihtiram Raza Khan
Jamia Hamdard
New Delhi, India

Sudhir Kumar
Inderprastha Engineering College
Ghaziabad, UP, India

V. Kumarakrishnan
Paavai Engineering College
Tamil Nadu, India

Manikandan Manoharan
VIT University
Vellore, Tamil Nadu, India

Kamar Mazloum
Marwadi University
Rajkot, India

Darshan Vishwasrao Medhane
Indian Institute of Information Technology Pune
Pune, Maharashtra, India

Sibabrata Mohanty
GIET University
Gunupur, India

Arnab Mitra
SRM University AP
Amaravati: Mangalagiri, Andhra Pradesh, India

Arivazhagan N.
VIT University
Vellore, Tamil Nadu, India

G. Nalinashini
RMD Engineering College
India

Lipika Nanda
KIIT University
Odisha, India

Van Chien Nguyen
Thu Dau Mot University
Thu Dau Mot City, Vietnam

Thu Thuy Nguyen
Thuongmai University
Hanoi, Vietnam

Babita Panda
KIIT University
Odisha, India

Dhiren Pandit
Nirma University
Ahmedabad, India

Arjyadhara Pradhan
KIIT University
Odisha, India

Priyaratnam
Amity University
Noida, Uttar Pradesh, India

R. Vani
SRM Institute of Science and Technology
India

Kali Charan Rath
GIET University
Gunupur, India

Renangi Sandeep
Vellore Institute of Technology
Vellore, India

Sourav Samanta
University of Burdwan
West Bengal, India

S. Tilva
S. V. National Institute of Technology
Surat, India

Yahia Said
University of Monastir
Monastir, Tunisia

Sarita Samal
KIIT University
Odisha, India

Dr. Amit Sata
Marwadi University
Rajkot, Gujarat, India

K. K. Senapati
Birla Institute of Technology Mesra
Ranchi, India

Dhairya Partap Singh
BSA Collage of Engineering & Technology
Mathura, UP, India

Pampa Sinha
KIIT University
Odisha, India

V. Srividhya
Meenakshi College of Engineering
Chennai, Tamil Nadu, India

K. Sujatha
Dr. MGR Educational & Research Institute
Chennai, India

Anita Ravi Tailor
Navyug Science College
Surat, India

Neelam Verma
Amity University
Noida, Uttar Pradesh, India

Editor Biographies

Om Prakash Jena is currently working as an Assistant Professor in the Department of Computer Science, Ravenshaw University, Cuttack, Odisha, India. He has 10 years of teaching and research experience in under graduate and post graduate levels. He has published several technical papers in international journals/conferences/edited book chapters of reputed publications. He is a member of IEEE, IETA, IAAC, IRED, IAENG, and WACAMLDS. His current research interest includes database, pattern recognition, cryptography, network security, artificial intelligence, machine learning, soft computing, natural language processing, data science, compiler design, data analytics, and machine automation. He has many edited books, published by Wiley, CRC Press, and Bentham Publication to his credit, and is also author of two textbooks under Kalyani Publishers. He also serves as review committee member and editor of many international journals.

Sudhansu Shekhar Patra is currently working as an Associate Professor in the School of Computer Applications, KIIT Deemed to be University, Bhubaneswar, India. He received his Master's degree in Computer Application from Motilal Nehru National Institute of Technology, Allahabad, India, M.Tech (Computer Science & Engg) from Utkal University, Bhubaneswar, India and PhD in Computer Science from KIIT University, Bhubaneswar, India. His research interests include grid computing, cloud computing, algorithms, blockchain, etc. He is a life-time member of the Indian Society for Technical Education.

Mrutyunjaya Panda holds a PhD degree in Computer Science from Berhampur University. He obtained his Master's in Engineering from Sambalpur University, MBA in HRM from IGNOU, New Delhi, and Bachelor in Engineering from Utkal University in 2002, 2009 and 1997 respectively. He has more than 20 years of teaching and research experience. He is presently working as Reader in P.G. Department of Computer Science and Applications, Utkal University, Bhubaneswar, Odisha, India. He is a member of MIR Labs (USA), KES (Australia),

IAENG (Hong Kong), ACEEE(I), IETE(I), CSI(I), and ISTE(I). He has published about 70 papers in international and national journals and conferences. He has also published seven book chapters. He has two textbooks and three edited books to his credit. He is a program committee member of various international conferences. He is acting as a reviewer for various international journals and conferences of repute. He is an Associate Editor of IJCINI Journal, IGI Global, USA, and an Editorial board member of IJKESDP Journal of Inderscience, UK. He is also a Special Issue Editor of the International Journal of Computational Intelligence Studies (IJCIStudies), Inderscience, UK. His active areas of research include data mining, image processing, intrusion detection and prevention, social networking, mobile communication, wireless sensor networks, natural language processing, internet of things, text mining, etc.

Zdzislaw Polkowski is Professor of UJW at Faculty of Technical Sciences and Rector's Representative for International Cooperation and Erasmus+ Program at the Jan Wyzykowski University Polkowice. Since 2019, he is also an Adjunct Professor, Department of Business Intelligence in Management, Wroclaw University of Economics and Business, Poland. Moreover, he is Visiting Professor at the University of Pitesti, Romania and Adjunct Professor at Marwadi University, India. He is the former Dean of the Technical Sciences Faculty during the period 2009–2012 at UZZM in Lubin. He holds a PhD degree in Computer Science and Management from Wroclaw University of Technology, a Post Graduate degree in Microcomputer Systems in Management from University of Economics in Wroclaw and a Post Graduate degree IT in Education from the Economics University in Katowice. He obtained his Engineering degree in Computer Systems in Industry from the Technical University of Zielona Gora. He has published more than 75 papers in journals, 25 conference proceedings, including more than 20 papers in journals indexed in the Web of Science, Scopus, and IEEE. He served as a member of the Technical Program Committee in many international conferences in Poland, India, China, Iran, Romania, and Bulgaria. To date he has delivered 24 invited talks at different international conferences across various countries. He is also a member of the Board of Studies and expert member of the doctoral research committee in many universities in India. He is also a member of the editorial board of several journals and has served as a reviewer in a wide range of international journals. His areas of interest include IT in business, IoT in business, and education technology. He has successfully completed a research project on developing the innovative methodology of teaching Business Informatics funded by the European Commission. He also owns an IT SME consultancy company in Polkowice and Lubin, Poland.

S. Balamurugan, ACM Distinguished Speaker, received his BTech Degree from P.S.G. College of Technology, Coimbatore, India, MTech and PhD Degrees from Anna University, India. He has published 57 books, 300+ international journals/conferences, and 79 patents. He is the recipient of Rashtriya Vidhya Gourav Gold Medal Award and The Best Educationalist Award from Hon. Justice O.P Saxena, Supreme Court, New Delhi, and the Former Chairman of the Minority Council, New Delhi, India. He is the recipient of two Lifetime Achievement Awards. He is the recipient of Dr. A.P.J. Abdul Kalam Sadhbhavana Award from Hon. Balmiki Prasad Singh, Former Governor of Sikkim, the Jewel of India Award from Mr. Gurpreet Singh, General Secretary, India, the Star of Asia Award from Mr. Korn Debbaransi, Former Dy. Prime Minister, Thailand, in an International Summit at (Bangkok) Thailand and the Pride of Asia Research Excellence Award from Hon. Anant V. Sheth, Deputy Speaker-Goa Legislative Assembly, Best Director Award. He also received the "Active Member CSI National Award". He was awarded the Prestigious Mahatma Gandhi Leadership Award at the House of Commons, British Parliament, London, UK. The book he authored on *Machine Learning and Deep Learning Algorithms using MATLAB and PYTHON* won the "Best MATLAB Book For Beginners" Award by BookAuthority. Dr. S. Balamurugan has won the CSI Young IT Professional Award, and is the winner of the "National CSI Youth Award 2020" by the Computer Society of India. He is also the recipient of the Best Researcher Award, Certificate of Exceptionalism, Young Scientist Award, Best Young Researcher Award, and Outstanding Scientist Award. His biography is listed in "Marquis WHO'S WHO", New Jersey, USA. His professional activities include roles as Editor-in-Chief/Associate Editor/Editorial Board Member for 500+ international journals/conferences of high repute and impact. He has been invited as chief guest/resource person/keynote plenary speaker to many reputed universities and colleges at national and international levels. His research interests include artificial intelligence, augmented reality, internet of things, big data analytics, cloud computing, and wearable computing. He is a life-time member of ACM, IEEE, ISTE, and CSI.

1 Computational Intelligence for Automation of Industrial Processes

K. Sujatha, G. Nalinashini, N. Kanimozhi, A. Kalaivani, N.P.G. Bhavani, V. Srividhya, and R. Vani

CONTENTS

DOI: 10.1201/9781003229018-1

1.1 INTRODUCTION

1.1.1 Introduction to Industrial Boilers

The first section of automation includes an intelligent scheme to monitor the temperature from furnace flame for an industrial boiler. These boilers in the industry are used to generate steam, where the feed water is superheated to produce steam. The water tube boilers have tubes which carry water and this water will be heated up by the furnace flame, so that it gets converted to steam. Industrial boilers play an important role in power generation plants where the pulverized coal is burnt in the presence of oxygen to facilitate combustion. Generally, power plant steam is produced and plays an important role in the case of steam power plant. It is a must that an appropriate air to fuel ratio is maintained to confirm the complete combustion in industrial boilers that are used for power generation. The law of conservation of energy suggests that in boilers, the potential energy of the water stored along with the chemical and light energy during burning of coal is converted to heat energy which converts water to steam, followed by conversion to mechanical energy and electrical energy. This combustion process takes place inside the combustion chamber. When the fuel is burnt inside the combustion chamber, heat energy is produced, and exhaust gases are liberated. The heat energy produced generates steam due to the combustion process in the boiler furnace within the boiler. The performance of the boiler can be optimized to reduce the cost. There are two methods for optimization of the performance of the boiler. Improvement in combustion quality can be obtained by controlling flame temperature, which offers energy savings and reduces the cost in a boiler. Another method commonly used for combustion improvement is injecting a correct combination of air and fuel; optimizing this air fuel mixture increases the boiler performance. Combustion efficiency is indicated by the combustion improvement which is carried out within the boiler combustion process. The flame colour indicates the combustion quality approach where statistical indicators are used to access the patterns.

The entire chapter is organized with six broader sections like Introduction, Material and Methods, Existing Method, Proposed Method, Results, its related Discussion and Conclusion. Some of these broader sections also include sub divisions to give the readers a better understanding.

1.1.2 Introduction to Bottling System

The bottling system focuses on mathematical modelling and automation by using control logic. The system is divided into three modules for estimation, controller design circuit and hardware for the controller. The controller parameters are estimated online, followed by designing a suitable controller where the controller settings are tuned using standard tuning rules on recursive basis. Once the optimal values are estimated, the simulation model is used to build a hardware model for the same purpose [1–3] using a recursive estimating technique; the parameter estimator estimates the parameters of a simplified model of the controlled system. The controller changes its actions to maintain its performance based on the estimated parameters and the approximation model. Hence an accurate model which is essential as compared with the conventional methods. Fuzzy logic is used as the parameter estimator in this project [4–6].

1.1.3 Introduction to Automated Inspection in Industries

The current trend as per Industry 4.0 is monitoring and transmission of industrial parameters using sophisticated manufacturing technologies. It comprises of cyber security, Internet of Things (IoT), cloud computing and intelligent computing techniques [7–14]. These kinds of Image based Automated Inspection (IAI) systems were used in fabricating electronic gadgets in early 1980s. The IAI systems will offer efficient monitoring replacing the human interference. When these systems were first introduced, there were some challenges with respect to speed, robustness, expenditure for installation and easiness in operating procedure [15–16]. The development of this technology took some time because of various reasons like prevalence of the image based system with AI technology, on-board complexity in execution of IP and AI algorithms, compatibility in hardware requirements and integration of these techniques with the Distributed Control System (DCS) [17–24]. There are numerous possible zones where IAI systems can be used, they include power plants, ore drying industry, food processing industries, pharmaceutical industry, printing industries, paper manufacturing, vehicle Assembly, etc.

1.1.4 Introduction to Quality Assessment in Industries

The fabrication of steel sheets involves efficient usage of the raw materials during the rolling process. Classically, flat, rectangular structures of steel sheets with required dimensions are available in the market. Consequently, steel sheets of required dimensions are being produced from larger uncut structures. Monitoring the fabrication of steel sheets is a significant process in the case of production industries. Many items have metal frames made by bending the sheet of metal. The range of thickness for steel sheets is 0.4 mm to 6 mm [25–32].

1.2 ALGORITHMS FOR AUTOMATION

1.2.1 Discriminant Function Analysis

The dimension reduction is carried out using FLDA. The multi-dimensional feature space is reduced to two-dimensional feature set. This type of feature reduction facilitates the reduction in computational complexity. The planar display is the result of these methods. Visualization of the two-dimensional visualization can be obtained from the unique features of the flame by applying FLDA. Linear mapping is achieved by using FLDA, where the clusters are formed among the three groups of flame images with optimal discrimination between the classes [1, 3].

The mapping is denoted by Equation (1.1)

$$Y = AX \tag{1.1}$$

where

$$A = \begin{bmatrix} \varphi_1 \\ \varphi_2 \end{bmatrix} \tag{1.2}$$

and φ_1 and φ_2 are the projection vectors in Equation (1.2).

The steps involved are:

Step 1: Compute φ_1 and φ_2
Step 2: Apply linear mapping to associate the vectors.

$$J(\varphi) = \frac{\varphi^T S_b \varphi}{\varphi^T S_w \varphi} \tag{1.3}$$

$$S_b = \Sigma\, p(\omega_i)(m_i - m_o)(m_i - m_o)^T \tag{1.4}$$

$$S_w = \Sigma\, p(\omega_i)E\left[X_i - m_o)(X_i - m_i)^T \omega_i\right] \tag{1.5}$$

It should be noted that Equations (1.3), (1.4) and (1.5) should be large enough to discriminate between the three groups of flame corresponding images to three varied temperatures.

$$S_b\varphi_1 = \lambda_{ml}\, S_w\varphi_1 \tag{1.6}$$

The eigenvalues are taken to be very large because this value will enhance the Euclidean distance measure when compared with Equation (1.6).

$$\varphi_2{}^T\varphi_1 = 0 \tag{1.7}$$

Orthogonality between the two vectors is given by Equation (1.7), which illustrates that the two vectors are perpendicular to each other.

$$Q_p S_b \varphi_2 = \lambda_{m2}\, S_w \varphi_2 \tag{1.8}$$

where

$$Q_p = I - \frac{\varphi_1\, \varphi_1^T S_W^{-1}}{\varphi_1^T\, S_W^{-1}\, \varphi_1} \tag{1.9}$$

I denotes an identity matrix (Equation 1.9).

Maximum eigenvalue corresponds to eigenvector of Equation (1.8) for φ_2. $S_w^{\,1}$ is calculated in Equation (1.10):

$$S_w^{\,1} = U * W * V^T \tag{1.10}$$

As per Equation (1.7), the inner product of φ_1 and φ_2 should be zero, but in the real scenario, it will not be zero and hence the two-dimensional mapping can be obtained in such a way as indicated in Equation (1.11).

$$y_i = \left(u_i, v_i\right) = \left(X_i^T \varphi_1, X_i^T \varphi_2\right) \tag{1.11}$$

1.2.2 Radial Basis Function (RBF)

The RBF trains a Feed Forward Neural Network (FFNN) consisting of input, hidden and output layers with the processing elements (PEs) interconnected in a particular fashion which depends on mapping. The PEs in the hidden and output layers consist of Gaussian and threshold activation function, respectively. In the training phase, the flame image [5] is acquired and FLDA is obtained followed by training RBF to obtain final weights. In the testing phase, the eigenvector of the new flame image is processed with φ_1 and φ_2 to obtain 2D vector. The final set of weights are used to obtain a single value from RBF. The obtained outcome is post processed to get the flame temperature.

1.2.3 Fundamentals of Fuzzy Logic

Fuzzy logic emulates human brain to assess various conditions pertaining to real-time applications where crisp values are to be inferred from the linguistic variables. The control algorithm for Fuzzy Logic Controller (FLC) can be realized using an embedded system with micro-controllers. Both simulation studies and a complex hardware model can also be built for specific industrial purpose with wide knowledge base about the system. This expert system can offer an optimal control action at a faster rate without being affected by the external disturbances with specific features [33–36].

1.2.4 Importance of Deep Learning

Computing machines mimic the intelligence of the human brain, which is called Artificial Intelligence (AI). AI system adopts Machine Learning, where machines learn by experience and acquire skills without any human involvement. Machine Learning algorithms take automatic decisions [1–3]. Deep Learning is a part of Machine Learning as shown in Figure 1.1.

Deep Learning is the next generation. Independent of human decisions, the Deep Learning algorithms can do prediction. For optimal performance (Figure 1.2), Machine Learning algorithms need human intervention whereas Deep Learning algorithms do not require them to make decisions during data analysis [37–38]. It mimics the human mind to take decisions by training the data set efficiently. These algorithms are applied for different applications based on the requirement and performance with different types of data [39–40].

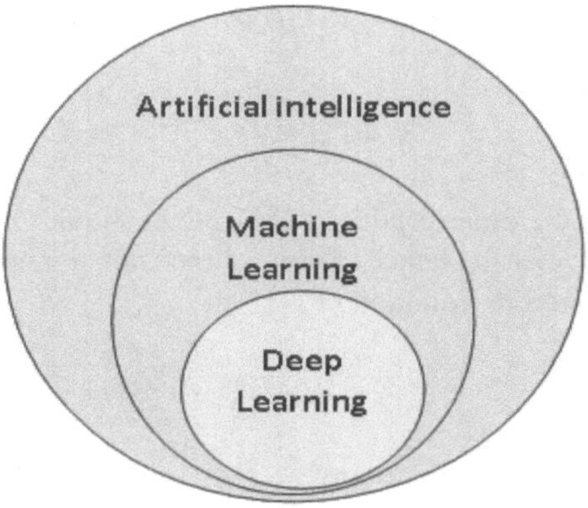

FIGURE 1.1 Interdependency diagram.

1.2.5 Recurrent Neural Network (RNN)

Conventional Back Propagation Algorithm (BPA) will not be useful to train the Deep neural networks because of the reduction in gradient which is slowly deceasing with respect to time. This may impact the predicted output. The objective function will be finalized by considering the Mean Squared Error (MSE). MSE is calculated by finding the difference between the actual and target values. The rate of change of error with respect to weights is used to calculate the new set of weights. Deep Learning algorithms perform best with problems with huge data set. The general architecture of RNN is shown in Figure 1.3. The algorithm for RNN is given below:

- Take the input images and pre-process them followed by feature extraction.
- Define the feature vectors as inputs along with desired target from the training and testing images.

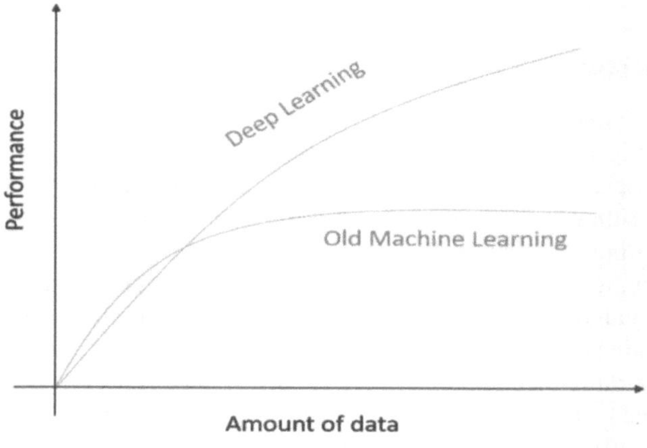

FIGURE 1.2 Machine Learning and Deep Learning performance.

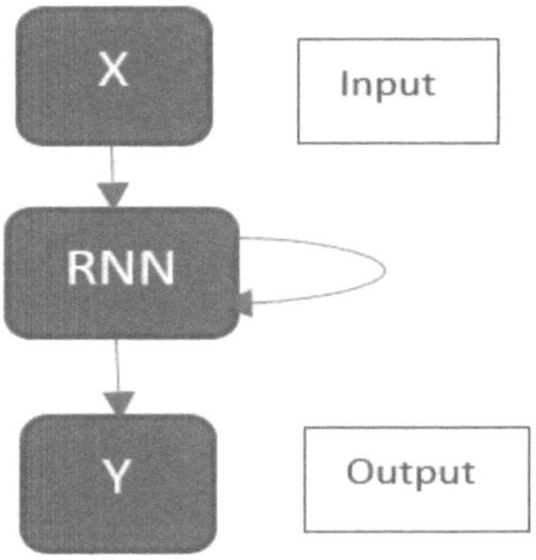

FIGURE 1.3 General architecture of recurrent neural network.

- Initialize the weights using randomize function between input and hidden layers and also between hidden and output layers.
- Calculate the value of outputs from hidden layer nodes and output layer nodes by using the sigmoid activation function.
- Calculate the actual value of network.
- Check for MSE. If it is within tolerance, stop training and proceed with testing. Else retrain the network.

1.3 EXISTING METHOD

1.3.1 Current System in Practice for Industrial Temperature Measurement in Combustion Chambers

The existing system states that the temperature measurement in the combustion chambers is done using non-contact type, J-K thermocouple. Generally, thermocouples consist of a pair of wires made up of different materials with different work function which is united at the end. This linkage point is called the measurement junction. The two wires are united at one point and it is ensured that there is no electromotive force existing at that point. The two junctions, namely the hot and the cold junction, are said to exist. This temperature difference generates a voltage. This is called as thermoelectric effect and is present along the length of the conductors present in the thermocouple. The variation in temperature difference is the physical quantity which gets converted to the voltage which is an electrical quantity. The conductors in the thermocouple need to be insulated to avoid short circuits so that the output voltage can be realized [36, 41–42].

1.3.2 Current System in Practice for Level Measurement in Bottling Industries

The major reasons to measure the level is to ensure safety, economy and monitoring of the process. The process variable that is to be measured and controlled is the level. The direct level measurement is done to satisfy the following objectives, which include the calculation of the inventories inside the tank, offering protective coverings for the equipment like compressors and turbines and providing the skilled labourers with safety against corrosion. Care must be taken to protect the environment from the hazards being released during the due course of level measurement. Finally, there should be possibility to separate the phases during loading operation [43–48].

1.3.3 Current System in Practice for Manufacturing PCB

The Printed Circuit Board (PCB) design needs to be a continuous one till the PCB is fabricated. The basic idea behind fabrication process is that there should be interconnectivity between the electronic components in the circuit with respect to final dimensions. Atmospheric temperature should be taken into account and must be capable to identify the required materials for PCB design. The chosen materials must be capable of performing effectively and should create a design which has a long duration. The circuit schematic drawing with its functional representation and electrical connectivity will be analysed and quality checked during the fabrication process [7–8, 49].

1.3.4 Current System in Practice for Monitoring of Steel Sheet Fabrication

For the past few years, the research community has been involved in automating the industrial processes. For having a quality check on the steel sheets being manufactured, barcode scanning and Radio Frequency Identification (RFID) tags are used. The performance of the process or project can be measured with accuracy using Building Information Modelling (BIM) and Virtual Reality (VR) techniques. Researchers have contributed in terms of quality, accuracy, time schedule and monitoring of the required process parameters, which is supposed to produce less deviation between the actual and the target values.

1.4 PROPOSED METHOD

1.4.1 An Intelligent Approach to Recognize Furnace Flame Temperature in an Industrial Boiler

The discriminant vectors from the FLDA computed based on the within class and between class mean facilitate the demarcation of the flame images for three different temperature ranges corresponding to 900°C, 1000°C and 1100°C. The block diagram for training and testing is shown in Figures 1.4 and 1.5, respectively. The furnace flame image captured from the boiler of the thermal power plant is illustrated in Figure 1.6. The RGB colour image is transformed to equivalent greyscale image. The intensity variation with respect to the temperature can be analysed by plotting the histogram for the flame images from every category. The output of FLDA, the discriminant vectors φ_1 and φ_2, are the inputs to the RBF. During the training process, based on the achievement of the objective function, the finalized weights are stored, which will facilitate the testing process when a new set of flame

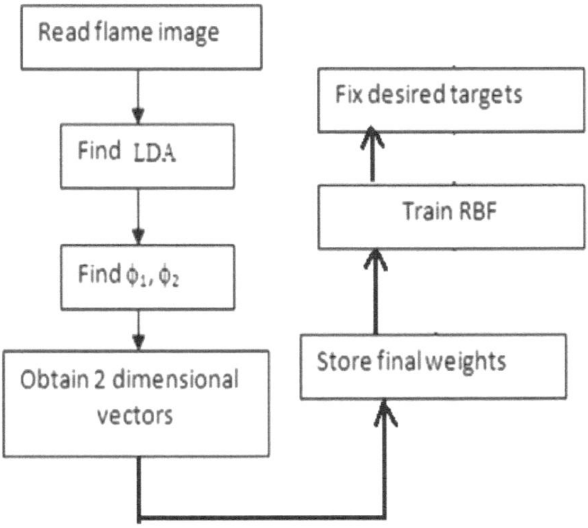

FIGURE 1.4 Training flame temperature using FLDA and RBF.

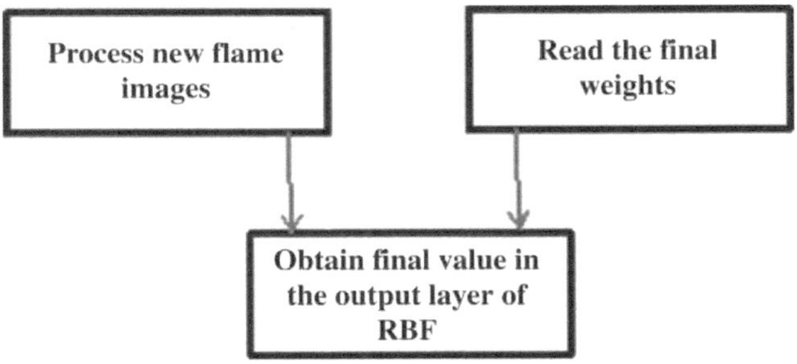

FIGURE 1.5 Testing flame temperature identification.

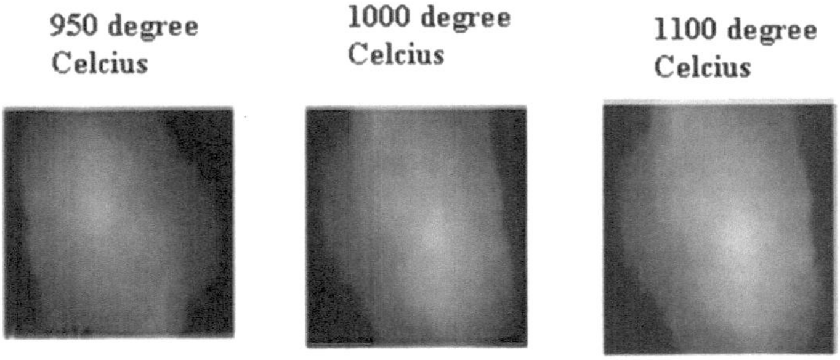

FIGURE 1.6 Flame images.

TABLE 1.1
Performance of the Algorithm

	Image 1	Image 2	Image 3
Actual temperature	900°C	1000°C	1100°C
Estimated temperature	850°C	1050°C	1090°C

images is given as inputs to the RBF network. Table 1.1 highlights the actual and estimated temperature values for the new images of the flame being presented to the neural network. It can be noted that the variation of the temperature is within ±10% as shown in Figure 1.6.

1.4.2 Design of Self Tuning Controller for Level Control using Fuzzy Logic in Bottling Aerated Drinks

The level of the aerated drinks during bottling is to be measured and monitored continuously. To perform the tests and collect the data, a prototype of the level process station was utilized. The level is monitored by converting this physical quantity to its equivalent current signal in the range of 4–20 mA.

1.4.3 Automated Testing of PCB using Sensor-Less Vision Based Technique

Indigenous Image based Automated Inspection (IAI) systems have four major components; they are the source of light, imaging system, probably a camera and an in-built source code to execute the program. IAI system captures the image of an object when light falls on that object and it is illuminated by a source of light. The image of the object captured is transferred to computing device and thereafter processed and analysed by the IP and AI algorithms [36, 41–45]. MATLAB simulation package is used for constructing the entire system so as to enable an efficient decision-making system to identify the high and low quality products being fabricated. The challenge involved is in the choice of suitable IP and AI algorithms. The schematic is shown in Figure 1.7.

The source of light can be a monochrome LED or white light or even coloured lighting can be used depending on the situation. Either single or an array of infra-red cameras or Charged Coupled Device cameras can be used to capture the image of the object in IAI systems [7–9, 44–49, 51].

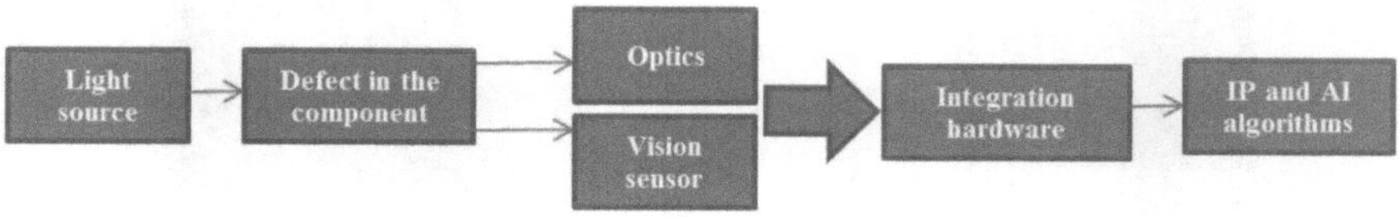

FIGURE 1.7 Schematic of an IAI system.

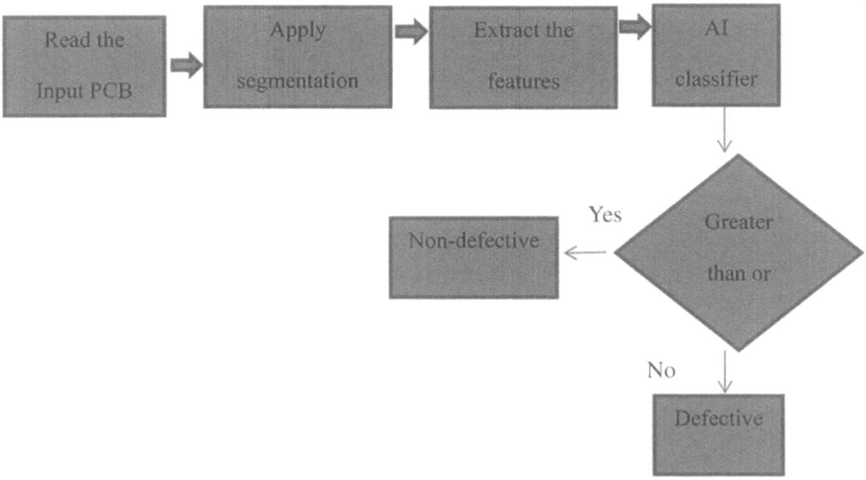

FIGURE 1.8 Flowchart for defect identification by IAI system.

The scope of this study is to assess the effectiveness of conventional IP algorithms and to identify a suitable one for the IAI system. The commonly used algorithms are boundary detection, point processing, feature extraction, etc. There are some other algorithms which include image segmentation like threshold, merge and split and region growing algorithms. After careful analysis, appropriate algorithms can be used to build the entire structure. The schematic representation is depicted in Figure 1.8. The algorithm is as follows:

1. Capture the image of the Printed Circuit Board which is free from flaws.
2. Segment the entire image into parts and find the Region of Interest (ROI).
3. Process the image of the pattern whose quality is to be found out using watershed segmentation algorithm.
4. Extract the features.
5. Implement the AI classifier so that it will classify the PCBs with and without defects.
6. Each time compute the Correlation Coefficient (CC) and compare with threshold.
7. If the value of CC exceeds or equals the threshold, infer that the PCB is non-defective, else identify the PCB to be defective.
8. Check for Quality Assurance.

1.4.4 Automatic Segmentation for Flaw detection in Steel Sheet fabrication using Deep Learning Neural Networks

A technique for segmenting network structure was presented in this section. Figure 1.8 depicts the recommended solution. The RNN is discussed further in this section. The faults are caused by early-stage defects caused by stress, a decreased load bearing

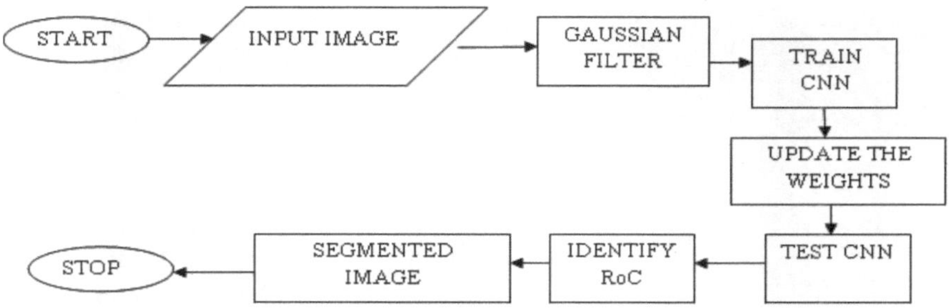

FIGURE 1.9 Generalize segmentation algorithm schematic.

portion of the product, or changes in the metallurgical structure. Some or all of these factors might cause the material to fail during fabrication or enter service with flaws that compromise its integrity.

The proposed method is capable of identifying three different sets: (1) Porosity, which causes formation of pores on the surface of the steel sheets during fabrication. (2) Cracks, in case of closed ingot forging between the die parts, material is pushed out. The term "flash" is used to describe this concept. After forging, the flash can be eliminated; however, excessive flash may break if there is a tension. (3) Surface defects: Due to impurities on the surface, high and low spots will be produced on the surface of the steel sheets. Because the three sets stated above are distinct, a fault cannot be found in both sets at the same time. The main phases in segmentation method are outlined in the flowchart in Figure 1.9.

1.4.4.1 Pre-Processing

The ROI should be used to resize chosen images. This reduces the amount of time required for computation. The scaled images were changed to greyscale to simplify processing because the difference in pixel intensities is increased when this is done. The speckle noise has unfavourable processing consequences in the future. To decrease noise and enhance the borders, low pass Gaussian filters are employed during the pre-processing.

1.4.4.2 Feature Extraction

The distinguishing characteristics are the different attributes that, when repeated in various orientations forms a complete image. The extraction of features is crucial since they serve as dataset for RNN-based Deep Learning.

1.4.4.3 Classification using RNN

RNN is used when the output needs to be sequential as in the case of image captioning and language translation. In a regular multilayer perceptron, each layer has its own weights and biases and hence cannot be combined. To combine these layers, same weights and biases are made in recurrent layer. This ensures that the neuron remembers the existing state and based on this state the next output is generated.

1.5 RESULTS AND DISCUSSION

1.5.1 Temperature Measurement of Furnace Flame

The flame image is split into three parts corresponding to RGB plane, where the images are acquired from the camera and closed circuit television. However, the types of fuel will influence the colour of the flame in the furnace. But the thermal power plants use heavy oil for firing and igniting. The complete combustion would yield flame images which are yellowish or whitish, deep yellow or orange for the partial combustion and blackish for incomplete combustion. Thus, the quality of combustion has intense effect on the colour of the flame images, and this association forms the basis for monitoring the exhaust gases' pollutants and combustion efficiency. The histogram and the distribution of the flame images on the 2D space is depicted in Figures 1.10 and 1.11, respectively. Table 1.1 depicts the actual values and desired values of the flame temperatures using the proposed non-contact type of intelligent control scheme.

1.5.2 Level Monitoring and Control using FLC in Bottling Industries

Conducting an open loop test and matching the output to a first order with dead time function determines the transfer function of the level process station. The procedure began with the tank being filled to a specific level. The reaction was given time to settle at this level. The reaction was allowed to settle at the new set point after the set point value was abruptly altered by a certain amount. Figure 1.12 was used to determine the values of time constant, dead time and gain.

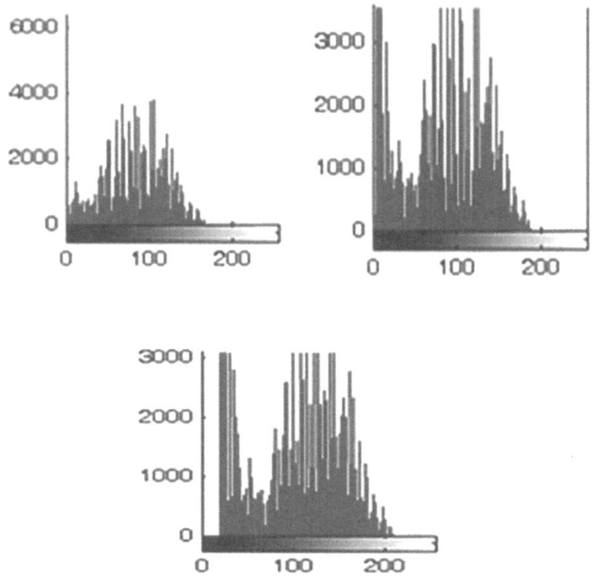

FIGURE 1.10 Flame images.

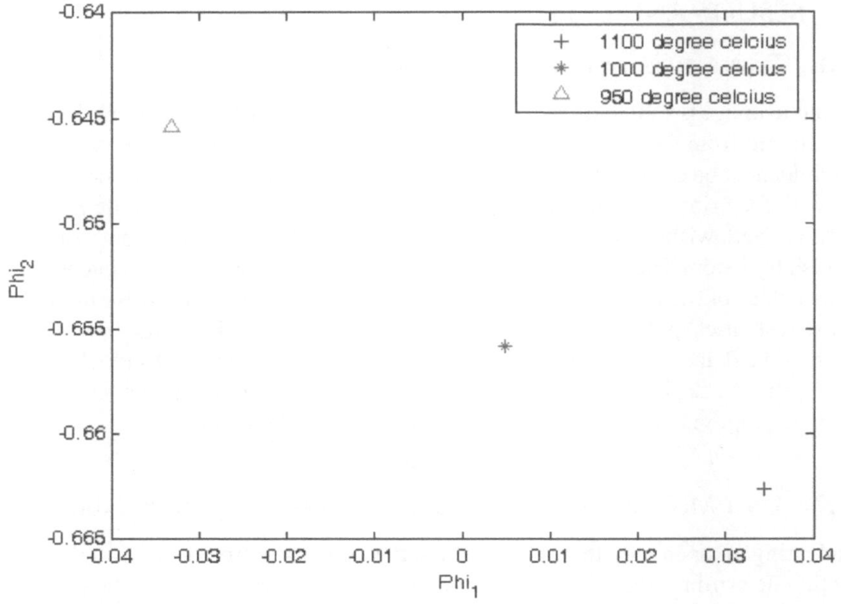

FIGURE 1.11 Distribution of trained images with extracted φ1 and φ2 vectors.

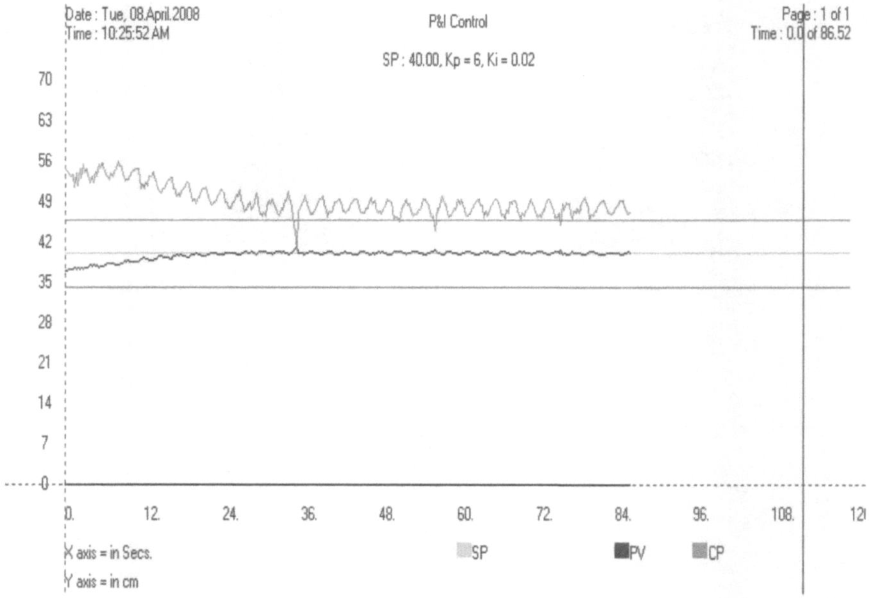

FIGURE 1.12 Output for the conventional system.

At the MATLAB prompt, enter "Fuzzy" to begin the program from the beginning. The standard untitled FIS Editor appears, with one labelled input 1 interface and one labelled output 1. With the aid of the Edit menu, we will create a two-input, two-output system for this project. Input 2 will show a second yellow box. Error and rate of change of error are the two inputs. Kp and Ki are the outputs.

1. Click once on the yellow input1 box on the left (the box will be highlighted in red).
2. Change input 1 to error in the white edit field on the right and press Return.
3. Click once on the input 2 box (yellow) (the box will be highlighted in red).
4. Change input 2 to rate of change of error in the white edit area on the right and press Return.
5. Click once on the output 1 box (blue) on the right.
6. Change output 1 to Kp in the white edit field on the right.
7. Click once on the output box (blue) on the right. Change output1 to Ki in the white edit field on the right.

The Integral Square Error (ISE) facilitates to measure the performance of the level process. The error's magnitude is squared, and integration is done throughout the range of zero to infinity. In MATLAB, the ISE values of a standard PI controller, a self-tuning PI controller and a fuzzy controller may be compared during simulation. This is accomplished using the math operation block, integrator and display. These concepts are illustrated in Figures 1.13 to 1.15, respectively.

1.5.3 Quality Check During PCB Manufacturing

The images of the PCB captured by the camera are divided into two parts. One set of PCB images are non-defective whereas the other set of images are defective. The PCB images are segmented using watershed segmentation algorithm to identify the ROI. Thereafter, the features are extracted from both the defective and non-defective PCB. The quality of PCB was ensured using watershed segmentation algorithm, which is used to isolate the objects that have contact with each other on an image. The "catchment basins" and "watershed ridge lines" in an image are considered where light and high value of pixels is less in number. The flowchart for watershed segmentation is shown in Figure 1.16.

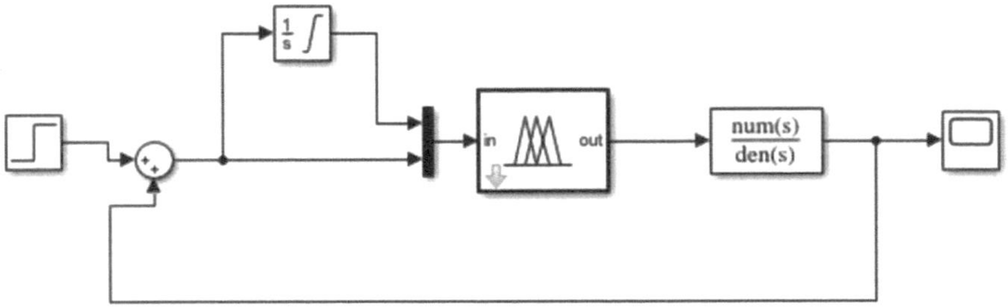

FIGURE 1.13 Simulink model for FLC.

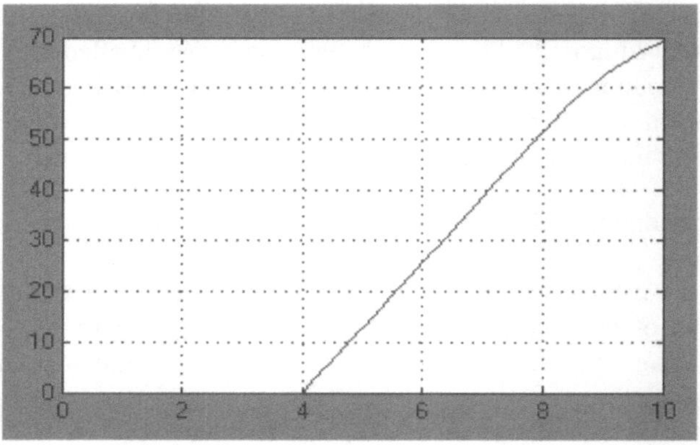

FIGURE 1.14 Output for FLC.

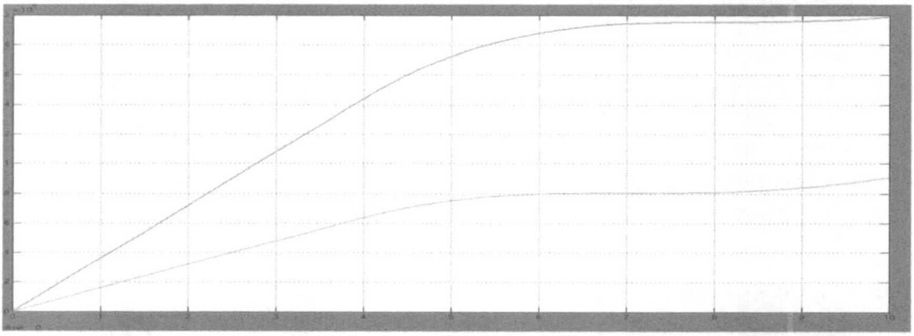

FIGURE 1.15 Simulink model for the bottling system with FLC.

The coefficients of watershed segmentation can be used as inputs to train the AI algorithms like the one which uses BPA-trained FFNN. Nearly 83 samples are used for training the FFNN, 22 image samples for testing and finally 23 images of the PCB for validation of the proposed FFNN. The above set of PCB images includes both defective and non-defective ones. The watershed coefficients are reduced to 21 by global averaging technique, which will serve as a feature reduction technique to minimize the number of inputs given to the FFNN. The challenge lies in implementation of FFNN trained with BPA. The number of nodes in the input layer corresponds to the number of watershed coefficients averaged using the global averaging method. The hidden layer's number of nodes will be determined by trial and error, but the output layer's number of output nodes will correspond to the two outputs, namely the faulty and non-defective PCB. During this process, the optimal least value of the MSE should be arrived at without compromising the time taken for training the FFNN. The example for FFNN architecture is shown in Figure 1.17. Also, the optimal parameters of the network like the learning factor and rate of momentum are also decided based on the optimal value of MSE which is achieved in less duration.

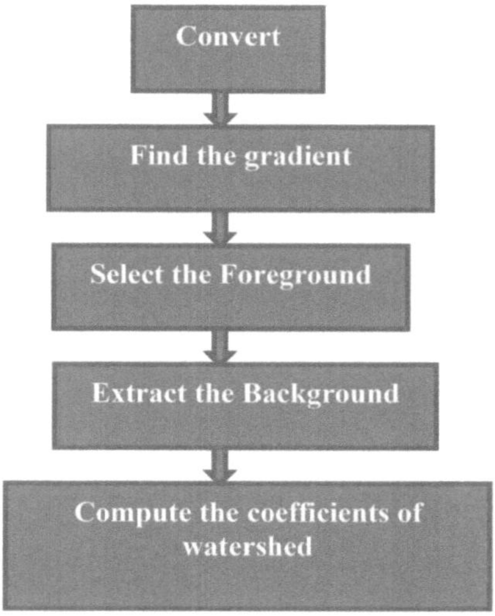

FIGURE 1.16 Flowchart for watershed segmentation algorithm.

The sample image of a PCB is shown in Figure 1.18. Simultaneously, the results for applying the watershed segmentation are also shown in the consecutive figures [Figures 1.19(a), 1.19(b) and 1.19(c)].

Figure 1.20 illustrates that the least value of the MSE is obtained during training, testing and validation with a very minimum deviation in the order of 1×10^{-3}. The number of correctly and incorrectly identified classified images is listed in Table 1.2. The classification efficiency was found to be satisfactory for testing the PCB design using an automated process.

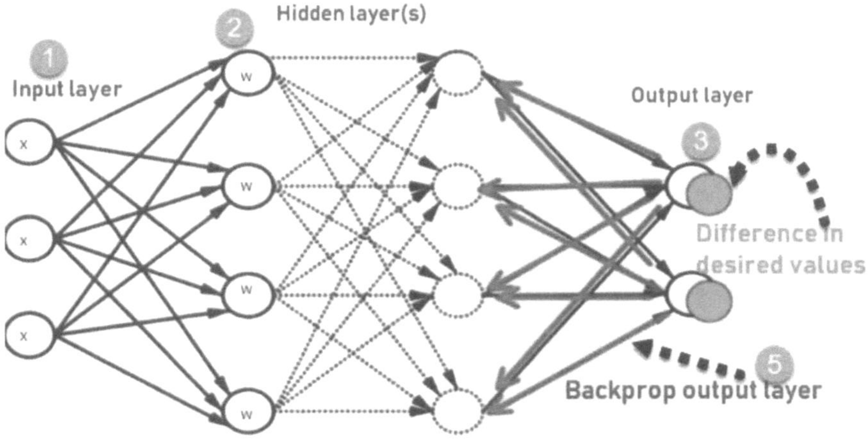

FIGURE 1.17 Sample architecture for FFNN.

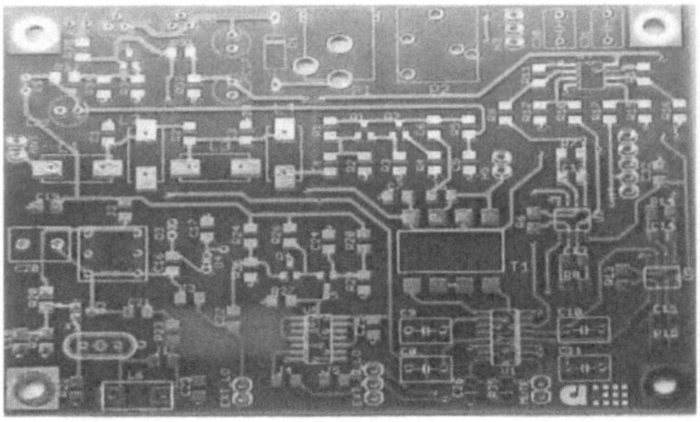

FIGURE 1.18 Original image of the PCB.

1.5.4 Quality Assessment During Steel Sheet Fabrication

The suggested approach is used to compare the findings with the results of segmentation algorithms, which have been utilized by many academics for quite some time. Table 1.3 indicates that the suggested technique's standard deviation and mean of the layers are 12 microns more precise, and that it takes 59 seconds to segment the layers against 72 seconds for the graph cut method.

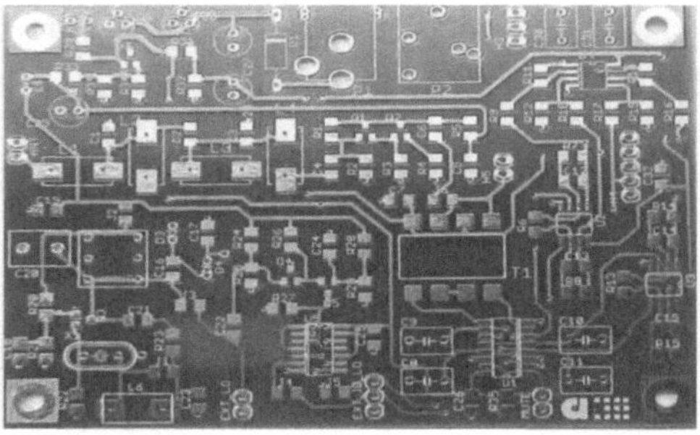

FIGURE 1.19 (a) Greyscale output for the PCB.

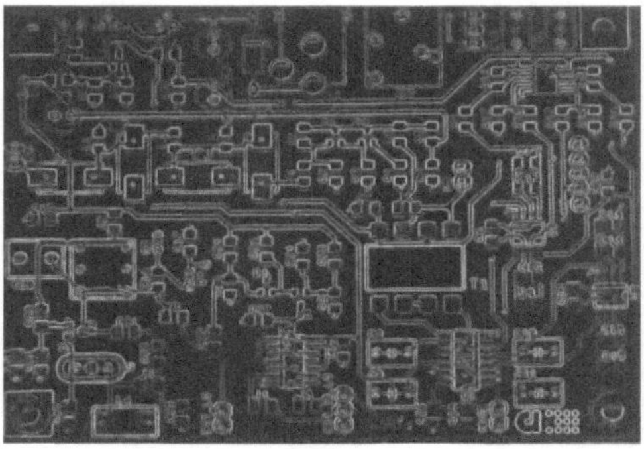

FIGURE 1.19 (b) Gradient image of the PCB.

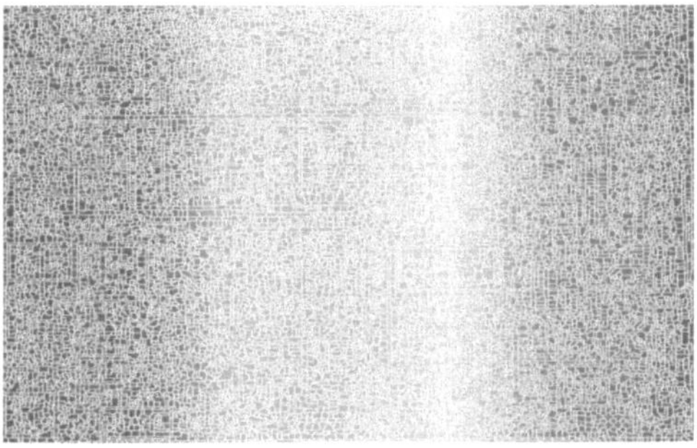

FIGURE 1.19 (c) Watershed segmented image.

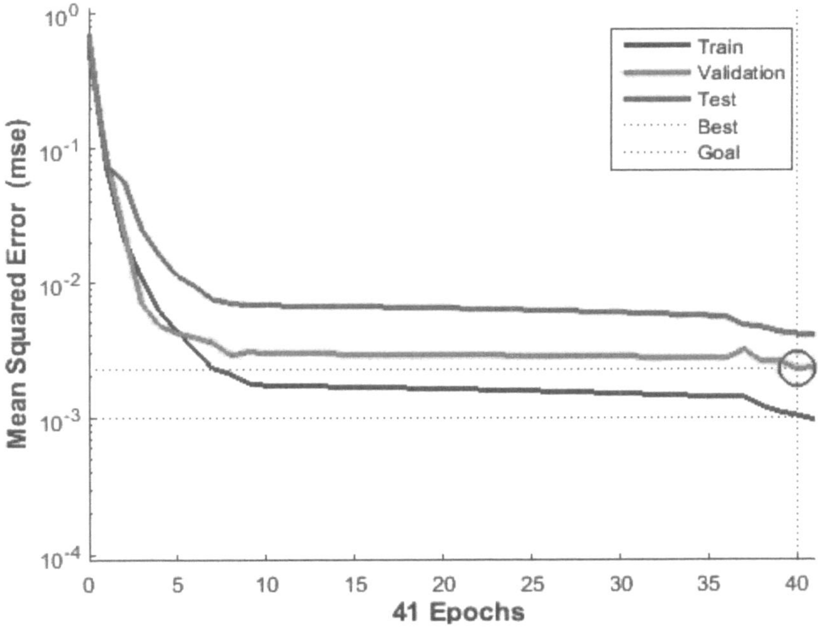

FIGURE 1.20 Tracking response for training the FFNN with BPA.

TABLE 1.2
Classification Efficiency for the Proposed Automated Testing of PCB

S.		Total Number of PCB Images			Correctly/Incorrectly Classified PCB Images						Classification Efficiency (%)		
No	Class	Training	Testing	Validation	Training		Testing		Validation		Training	Testing	Validation
1.	Defective	42	11	12	41	1	9	2	10	2	97.6	81.8	83.3
2.	Non-defective	41	11	11	40	1	9	2	9	2	97.5	81.8	81.8

TABLE 1.3
Comparison Chart for Standard Deviation and Mean Calculation

Algorithm	Mean for Proposed Work	SD for Proposed Work
RNN	25.24	11.22
Merge and Split	26.18	22.71
Region based Segmentation	52.75	11.39

1.6 CONCLUSION

In this work, firstly, three flame images were processed using FLDA and RBF to obtain final weights. New flame images were tested. The temperature obtained is in the range of $\pm10\%$. However, many images should be considered to identify the exact performance of the algorithm for temperature detection.

Second, as compared to a standard PI controller, the value of ISE for a fuzzy logic based self-tuning controller is the smallest. This project has demonstrated a practical method of regulating any process. It employs fuzzy logic, a relatively new technology. The settings of a self-tuning controller are automatically changed so that it can function efficiently even when there are process disruptions. By carefully constructing the database, sketching out the rules and employing the fuzzy logic control as a parameter estimator, it may be created for any operation. The performance of the fuzzy logic control is superior than that of traditional PI, as evidenced by the performance study.

Thirdly, it is observed that automation of any industrial process is mandatory as per Industry 4.0, as it is of very great use to the manufactures and customers for obtaining a very high-quality end product. The implementation of such indigenous automated systems will facilitate all the industries in India to be the trend setters in the field of manufacturing technology. Automation of the PCB testing process will replace the human intervention, thereby reducing the operating hours and increasing the profit. The image of the PCB is captured from all directions to extract all possible features. Before the features are extracted, the RGB image is to be converted into greyscale to obtain some important information. These features are classified by the intelligent classifier using FFNN trained with BPA, so as to identify the defects during fabrication. The classification efficiency is found to be 83% efficient during validation process. This efficiency can be further improved during validation process by considering more number of images for validation.

Lastly, a new technique is opted to identify the flaws during the fabrication of sheets from the images captured by the infra-red camera. With the help of RNN, online monitoring is possible by the method of real-time segmentation process. In future, this method can also be used to obtain more accuracy in identification of other flaws created during fabrication of steel sheets.

REFERENCES

[1] R.A. Fisher, "The Use of Multiple Measurements in Taxonomic Problems," Annals of Eugenics, vol. 7, pp. 178–188, 1936.
[2] Z.Q. Hong, Y.J. Yang, "Optimal Discriminate Plane for a Small Number of Samples and Design Method of Classifier on the Plane," Pattern Recognition, vol. 24, pp. 317–324, 1991.
[3] D.H. Foley, "Consideration of Sample and Feature Size," IEEE Transactions on Information Theory, vol. 18, no. 5, pp. 626–681, 1972.
[4] R.P. Lippmann, "An Introduction to Computing with Neural Nets," IEEE Transactions on Acoustics, Speech and Signal Processing Magazine, vol. 35, no. 4, pp. 4–22, 1987.
[5] S. Purushothaman, Y.S. Srinivasa, "A Back-Propagation Algorithm Applied to Tool Wear Monitoring," International Journal of M/C Tools & Manufacturing, vol. 34, no. 5, pp. 625–631, 1994.
[6] K. Sujatha, N.P.G. Bhavani, S.-Q. Cao, K.S Ram Kumar, "Soft Sensor for Flame Temperature Measurement and IoT based Monitoring in Power Plants," Materials Today: Proceedings, Elsevier, 2018.
[7] M. Janóczki, Á. Becker, L. Jakab, R. Gróf, T. Takács, "Automatic Optical Inspection of Soldering," In Y. Mastai (Ed.), Materials Science - Advanced Topics, Chapter 16, IntechOpen, https://www.intechopen.com/books/materials-science-advanced-topics/automatic-optical-inspection-of-soldering, 2013. DOI: 10.5772/51699
[8] C. Marques, N. Lopes, G. Santos, I. Delgado, P. Delgado, "Improving Operator Evaluation Skills for Defect Classification Using Training Strategy Supported by Attribute Agreement Analysis," Measurement, vol. 119, pp. 129–141, 2018.
[9] A. Korodi, D. Anitei, A. Boitor, I. Sile, "Image-Processing-Based Low-Cost Fault Detection Solution for End-of-Line ECUs in Automotive Manufacturing," Sensors (Basel), vol. 20, no. 12, p. PMC7349585, 2020.
[10] Abd Al Rahman, M., A. Ebayyeh, et al., "A Review and Analysis of Automatic Optical Inspection and Quality Monitoring Methods in Electronics Industry," IEEE Access, vol. 8, pp. 183192–183271, 2020. DOI: 10.1109/ACCESS.2020.3029127
[11] M. Guo, R. Wang, "The introduction of AOI in PCB defect detection based on linear array camera," International Forum on Management, Education and Information Technology Application (IFMEITA 2016), Atlantis Press, 2016.
[12] S. Singh, M. Bharti, "Image Processing Based Automatic Visual Inspection System for PCBs," IOSR Journal of Engineering, vol. 2, no. 6, pp. 1451–1455, 2012.
[13] T. Czimmermann, G. Ciuti, M. Milazzo, M. Chiurazzi, S. Roccella, C.M. Oddo, P. Dario, "Visual-Based Defect Detection and Classification Approaches for Industrial Applications—A SURVEY," Sensors, vol. 20, 1459, 2020. DOI: 10.3390/s20051459
[14] N. Rahmatov, A. Paul, F. Saeed, et al., "Machine Learning–Based Automated Image Processing for Quality Management in Industrial Internet of Things," International Journal of Distributed Sensor Networks, vol. 15, no. 10, 2019. DOI: 10.1177/1550147719883551
[15] P. Perner, "A Knowledge-Based Image-Inspection System for Automatic Defect Recognition, Classification, and Process Diagnosis," Machine Vision and Applications, vol. 7, pp. 135–147, 1994. DOI: 10.1007/BF01211659
[16] C. Sacco, A.B. Radwan, R. Harik, M. Van Tooren, "Automated Fiber Placement Defects: Automated Inspection and Characterization," Proceedings of the SAMPE 2018 Conference and Exhibition, Long Beach, CA, USA, pp. 21–24, 2018.
[17] Y. Yang, L. Pan, J. Ma, R. Yang, Y. Zhu, Y. Yang, L. Zhang, "A High-Performance Deep Learning Algorithm for the Automated Optical Inspection of Laser Welding," Applied Sciences, vol. 10, p. 933, 2020.
[18] R. Ren, T. Hung, K.C. Tan, "A Generic Deep-Learning-Based Approach for Automated Surface Inspection," IEEE Transactions on Cybernetics, vol. 48, no. 3, pp. 929–940, 2018.

[19] Y.J. Cha, W. Choi, O. Büyüköztürk, "Deep Learning-Based Crack Damage Detection Using Convolutional Neural Networks," Computer Aided Civil and Infrastructure Engineering, vol. 32, pp. 361–378, 2017. DOI: 10.1111/mice.12263
[20] X. Tao, D. Zhang, W. Ma, X. Liu, "Automatic Metallic Surface Defect Detection and Recognition with Convolutional Neural Networks," Applied Sciences, vol. 8, no. 9, p. 1575, 2018. DOI: 10.3390/app8091575
[21] H. Lin, B. Li, X. Wang, Y. Shu, S. Niu, "Automated Defect Inspection of LED Chip Using Deep Convolutional Neural Network," Journal of Intelligent Manufacturing, vol. 30, pp. 2525–2534, 2019.
[22] D. Tabernik, S. Šela, J. Skvarč, D. Skočaj, "Segmentation-based Deep-Learning Approach for Surface-Defect Detection," Journal of Intelligent Manufacturing, vol. 31, pp. 759–776, 2020.
[23] X. Fang, Q. Luo, B. Zhou, C. Li, L. Tian, "Research Progress of Automated Visual Surface Defect Detection for Industrial Metal Planar Materials," Sensors, vol. 20, p. 5136, 2020. DOI: 10.3390/s20185136
[24] H. Yang, T. Haist, M. Gronle, W. Osten, "Realistic Simulation of Camera Images of Micro-scale Defects for Automated Defect Inspection," Bildverarbeitung, KIT Scientific Publishing: Karlsruhe, Germany, vol. 84, p. 63, 2016.
[25] T. Lindeberg, "Edge Detection and Ridge Detection with Automatic Scale Selection," International Journal of Computer Vision, vol. 30, no. 2, pp. 117–154, 1998.
[26] J. Ker, L. Wang, J. Rao, T. Lim, "Deep Learning Applications in Medical Image Analysis," IEEE Access, vol. 6, pp. 2169–3536, 2018. DOI: 10.1109/ACCESS.2017.2788044
[27] A.S. Lundervold, A. Lundervold, "An Overview of Deep Learning in Medical Imaging Focusing on MRI," Zeitschrift für Medizinische Physik, vol. 29, no. 2, pp. 102–127, 2019.
[28] R. Sizyakin, V. Voronin, N. Gapon, A. Nadykto, A. Pižurica, A. Zelensky, "Automated visual inspection algorithm for the reflection detection and removing in image sequences," Proc. SPIE 11433, Twelfth International Conference on Machine Vision (ICMV 2019), 114332B, 2020. DOI: 10.1117/12.2559362
[29] S. Dupond,"A Thorough Review on the Current Advance of Neural Network Structures," Annual Reviews in Control, vol. 14, pp. 200–230, 2019.
[30] O.I. Abiodun, A. Jantan, A.E. Omolara, K.V. Dada, N.A. Mohamed, H. Arshad, "State-of-the-Art in Artificial Neural Network Applications: A Survey," Heliyon, vol. 4, no. 11, p. e00938, 2018. DOI: 10.1016/j.heliyon.2018.e00938
[31] A. Tealab,"Time Series Forecasting using Artificial Neural Networks Methodologies: A Systematic Review," Future Computing and Informatics Journal, vol. 3, no. 2, pp. 334–340, 2018.
[32] S. Niu, et al., "Automated Retinal Layers Segmentation in SD-OCT Images using Dual-Gradient and Spatial Correlation Smoothness Constraint," Computers in Biology and Medicine, vol. 54, pp. 116–128, 2014.
[33] K. Sujatha, N.P.G. Bhavani, V. Srividhya, V. Karthikeyan, N. Jayachitra, "Soft Sensor with Shape Descriptors for Flame Quality, Prediction Based on LSTM Regression," In H. Das, N. Dey, V.E. Balas (Eds.), Real-Time Data Analytics for Large Scale Sensor Data, pp. 115–138, Academic Press, 2020. DOI: 10.1016/B978-0-12-818014-3.00006-1
[34] K. Sujatha, N.P.G. Bhavani, R.S. Ponmagal, "Impact of NOx emissions on climate and monitoring using smart sensor technology," 2017 International Conference on Energy, Communication, Data Analytics and Soft Computing, ICECDS 2017, 2018.
[35] N.P.G. Bhavani, K. Sujatha, R.S. Ponmagal, T.K. Reddy, "Monitoring of SO2emissions in power plants using Internet of Things," International Conference on Energy, Communication, Data Analytics and Soft Computing, ICECDS 2017, 2018.
[36] Government of India, Economic Survey, 2014-15.
[37] L. Zhang, Y. Gao, Y. Xia, et al., "Representative Discovery of Structure Cues for Weakly-Supervised Image Segmentation," IEEE Transactions on Multimedia, vol. 16, no. 2, pp. 470–479, 2014.

[38] L.C. Chen, G. Papandreou, I. Kokkinos, et al., Deeplab: Semantic image segmentation with deep convolutional nets, atrous convolution, and fully connected crfs[J]. arXiv preprint arXiv:1606.00915, 2016.
[39] J. Lancaster. Handbook of Structural Welding: Processes, materials and methods used in the welding of major structures, pipelines and process plant. Abington Publishing, 1992.
[40] E.P. Polushkin. Defects and failures in metals. Elsevier Publishing, 1956.
[41] Indian Manufacturing Industry: Technology Status and Prospects, 2016.
[42] Wikipedia. https://en.wikipedia.org/wiki/Industry¬_4.0
[43] Industry 4.0 India Inc. gearing up for change, KPMG, March 2018.
[44] Top 5 IoT initiatives by Government of India, IOT magazine, 15th December 2016.
[45] 2016 Global Manufacturing Competitiveness Index, by Deloitte, 2016.
[46] TechSci report, "India Internet of Things (IoT) market opportunities and forecast, 2020," 2016.
[47] A. Iyer, "Moving from Industry 2.0 to 4.0: a case study of India a leapfrogging in smart manufacturing," Procedia Manufacturing, vol. 21, pp. 663–670, 2018.
[48] D. Sarma, "Case studies of how Industry 4.0 is changing the country," November 2017.
[49] W.-C. Wang, S.-L. Chen, L.-B. Chen, W.-J. Chang, "A Machine Vision Based Automatic Optical Inspection System for Measuring Drilling Quality of Printed Circuit Boards," IEEE Access, vol. 5, no. 1, pp. 10817–10833, 2017. DOI: 10.1109/ACCESS.2016.2631658
[50] Centre of Excellence (CoE) on IT for Industry 4.0, Press Information Bureau, Government of India, 5th May 2017.
[51] Y.A. Furman, "Modern Problems of Brain-Signal Analysis and Approaches to Their Solution," Pattern Recognition and Image Analysis, vol. 29, no. 1, pp. 99–119, 2019.

2 Machine Learning Applications for Additive Manufacturing

State-of-the-Art and Future Perspectives

Renangi Sandeep, Bibin Jose, K. Gokul Kumar, Manikandan Manoharan, and Arivazhagan N.

CONTENTS

2.1 INTRODUCTION

Additive Manufacturing (AM) is also referred to as 3D printing technology. It builds components by adding a similar or dissimilar combination of material, usually in layers[1]. AM technology was initially established in 1987 and gradually increased ever since, with more significant leaps and bounds in the recent past. Compared to conventional manufacturing processes, it can produce complex geometrical components with unique microstructures and properties to reduce lead time and cost.

DOI: 10.1201/9781003229018-2

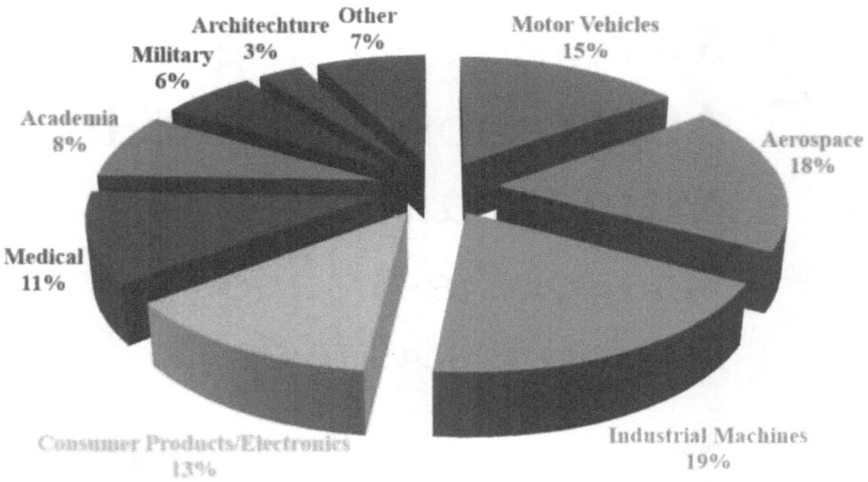

FIGURE 2.1 Industrial market share of AM in 2017 (Reprinted from Altıparmak and Xiao[3] with permission from Elsevier 2021).

As a result, AM has garnered considerable attention in recent years, both in academic research and industrial applications globally. AM finds its applications in sectors like aerospace, automobile, military, medical, and much more. Figure 2.1 shows the industrial market share of AM in 2017. Since 2015, the aerospace industry has risen to prominence as a vital segment of the global AM market. According to a recent market study, the worldwide AM market will exceed $3000 million by 2025[2]. This clearly shows the pace at which AM technology is growing.

Presently, AM methods can print small to large complex structures with high resolution, considerably decreased failure rates, and improved mechanical characteristics. However, a large amount of data is being generated during different stages of AM processes. This has led to the emergence of ML technology in AM. Therefore, ML has been targeted to accelerate AM innovations and products in the upcoming era. Also, ML and AM are the two most essential components of the upcoming industrial revolution, i.e., Industry 4.0[4].

2.1.1 Types of AM

Based on ASTM and ISO standards, AM techniques are classified into seven types, namely "binder jetting (BJ), directed energy deposition (DED), powder bed fusion (PBF), sheet lamination (SLA), material extrusion (ME), material jetting (MJ), and vat photopolymerization (VPP)." Figure 2.2 schematically represents different AM processes[5].

BJ is a unique AM technique in which the fusion of the material does not require any thermal energy. Here a binder is selectively deposited to bind the powder material together to form the final component. The procedure begins with the powder material being spread throughout the construction platform with a roller, and the binder is then deposited as needed by the print head. The construction platform

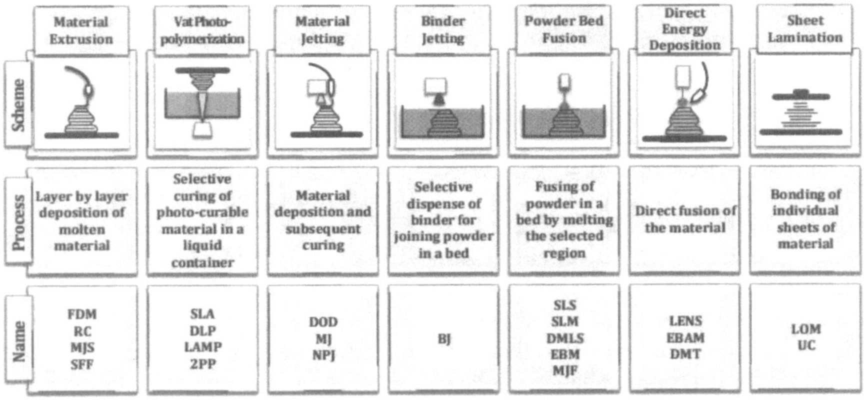

FIGURE 2.2 Different AM processes (Reprinted from Dilberoglu et al.[6] with permission from Elsevier 2017).

descends to accommodate the subsequent layer, and the procedure is repeated until the item is completed. The DED process utilizes focused energy sources, such as lasers, electron beams, or plasma arcs to melt and deposit powdered or wire-based materials to build the 3D objects. The DED method is commonly used to manufacture metal components with complicated geometries. However, this method can also produce ceramic and polymer components as well. Finally, because of the nature of the process, it is mainly used to repair existing components by adding material.

The PBF method uses an energy source (laser or electron based) to melt the material (metal or plastic), solidifying and fusing to form the desired component. The main components of the PBF system include a build chamber, a powder chamber, and a coating roller. The coating roller travels across the build chamber to construct the items, spreading the powder material and depositing a thin layer of powder. Following that, the energy source melts the metal powder base's deposited top layer. After scanning and fusing that layer, the build platform is progressively lowered. At the same time, the same amount elevates the powder chamber, and the process is repeated until the component is fully built. SLA constructs 3D structures by bonding, ultrasonic welding, or brazing thin sheets of material together. The process is followed by CNC machining or laser cutting to generate the final form of the component. SL creates components with the least additive resolution or detail of the different AM processes but at a lower cost and shorter production time. This is making it ideal for rapid prototyping at the lowest material cost. ME is the most widely used AM process in terms of accessibility and quality. This process utilizes a continuous filament of composite or thermoplastic material to build 3D objects. The material will be heated in the extruder nozzle. The nozzle moves horizontally. During the movement of the nozzle, the heated material deposits as layer-by-layer onto the build platform.

MJ method involves selective deposition of droplets of wax-like materials on a construction platform. Further, the substance cools and hardens, allowing for the layering of components. After the construction of the component, the support structures are removed by melting or by mechanical means. VPP employs an ultraviolet laser to cure a liquid resin in a vat, resulting in hard plastic formation. Digital Light Processing and stereolithography are a few of the most prevalent kinds of this technique.

2.2 MACHINE LEARNING (ML)

"ML is a subset of artificial intelligence technology that enables a machine to learn from datasets and takes decisions without being explicitly programmed"[7]. ML has recently gained traction in various fields, including medical monitoring, material property prediction, smart manufacturing, self-driving automobiles, and picture recognition. ML algorithms are typically classified into three types: reinforcement learning (RL), unsupervised learning (UL), and supervised learning (SL)[8].

SL enables software to accurately recognize unlabeled information in a test set by learning from a training batch of labeled information. Photographs, audio recordings, or written documents may be included in the datasets. The difference between planned and actual output values is calculated using an objective function called the cost function. After each repetition, the training approach updates the settings among neurons in neighboring layers to decrease the cost function. The testing phase includes previously unknown additional data, dubbed the test set, to offer an objective assessment of the model's perfection. UL makes inferences about unlabeled data. It is a data-driven ML strategy that enables the finding of hidden patterns or grouping of comparable data within a given random sample. UL is frequently used in the detection of anomalies, recommendation systems, and market segmentation. RL is a semi-supervised ML model that lets the model to interrelate through its environment and learn which behaviors are most advantageous. This is a model that is self-learning and does not require a training dataset. Reinforcement learning is used in robotic arms, autonomous vehicles, and AlphaGo.

The purpose of this chapter is to offer a sound understanding of the various ML approaches used in AM, as disclosed by the current literature. To demonstrate the benefits of ML in AM, we categorize the applications into four groups. They are design for AM using ML, ML in AM production, ML in inspection, testing and validation, and repair and restoration of AM components using ML. Finally, it accomplishes with a section summarizing major observations from the previous works and offering insights into the future prospects of ML applications in AM research and development.

2.3 DESIGN FOR AM USING ML

The AM design process can be divided into various stages, each having a distinct function. At the moment, these stages are being accomplished utilizing two ML techniques, namely design recommendations and topology optimization.

2.3.1 Design Recommendations

To assist AM designers, ML-based design recommendation systems were established. Yao et al.[9] developed a hybrid ML method for recommending design features. The UL is accomplished via hierarchical clustering, while SL is accomplished using a support vector machine algorithm (SVM). Additionally, the evidence shows that the corporation intends to use expert systems based on ontologies to define AM design expertise, which is more sophisticated.

Recently, several initiatives have been taken to aid beginner designers in estimating design characteristics such as essential support structures and build time for additive manufacturing (DfAM). McComb et al.[10] developed intelligent design compression and reconstruction system by training an auto-encoder neural network to compact and reconstruct component design. It was followed by analytical neural networks to estimate component weight, build time, and support material. Their previous attempts had inadequate predictive power; as a result, they want to use CNNs in the impending to improve predictive power by accurately and comprehensively detecting and expressing both local geometric patterns such as lattices and non-local structures such as lagoons.

The NN technique was employed by Munguía et al.[11] to simulate the required time to build components through the Laser Powder Bed Fusion (L-PBF) process. It is employed for two different purposes. Most importantly, it is capable of learning and adapting to diverse cases. Secondly, it predicts the costs for all machine models independent of their type. For this purpose, three parameters were used: bounding-box volume, part volume, and Z-height. When it comes to prediction error rates, analytic and parametric time estimators have error rates between 20% and 25%, while the NN yielded error rates between 2% and 15%.

Another method utilized by Di Angelo and Di Stefano[12] was to create a build time estimator based on NN. While they collected more comprehensive build-time parameters using a parametric approach, they also included dimensional and geometric aspects in their dataset. The authors found that NN required eight build time drive parameters to produce favorable results.

To conduct a more accurate cost prediction of AM, researchers are applying ML techniques to access a significant amount of available product and production-related data. For example, Chan et al.[13] used historical data from similar parts to forecast the cost of a new print job. They extracted critical features from the part geometry using commonalities in the parts' 3D geometry and printing processes. On feature vectors, ML algorithms such as dynamic clustering, most minor absolute shrinkage, selection operator (ASSO), and elastic net regression are employed to estimate the cost using historical data.

2.3.2 Topology Optimization

AM procedures have far more available customization than traditional, subtractive methods when it comes to topology optimization. Minimizing the total mass of the structure usually involves picking the topology that has the lowest total mass. A wide range of gradient-based, stochastic, and NN techniques has been investigated for topology optimization in AM[14].

Chowdhury and Anand[15] created a method to reduce the thermal deformation caused by temperature variations while fabricating AM parts. Their approach predicts the manufactured part's surface using a backpropagation neural network trained on the CAD model's surface data. When the model data for a new part indicates that the final part will have a poor surface quality, the trained network can modify the stereolithography (STL) file. By utilizing their NN findings, the authors effectively claimed a decrease in error in manufactured parts' compliance to CAD design.

2.4 ML IN AM PRODUCTION

2.4.1 ML IN PLANNING

Since AM is still regarded as an expensive production technique, higher productivity is crucial for the consumers. A sophisticated pre-manufacturing strategy for AM production is required, from CAD design to final product quality. As a result, several researchers have used ML to aid with AM planning[16].

ML can be used to identify the manufacturability of a part during pre-manufacturing, as shown by Tang et al.[17] for testing structures created by FDM printing. Further, a multimodal learning strategy has been presented to forecast if a metal item may be successfully produced using AM. The strategy includes CNN and MLP[18]. Similarly, Lu[19] employed SVM to increase the accuracy of a 3D printability checker software that determines whether the AM method is appropriate for a given design. Additionally, time estimation accuracy was improved by implementing and training an MLP model on a simulated SLS machine to lower the error rate from 35% to 15%[20].

2.4.2 ML IN AM QUALITY CONTROL

One crucial obstacle impeding certification of AM items is variation in product quality between machines in the same process or even between builds of the same machine. The incoherence may cause changes in geometrical precision, density, stability of processes, and mechanical characteristics. Thus, comprehensive research has sought to use ML approaches to achieve AM components quality assurance. Three techniques, namely altering the original CAD, rescaling the whole part, and executing the process control, can minimize the geometry errors. To change the overall size of components before production, the scaling ratio can be predicted using CNN or MLP[21]. Chowdhury et al.[22] demonstrated that MLP was particularly useful to compensate for geometric distortion caused by SLM processing. The authors trained FEM simulation information to forecast deformed regions (Figure 2.3).

Noriega et al.[23] used a similar strategy in FDM printing, but experimental values were used to train rather than simulation data. SOM can associate various sorts

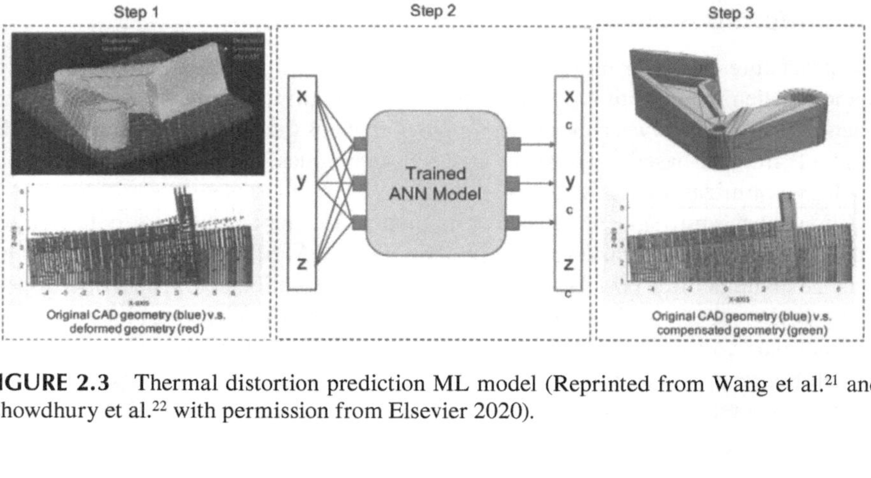

FIGURE 2.3 Thermal distortion prediction ML model (Reprinted from Wang et al.[21] and Chowdhury et al.[22] with permission from Elsevier 2020).

of geometric deviations with specific process conditions to accomplish process control[24]. Additionally, when evaluating the geometric accuracy of AM components using a laser scanner, this approach dramatically reduces the amount of 3D point cloud data required compared to several current supervised ML approaches[25]. Additionally, individual tracks can be altered by manipulating the DED process settings to minimize geometric mistakes at the macro scale[26]. As part of the PBF process, surface layer photos are captured before powder coating. These images are utilized for training ML software for the early detection of warped sections[27].

In addition to capturing images during processing, few authors used sensors to increase the density, process stability, and mechanical properties of AM-built parts. The signal emissions, mostly visual or audio, are gathered and analyzed to train various ML algorithms for monitoring the printing process, such as surface roughness prediction[28], melting condition[29], tensile property prediction[30], porosity detection[31], and printing status[32].

2.5 CLOUD-BASED AM

Cloud solutions in AM play an essential role in the widespread adoption of 3D printing and the advancement of Industry 4.0. It is server-based computing (CBC) that utilizes both software and hardware resources. It facilitates sharing 3D models or printing services to a public repository and integrates those to produce a complete pool of resources (Figure 2.4)[33].

ML algorithms can be trained to evaluate service requirements to conduct a complete analysis of terminal printers and manage resources intelligently based on

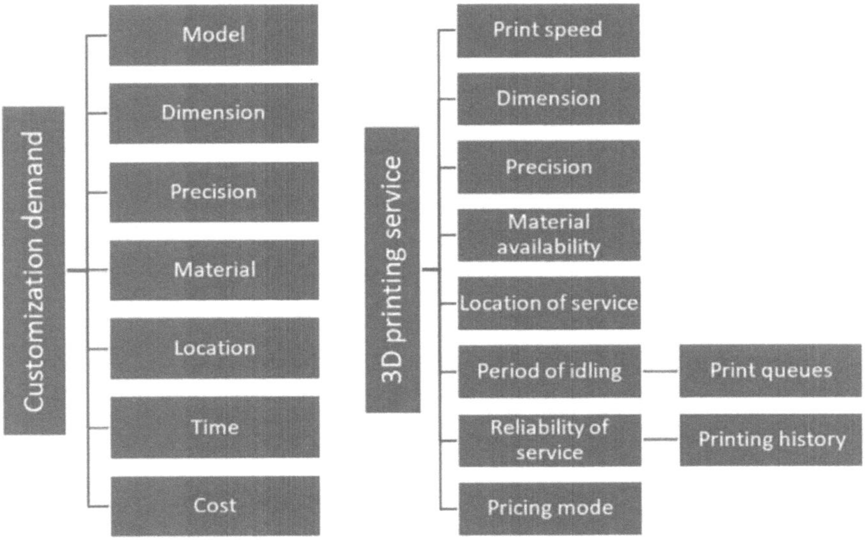

FIGURE 2.4 Variables of customization demand and 3D printing service (Reprinted from Goh et al.[34] with permission from Springer 2021).

time, cost, and print accuracy[35]. Additionally, ML offers feature recommendations for designs, enabling innovative product customization[36]. This removes a restriction to access for public users. Resource algorithms for collective and adaptive resource management have been applied for development[37]. Wu et al.[38] measured the quality of service (QoS) and improved the accuracy of service selection with a fuzzy-based hamming distance-based algorithm. Dong et al.[39] used a genetic algorithm to develop a mechanism for acquiring the quality of service and developed a cloud production trust evaluation model.

2.6 ML IN AM DATA SECURITY

In many fields, the protection of intellectual property (IP) is a top priority. There are two major components, cyber domain and physical domain, in digital production. While data violation or IP leakage is usually the result of a cyber domain, it may occur via a physical domain (sometimes referred to as side channels) because AM systems can emit different signals in creating 3D objects. To indirectly rebuild the CAD data, IP espionage can use ML algorithms to process emission signals[40].

To identify malicious attacks in fused filament fabrication (FFF) systems, Wu et al.[41] used various SL methods: the k Nearest Neighbors (kNN) approach, unsupervised anomaly detection system (USAD), and the random forest algorithm (RFA). The images from the optical camera are transformed into a grayscale plot. The grayscale value distribution is used to extract the standard deviation, mean of grayscale, and the pixel numbers with a significance greater than or equal to the threshold value. The authors also demonstrated that the USAD system achieved a greater accuracy (96%) than the kNN (87%) and random forest learning algorithms (95%).

Faruque[42] explored possibilities with SL, and UL K-Means ML approaches to analyze the design specifications of fabricated components and the thermal history of AM (Figure 2.5). While these limitations constrain the capabilities of assault detection, low sample frequency, resolution, and the absence of dynamic focus all contribute to this[43]. For the average user, file sharing has simplified the process of obtaining and producing various parts via 3D printing. These items could endanger the community if shared online and created 3D-printed replicas of real-world weapons.

Pham et al.[44] designed an anti-weapon model recognition method to prevent others from sharing and printing the restrictions. The suggested approach generates pairs of random points from the 3D mesh using facets and vertices (Figure 2.6). Once the pairs of two points are selected, the distances between them are calculated. To determine the D2 vector, the D2 shape distribution is calculated. To help train the CNN to detect firearm and knife models, the D2 vectors are applied. According to an experiment CNN performed, their method has an accuracy of 98.03%, which is higher than other methods that used depth images[45].

2.7 ML IN INSPECTION, TESTING, AND VALIDATION

The final AM part inspection and validation can be done using ML techniques. This section focuses on surface metrology, which uses ex-situ measurements, such as X-CT data, to identify defects and make estimations.

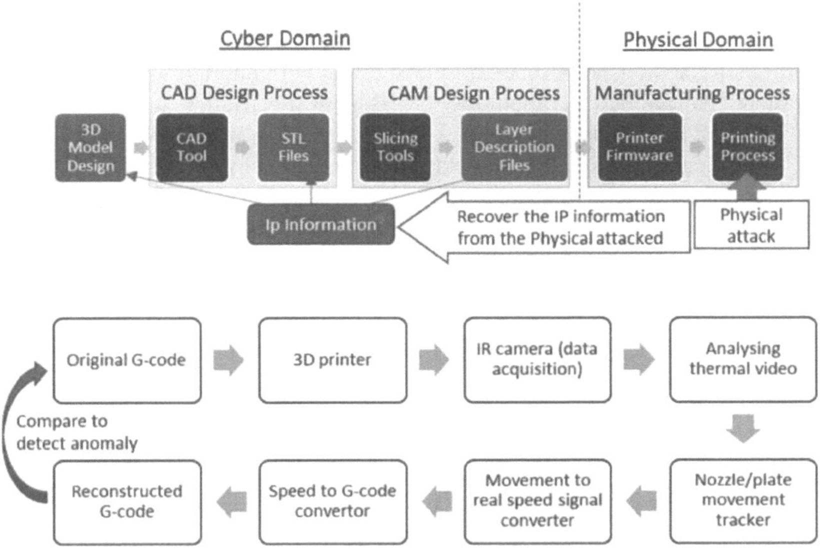

FIGURE 2.5 Cyber security workflow using ML (Reprinted from Goh et al.[43] with permission from Springer 2021).

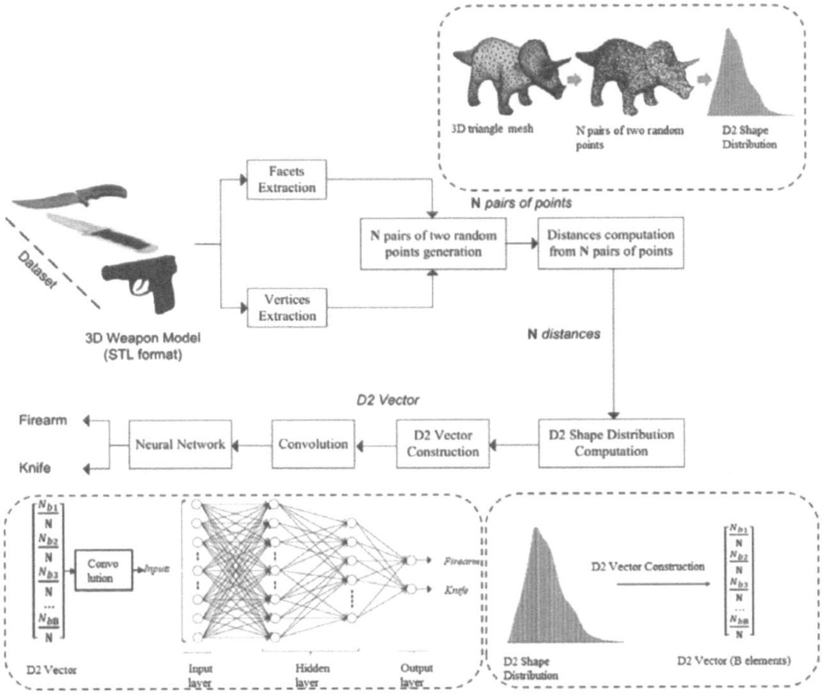

FIGURE 2.6 3D Weapon workflow using ML[44].

New spectral graph theory classifications for dimensional variations were established by Tootooni et al.[46] with FDM components. To do so, they used 3D laser scanning data for their point cloud, processed this with an algorithm that extracted eigenvalues, and applied SL methods to classify the variation in the dimensions, which included sparse representation, k-NN, NN, naïve Bayes, SVM, and decision tree. This strategy provided the best classification accuracy (F-score > 95%) by utilizing a sparse representation. Since the training approach uses prior knowledge of the part, it can only be used on other parts.

In addition to assessing the distribution of pores inside each layer of an AM component, a layer-wise spatiotemporal log Gaussian Cox process (ALS-LGCP) model was presented by Liu et al.[47] to quantify the sequential evolution of pores within layers. To make binder-jetted parts, they first applied the ALS-LGCP to the binder. They utilized Bayesian predictive analytics to forecast the existence of pores inside successive layers, ultimately attaining a level of statistical fidelity nearing 85%.

A multi-sensor data fusion and ML-based information-rich surface metrology technique were developed by Senin and Leach[48]. Since there are many complex geometries and different material properties, AM serves as an example of the requirement for improved measurement techniques.

2.8 REPAIR AND RESTORATION OF AM COMPONENTS USING ML

The product life cycle demands repair and restoration, enabling reusable components to be returned to an operational state. To justify its status as the economy's most resource-efficient approach, the process must be successful and efficient. With the advancement of AM technologies, most industries are now able to automate their repair and restoration procedures. Automated repair and restoration should alleviate some of the difficulties of manual repair and restoration performed by qualified professionals.

Den Hollander et al.[49] defined CE repair as a preventive strategy for resolving component failures and variances. Natural resource depletion and landfill accumulation intensify the necessity of product repair and restoration. Authors indicate that one technique for recovering objects is to design for repair, as depicted in Figure 2.7. Additionally, designing goods to be repairable and restorable quickly simplifies the process of recovering EoL components, resulting in increased product durability.

Recovery (design approaches for recovery)	Design for Recontextualising
	Design for Refurbishment
	Design for Repair
	Design for Remanufacture

FIGURE 2.7 Design strategy for recovering (Reproduced from den Hollander et al.[49] with permission from Wiley Materials 2017).

Researchers have shown interest in continual innovation to repair and restore. In their studies, Houston et al.[50] emphasized the benefits of repair and restoration, focusing on a variety of themes, including the influence of repair procedures on communities and the extension of product lifespan. While the available literature addresses repair and restoration difficulties, most studies have focused on the repair technique itself. However, the need for repair can be anticipated and felt well in advance of the onset of damage, for example, through the product's design that facilitates and expedites repair and restoration[51].

Harrison et al.[52] emphasized that rather than focusing on improved reparability once a product fails, the design phase should prioritize an ideal design that combines sustainable features. Durable reparability characteristics include simplicity of disassembly without destroying the product, ease of inspection, and accessibility for component replacement; these characteristics necessitate careful product design and setup[53].

Automated restoration with AM is more efficient and versatile in terms of materials and construction techniques. However, repair and restoration are much easier to accomplish if the component design aspect is prioritized and strategic throughout the design stage. Based on the literature, it was noticed that both DMLS and DED processes could repair products in the failure stage, including automobile components. Whenever a component is reconstituted to its original state, it benefits both organizations and their end-users. The majority of studies have focused on process optimization for AM component manufacturing. Considering the increasing need for automatic restoration by AM, parts of component design optimization for repairability have remained unaddressed. Design optimization will improve the efficiency of AM repair and restoration, outperforming the traditional restoration approach. When combined with ML technologies, the optimization technique may ease the burden on designers and manufacturers of analyzing various types of problems prior to sending the component to the repair shop. Thus, it is necessary to investigate the integration of ML optimization approaches into AM repair to meet the economy's demand.

2.9 SUMMARY AND FUTURE PROSPECTS

Despite the fact that ML has been around for decades, the emergence of applications in the field of AM is only a few years old. This chapter presents a detailed review of ML applications throughout the AM product life cycle. Starting from the design stage, process planning, production, quality control, data storage, data security, and finally, testing and validation. The chapter also discusses the repair and restoration of AM components using ML, which has gained popularity in recent years.

In the design stage, ML-based design recommendation and topology optimization systems were established to assist AM designers. In general, UL is achieved via hierarchical clustering, while SL is achieved through the use of the SVM algorithm. NN is effective in estimating component weight, build time, and support material. NN also plays a crucial role in topology optimization in AM. Researchers are using ML to access vast amounts of product and manufacturing data to improve cost forecast accuracy. ML algorithms such as dynamic clustering, ASSO, and elastic net regression are employed to estimate the cost of new jobs from the available product data.

ML is an extremely useful technique as far as AM production is concerned. AM production involves different stages like planning, quality control, data storage, and data security. ML can be used to identify the manufacturability of a part and forecast if it can be produced using AM. Hence, the strategy includes CNN and MLP, which are widely used in process planning. To ensure the quality of AM components, ML approaches have been extensively investigated. Generally, geometry errors are reduced by altering the source CAD, rescaling the part, and executing the process control. The use of CNN or MLP can forecast the scaling ratio and alter component sizes before production.

In addition, few authors used sensors to increase the density, process stability, and mechanical properties of AM-built parts. The signal emissions are gathered and analyzed to train various ML algorithms for monitoring the printing process, such as surface roughness prediction, melting condition, tensile property prediction, porosity detection, and printing status. Cloud solutions in AM play an essential role in the widespread adoption of 3D printing and the advancement of Industry 4.0. It allows users to upload 3D models to a public repository and combine them. ML algorithms can be trained to evaluate service requirements for 3D printers. They can manage resources based on time, cost, and print accuracy. This removes a restriction to access for public users in the digital manufacturing industry. It also offers feature recommendations for designs, enabling innovative product customization. Data security is gaining attention recently as AM processes utilize file sharing and cloud manufacturing. The malicious changes to process parameters can lead to unintended product flaws. ML can help to avoid such problems by detecting anomalies in the AM process utilizing kNN, random forest, and anomaly detection algorithms.

After production, the final AM part inspection and validation can also be done using ML techniques. The commonly used ML algorithms for inspection are k-NN, NN, naïve Bayes, SVM, and decision tree. Researchers are also coming up with innovative models for inspections; the ALS-LGCP model for assessing the distribution of pores inside each layer of an AM component is one among many.

The product life cycle demands repair and restoration, enabling reusable components to be returned to an operational state. With the advancement of AM technologies, most industries are now able to automate their repair and restoration procedures. Repair and restoration are much easier to accomplish if the component design aspect is prioritized and strategic throughout the design stage. Based on the literature, it was noticed that both DMLS and DED processes could repair products in the failure stage, including automobile components. When combined with ML technologies, the optimization technique may ease the burden on designers and manufacturers of analyzing various types of problems prior to sending the component to the repair shop.

However, in every AM process, there are an extensive set of underutilized data. In this sense, expanding data gathering methods, researching other ML approaches, and developing more advanced algorithms will be the primary research directions in this emerging field of study. To accommodate the enormous dataset generated by the sensors, more advanced data compression techniques would be necessary.

Future research should focus on multi-task learning, which dramatically improves the reliability of build models and enables designers to evaluate the functioning of AM goods prior to mass production. This prediction approach will expedite the process of developing digital twins for AM. This opens up an attractive opportunity for

ML to grow and be employed in AM applications. Additionally, the ML use in AM examined their capacity to tackle crucial problems at various manufacturing phases. Algorithmically driven methods are regarded to have enormous potential and promise in Industry 4.0.

ABBREVIATIONS

AM	Additive manufacturing
ML	Machine learning
BJ	Binder jetting
DED	Directed energy deposition
PBF	Powder bed fusion
SLA	Sheet lamination
ME	Material extrusion
MJ	Material jetting
VPP	Vat photopolymerization
AI	Artificial Intelligence
RL	Reinforcement learning
UL	Unsupervised learning
SL	Supervised learning
kNN	k Nearest Neighbors
USAD	Unsupervised anomaly detection system
FFF	Fused filament fabrication
RFA	Random forest algorithm
L-PBF	Laser Powder Bed Fusion
ASSO	Absolute shrinkage and selection operator
STL	Stereolithography
ASTM	American Society for Testing and Materials
ISO	International Organization for Standardization
SVM	Support vector machine algorithm
NN	Neural network
CAD	Computer-aided design
SLS	Selective laser sintering
CNN	Convolutional neural network
MLP	Multilayer perceptron

NOTES

1 Johnson et al., "Invited Review: Machine Learning for Materials Developments in Metals Additive Manufacturing."
2 Altıparmak and Xiao, "A Market Assessment of Additive Manufacturing Potential for the Aerospace Industry."
3 Altıparmak and Xiao.
4 Ahuett-Garza and Kurfess, "A Brief Discussion on the Trends of Habilitating Technologies for Industry 4.0 and Smart Manufacturing."
5 Dilberoglu et al., "The Role of Additive Manufacturing in the Era of Industry 4.0."
6 Dilberoglu et al.

7 Wang et al., "Machine Learning in Additive Manufacturing: State-of-the-Art and Perspectives."
8 Meng et al., "Machine Learning in Additive Manufacturing: A Review."
9 Yao, Moon, and Bi, "A Hybrid Machine Learning Approach for Additive Manufacturing Design Feature Recommendation."
10 McComb et al., "Predicting Part Mass, Required Support Material, and Build Time via Autoencoded Voxel Patterns."
11 Munguía, Ciurana, and Riba, "Neural-Network-Based Model for Build-Time Estimation in Selective Laser Sintering."
12 Di Angelo and Di Stefano, "A Neural Network-Based Build Time Estimator for Layer Manufactured Objects."
13 Chan, Lu, and Wang, "Data-Driven Cost Estimation for Additive Manufacturing in Cybermanufacturing."
14 Gaynor, "Topology Optimization Algorithms for Additive Manufacturing"; Razvi et al., "A Review of Machine Learning Applications in Additive Manufacturing."
15 Chowdhury and Anand, "Artificial Neural Network Based Geometric Compensation for Thermal Deformation in Additive Manufacturing Processes."
16 Wang et al., "Machine Learning in Additive Manufacturing: State-of-the-Art and Perspectives."
17 Tang et al., "Halomonas Alkalicola Sp. Nov., Isolated from a Household Product Plant."
18 Zhang et al., "Machine Learning Assisted Prediction of the Manufacturability of Laser-Based Powder Bed Fusion Process."
19 Lu, "Towards a Fully Automated 3D Printability Checker."
20 Munguía, Ciurana, and Riba, "Neural-Network-Based Model for Build-Time Estimation in Selective Laser Sintering."
21 Wang et al., "Machine Learning in Additive Manufacturing: State-of-the-Art and Perspectives."
22 Chowdhury, Mhapsekar, and Anand, "Part Build Orientation Optimization and Neural Network-Based Geometry Compensation for Additive Manufacturing Process."
23 Noriega et al., "Dimensional Accuracy Improvement of FDM Square Cross-Section Parts Using Artificial Neural Networks and an Optimization Algorithm."
24 Khanzadeh et al., "Quantifying Geometric Accuracy with Unsupervised Machine Learning: Using Self-Organizing Map on Fused Filament Fabrication Additive Manufacturing Parts."
25 Samie Tootooni et al., "Classifying the Dimensional Variation in Additive Manufactured Parts From Laser-Scanned Three-Dimensional Point Cloud Data Using Machine Learning Approaches."
26 Lu et al., "The Prediction of the Building Precision in the Laser Engineered Net Shaping Process Using Advanced Networks"; Xiong et al., "Bead Geometry Prediction for Robotic GMAW-Based Rapid Manufacturing through a Neural Network and a Second-Order Regression Analysis"; Caiazzo and Caggiano, "Laser Direct Metal Deposition of 2024 Al Alloy: Trace Geometry Prediction via Machine Learning."
27 zur Jacobsmühlen et al., "Detection of Elevated Regions in Surface Images from Laser Beam Melting Processes."
28 Li et al., "Prediction of Surface Roughness in Extrusion-Based Additive Manufacturing with Machine Learning."
29 Gobert et al., "Application of Supervised Machine Learning for Defect Detection during Metallic Powder Bed Fusion Additive Manufacturing Using High Resolution Imaging"; Ye et al., "Defect Detection in Selective Laser Melting Technology by Acoustic Signals with Deep Belief Networks."
30 Okaro et al., "Automatic Fault Detection for Laser Powder-Bed Fusion Using Semi-Supervised Machine Learning."
31 Shevchik et al., "Acoustic Emission for in Situ Quality Monitoring in Additive Manufacturing Using Spectral Convolutional Neural Networks"; Jafari-Marandi et al., "From In-Situ Monitoring toward High-Throughput Process Control: Cost-Driven Decision-Making Framework for Laser-Based Additive Manufacturing."

32 Rao et al., "Online Real-Time Quality Monitoring in Additive Manufacturing Processes Using Heterogeneous Sensors"; Delli and Chang, "Automated Process Monitoring in 3D Printing Using Supervised Machine Learning"; Uhlmann et al., "Intelligent Pattern Recognition of a SLM Machine Process and Sensor Data"; Wu, Yu, and Wang, "Real-Time FDM Machine Condition Monitoring and Diagnosis Based on Acoustic Emission and Hidden Semi-Markov Model."
33 Wang et al., "Secure Cloud Manufacturing: Research Challenges and a Case Study."
34 Goh, Sing, and Yeong, "A Review on Machine Learning in 3D Printing: Applications, Potential, and Challenges."
35 Wu et al., "Detecting Malicious Defects in 3D Printing Process Using Machine Learning and Image Classification"; Wu et al., "Service Architecture and Evaluation Model of Distributed 3D Printing Based on Cloud Manufacturing."
36 Yao, Moon, and Bi, "A Hybrid Machine Learning Approach for Additive Manufacturing Design Feature Recommendation."
37 Wang et al., "Secure Cloud Manufacturing : Research Challenges and a Case Study."
38 Wu et al., "Detecting Malicious Defects in 3D Printing Process Using Machine Learning and Image Classification"; Wu et al., "Service Architecture and Evaluation Model of Distributed 3D Printing Based on Cloud Manufacturing."
39 Yuan-f, "Evaluation and Selection Approach for Cloud Manufacturing Service Based on Template and Global Trust Degree."
40 Wang et al., "Machine Learning in Additive Manufacturing: State-of-the-Art and Perspectives."
41 Wu, Song, and Moon, "Detecting Cyber-Physical Attacks in CyberManufacturing Systems with Machine Learning Methods."
42 Faruque, "Forensics of Thermal Side-Channel in Additive Manufacturing Systems."
43 Goh, Sing, and Yeong, "A Review on Machine Learning in 3D Printing: Applications, Potential, and Challenges."
44 Pham et al., "Anti-3D Weapon Model Detection for Safe 3D Printing Based on Convolutional Neural Networks and D2 Shape Distribution."
45 Yao, Moon, and Bi, "A Hybrid Machine Learning Approach for Additive Manufacturing Design Feature Recommendation."
46 Samie Tootooni et al., "Classifying the Dimensional Variation in Additive Manufactured Parts from Laser-Scanned Three-Dimensional Point Cloud Data Using Machine Learning Approaches."
47 Liu et al., "Layer-Wise Spatial Modeling of Porosity in Additive Manufacturing."
48 Senin and Leach, "Information-Rich Surface Metrology."
49 den Hollander, Bakker, and Hultink, "Product Design in a Circular Economy: Development of a Typology of Key Concepts and Terms."
50 Houston et al., "Values in Repair."
51 van Nes and Cramer, "Influencing Product Lifetime through Product Design."
52 Harrison, "Design for Service: Harmonising Product Design with a Services Strategy."
53 Shahbazi and Jönbrink, "Design Guidelines to Develop Circular Products: Action Research on Nordic Industry."

REFERENCES

Ahuett-Garza, H, and T Kurfess. "A Brief Discussion on the Trends of Habilitating Technologies for Industry 4.0 and Smart Manufacturing." *Manufacturing Letters* 15 (2018): 60–63. https://doi.org/10.1016/j.mfglet.2018.02.011

Altıparmak, Sadettin Cem, and Bowen Xiao. "A Market Assessment of Additive Manufacturing Potential for the Aerospace Industry." *Journal of Manufacturing Processes* 68 (2021): 728–38. https://doi.org/10.1016/j.jmapro.2021.05.072

Angelo, Luca Di, and Paolo Di Stefano. "A Neural Network-Based Build Time Estimator for Layer Manufactured Objects." *The International Journal of Advanced Manufacturing Technology* 57, no. 1 (2011): 215–24. https://doi.org/10.1007/s00170-011-3284-8

Caiazzo, Fabrizia, and Alessandra Caggiano. "Laser Direct Metal Deposition of 2024 Al Alloy: Trace Geometry Prediction via Machine Learning." *Materials* 11, no. 3 (2018): 444. https://doi.org/10.3390/ma11030444

Chan, Siu L, Yanglong Lu, and Yan Wang. "Data-Driven Cost Estimation for Additive Manufacturing in Cybermanufacturing." *Journal of Manufacturing Systems* 46 (2018): 115–26. https://doi.org/10.1016/j.jmsy.2017.12.001

Chowdhury, Sushmit, and Sam Anand. "Artificial Neural Network Based Geometric Compensation for Thermal Deformation in Additive Manufacturing Processes," In *ASME 2016 11th International Manufacturing Science and Engineering Conference*, Blacksburg, Virginia, USA, June 27, 2016. https://doi.org/10.1115/MSEC2016-8784

Chowdhury, Sushmit, Kunal Mhapsekar, and Sam Anand. "Part Build Orientation Optimization and Neural Network-Based Geometry Compensation for Additive Manufacturing Process." *Journal of Manufacturing Science and Engineering* 140, no. 3 (2017): 031009. https://doi.org/10.1115/1.4038293

Delli, Ugandhar, and Shing Chang. "Automated Process Monitoring in 3D Printing Using Supervised Machine Learning." *Procedia Manufacturing* 26 (2018): 865–70. https://doi.org/10.1016/j.promfg.2018.07.111

Dilberoglu, Ugur M, Bahar Gharehpapagh, Ulas Yaman, and Melik Dolen. "The Role of Additive Manufacturing in the Era of Industry 4.0." *Procedia Manufacturing* 11 (2017): 545–54. https://doi.org/10.1016/j.promfg.2017.07.148

Faruque, M A. *Forensics of Thermal Side-Channel in Additive Manufacturing Systems*, University of California, Irvine, 12 (2016): 13.

Gaynor, Andrew T. "Topology Optimization Algorithms for Additive Manufacturing." Ph.D. Dissertation; The Johns Hopkins University, 2015.

Gobert, Christian, Edward W Reutzel, Jan Petrich, Abdalla R Nassar, and Shashi Phoha. "Application of Supervised Machine Learning for Defect Detection during Metallic Powder Bed Fusion Additive Manufacturing Using High Resolution Imaging." *Additive Manufacturing* 21 (2018): 517–28. https://doi.org/https://doi.org/10.1016/j.addma.2018.04.005

Goh, G D, S L Sing, and W Y Yeong. "A Review on Machine Learning in 3D Printing: Applications, Potential, and Challenges." *Artificial Intelligence Review* 54, no. 1 (2021): 63–94. https://doi.org/10.1007/s10462-020-09876-9

Harrison, Andrew. "Design for Service: Harmonising Product Design with a Services Strategy." In *ASME Turbo Expo: Power for Land, Sea, and Air*, 135–43. Barcelona, Spain, May 8–11, 2006. https://doi.org/10.1115/GT2006-90570

Hollander, Marcel C den, Conny A Bakker, and Erik Jan Hultink. "Product Design in a Circular Economy: Development of a Typology of Key Concepts and Terms." *Journal of Industrial Ecology* 21, no. 3 (2017): 517–25. https://doi.org/https://doi.org/10.1111/jiec.12610

Houston, Lara, Steven J Jackson, Daniela K Rosner, Syed Ishtiaque Ahmed, Meg Young, and Laewoo Kang. "Values in Repair." In *Proceedings of the 2016 CHI Conference on Human Factors in Computing Systems*, 1403–14. New York, NY, USA: Association for Computing Machinery, 2016. https://doi.org/10.1145/2858036.2858470

Jacobsmühlen, Joschka zur, Stefan Kleszczynski, Gerd Witt, and Dorit Merhof. "Detection of Elevated Regions in Surface Images from Laser Beam Melting Processes." In *IECON 2015 - 41st Annual Conference of the IEEE Industrial Electronics Society*, 1270–75, 2015. https://doi.org/10.1109/IECON.2015.7392275

Jafari-Marandi, Ruholla, Mojtaba Khanzadeh, Wenmeng Tian, Brian Smith, and Linkan Bian. "From In-Situ Monitoring toward High-Throughput Process Control: Cost-Driven Decision-Making Framework for Laser-Based Additive Manufacturing." *Journal of Manufacturing Systems* 51 (2019): 29–41. https://doi.org/10.1016/j.jmsy.2019.02.005

Johnson, N S, P S Vulimiri, A C To, X Zhang, C A Brice, B B Kappes, and A P Stebner. "Invited Review: Machine Learning for Materials Developments in Metals Additive Manufacturing." *Additive Manufacturing* 36 (2020). https://doi.org/10.1016/j.addma.2020.101641

Khanzadeh, Mojtaba, Prahalada Rao, Ruholla Jafari-Marandi, Brian K Smith, Mark A Tschopp, and Linkan Bian. "Quantifying Geometric Accuracy with Unsupervised Machine Learning: Using Self-Organizing Map on Fused Filament Fabrication Additive Manufacturing Parts." *Journal of Manufacturing Science and Engineering* 140, no. 3 (2018): 031011. https://doi.org/10.1115/1.4038598

Li, Zhixiong, Ziyang Zhang, Junchuan Shi, and Dazhong Wu. "Prediction of Surface Roughness in Extrusion-Based Additive Manufacturing with Machine Learning." *Robotics and Computer-Integrated Manufacturing* 57 (2019): 488–95. https://doi.org/10.1016/j.rcim.2019.01.004

Liu, Jia (Peter), Chenang Liu, Yun Bai, Prahalada Rao, Christopher B Williams, and Zhenyu (James) Kong. "Layer-Wise Spatial Modeling of Porosity in Additive Manufacturing." *IISE Transactions* 51, no. 2 (2019): 109–23. https://doi.org/10.1080/24725854.2018.1478169

Lu, Tianxiang. "Towards a Fully Automated 3D Printability Checker." In *2016 IEEE International Conference on Industrial Technology (ICIT)*, 922–27, 2016. https://doi.org/10.1109/ICIT.2016.7474875

Lu, Z L, D C Li, B H Lu, A F Zhang, G X Zhu, and G Pi. "The Prediction of the Building Precision in the Laser Engineered Net Shaping Process Using Advanced Networks." *Optics and Lasers in Engineering* 48, no. 5 (2010): 519–25. https://doi.org/https://doi.org/10.1016/j.optlaseng.2010.01.002

McComb, Christopher, Nicholas Meisel, T Simpson, and Christian Murphy. "Predicting Part Mass, Required Support Material, and Build Time via Autoencoded Voxel Patterns," In *International Solid Freeform Fabrication Symposium*, The University of Texas at Austin (2018): 1–15.. https://doi.org/10.31224/osf.io/8kne7

Meng, Lingbin, Brandon McWilliams, William Jarosinski, Hye Yeong Park, Yeon Gil Jung, Jehyun Lee, and Jing Zhang. "Machine Learning in Additive Manufacturing: A Review." *JOM* 72, no. 6 (2020): 2363–77. https://doi.org/10.1007/s11837-020-04155-y

Munguía, J, J Ciurana, and C Riba. "Neural-Network-Based Model for Build-Time Estimation in Selective Laser Sintering." *Proceedings of the Institution of Mechanical Engineers, Part B: Journal of Engineering Manufacture* 223, no. 8 (2009): 995–1003. https://doi.org/10.1243/09544054JEM1324

Nes, Nicole van, and Jacqueline Cramer. "Influencing Product Lifetime through Product Design." *Business Strategy and the Environment* 14, no. 5 (2005): 286–99. https://doi.org/https://doi.org/10.1002/bse.491

Noriega, A, D Blanco, B J Alvarez, and A Garcia. "Dimensional Accuracy Improvement of FDM Square Cross-Section Parts Using Artificial Neural Networks and an Optimization Algorithm." *The International Journal of Advanced Manufacturing Technology* 69, no. 9 (2013): 2301–13. https://doi.org/10.1007/s00170-013-5196-2

Okaro, Ikenna A, Sarini Jayasinghe, Chris Sutcliffe, Kate Black, Paolo Paoletti, and Peter L Green. "Automatic Fault Detection for Laser Powder-Bed Fusion Using Semi-Supervised Machine Learning." *Additive Manufacturing* 27 (2019): 42–53. https://doi.org/10.1016/j.addma.2019.01.006

Pham, Giao N, Suk-Hwan Lee, Oh-Heum Kwon, and Ki-Ryong Kwon. "Anti-3D Weapon Model Detection for Safe 3D Printing Based on Convolutional Neural Networks and D2 Shape Distribution." *Symmetry* 10, no. 4 (2018): 90. https://doi.org/10.3390/sym10040090

Rao, Prahalad K, Jia (Peter) Liu, David Roberson, Zhenyu (James) Kong, and Christopher Williams. "Online Real-Time Quality Monitoring in Additive Manufacturing Processes Using Heterogeneous Sensors." *Journal of Manufacturing Science and Engineering* 137, no. 6 (2015): 061007. https://doi.org/10.1115/1.4029823

Razvi, Sayyeda Saadia, Shaw Feng, Anantha Narayanan, Yung Tsun Tina Lee, and Paul Witherell. "A Review of Machine Learning Applications in Additive Manufacturing." In *Proceedings of the ASME Design Engineering Technical Conference* 1, no. August (2019). https://doi.org/10.1115/DETC2019-98415

Samie Tootooni, M, Ashley Dsouza, Ryan Donovan, Prahalad K Rao, Zhenyu (James) Kong, and Peter Borgesen. "Classifying the Dimensional Variation in Additive Manufactured Parts from Laser-Scanned Three-Dimensional Point Cloud Data Using Machine Learning Approaches." *Journal of Manufacturing Science and Engineering* 139, no. 9 (2017): 091005. https://doi.org/10.1115/1.4036641

Senin, Nicola, and Richard Leach. "Information-Rich Surface Metrology." *Procedia CIRP* 75 (2018): 19–26. https://doi.org/10.1016/j.procir.2018.05.003

Shahbazi, Sasha, and Anna Karin Jönbrink. "Design Guidelines to Develop Circular Products: Action Research on Nordic Industry." *Sustainability* 12, no. 9 (2020): 3679. https://doi.org/10.3390/su12093679

Shevchik, S A, C Kenel, C Leinenbach, and K Wasmer. "Acoustic Emission for in Situ Quality Monitoring in Additive Manufacturing Using Spectral Convolutional Neural Networks." *Additive Manufacturing* 21 (2018): 598–604. https://doi.org/10.1016/j.addma.2017.11.012

Tang, Xiaoli, Lei Zhai, Yafang Lin, Su Yao, Lijiang Wang, Yuanyuan Ge, Yang Liu, et al. "Halomonas Alkalicola Sp. Nov., Isolated from a Household Product Plant." *International Journal of Systematic and Evolutionary Microbiology* 67, no. 5 (2017): 1546–50. https://doi.org/10.1099/ijsem.0.001757

Uhlmann, Eckart, Rodrigo Pastl Pontes, Abdelhakim Laghmouchi, and André Bergmann. "Intelligent Pattern Recognition of a SLM Machine Process and Sensor Data." *Procedia CIRP* 62 (2017): 464–69. https://doi.org/10.1016/j.procir.2016.06.060

Wang, C, X P Tan, S B Tor, and C S Lim. "Machine Learning in Additive Manufacturing: State-of-the-Art and Perspectives." *Additive Manufacturing* 36 (2020): 101538. https://doi.org/https://doi.org/10.1016/j.addma.2020.101538

Wang, Weichao, Y Wang, Wesley B Williams, and Aidan Browne. "Secure Cloud Manufacturing: Research Challenges and a Case Study," In IFIP Workshop on Emerging Ideas and Trends in Engineering of Cyber-Physical Systems (EITEC), in CPS Week, Seattle, WA, April 2015: 1–12...

Wu, Haixi, Zhonghua Yu, and Yan Wang. "Real-Time FDM Machine Condition Monitoring and Diagnosis Based on Acoustic Emission and Hidden Semi-Markov Model." *The International Journal of Advanced Manufacturing Technology* 90, no. 5 (2017): 2027–36. https://doi.org/10.1007/s00170-016-9548-6

Wu, Mingtao, Vir V Phoha, Young B Moon, and Amith K Belman. "Detecting Malicious Defects in 3D Printing Process Using Machine Learning and Image Classification," In ASME 2016 International Mechanical Engineering Congress and Exposition, IMECE2016-67641, November 11–17, 2016, Phoenix, Arizona, USA. https://doi.org/10.1115/IMECE2016-67641

Wu, Mingtao, Zhengyi Song, and Young B Moon. "Detecting Cyber-Physical Attacks in CyberManufacturing Systems with Machine Learning Methods." *Journal of Intelligent Manufacturing* 30, no. 3 (2019): 1111–23. https://doi.org/10.1007/s10845-017-1315-5

Wu, Yinan, Gongzhuang Peng, Lu Chen, and Heming Zhang. "Service Architecture and Evaluation Model of Distributed 3D Printing Based on Cloud Manufacturing." In *2016 IEEE International Conference on Systems, Man, and Cybernetics (SMC)*, 2762–67, 2016. https://doi.org/10.1109/SMC.2016.7844657

Xiong, Jun, Guangjun Zhang, Jianwen Hu, and Lin Wu. "Bead Geometry Prediction for Robotic GMAW-Based Rapid Manufacturing through a Neural Network and a Second-Order Regression Analysis." *Journal of Intelligent Manufacturing* 25, no. 1 (2014): 157–63. https://doi.org/10.1007/s10845-012-0682-1

Yao, Xiling, Seung Ki Moon, and Guijun Bi. "A Hybrid Machine Learning Approach for Additive Manufacturing Design Feature Recommendation." *Rapid Prototyping Journal* 23, no. 6 (2017): 983–97. https://doi.org/10.1108/RPJ-03-2016-0041

Ye, Dongsen, Geok Soon Hong, Yingjie Zhang, Kunpeng Zhu, and Jerry Ying Hsi Fuh. "Defect Detection in Selective Laser Melting Technology by Acoustic Signals with Deep Belief Networks." *The International Journal of Advanced Manufacturing Technology* 96, no. 5 (2018): 2791–2801. https://doi.org/10.1007/s00170-018-1728-0

Yuan-f, Dong. "Evaluation and Selection Approach for Cloud Manufacturing Service Based on Template and Global Trust Degree." *Computer Integrated Manufacturing Systems*, 20, no. 1 (2014): 207–14.

Zhang, Ying, Guoying Dong, Sheng Yang, and Yaoyao Fiona Zhao. "Machine Learning Assisted Prediction of the Manufacturability of Laser-Based Powder Bed Fusion Process," Vol. 1: International Design Engineering Technical Conferences and Computers and Information in Engineering Conference, 2019. https://doi.org/10.1115/DETC2019-97610

3 Implementation of Total Productive Maintenance (TPM) in the Manufacturing Industry for Improving Production Effectiveness

*Sibabrata Mohanty, Kali Charan Rath, and Om Prakash Jena**

CONTENTS

DOI: 10.1201/9781003229018-3

3.1 INTRODUCTION AND LITERATURE REVIEWS ON MANUFACTURING INDUSTRY

Production is the procedure of making gadgets with the aid of combining people, machines, equipment, and the system. The process involved in the manufacturing industry is well defined for its specified product. Manufacturing process involves the steps of conversion of raw material into finished product. Maintenance is a function that is directly mapping to production [Ahuja and Khamba (2008), Al Mannai et al. (2016), Chana et al. (2005)]. The preferred product is the primary output of manufacturing whilst protection is the secondary output. Maintenance is the primary action responsible to look after all the equipment and machines. Preservation, according to the definition, is "the work of keeping something in proper condition, upkeep". This statement explains that the preventive action is to be taken seriously before a device or equipment fails. Maintenance's key purpose is to make the equipment or device ready for any industrial operation while also minimising the machine's ideal time.

As breakdown maintenance takes a longer time, initial preventative maintenance is the best option to reduce the ideal time of machine.It also helps in increasing the productivity and gives more profit to the manufacturing industries [Eti et al. (2004), Flynn et al. (1995), Green et al. (2011)].

The primary purpose of productive maintenance is to maximize the efficiency of plants and equipment in order to achieve the lowest possible production costs. Jeon et al. (2011), Singh and Ahuja (2020), and Kutucuoglu et al. (2001) explained the meaning of maintenance to industry. Here maintenance is divided into four categories.

Kumar et al. (2004), Khanna et al. (2010), and Kigsirisin et al. (2016) approached total productive maintenance (TPM) with an objective of maintenance and concluded that the machine should be kept in good condition to increase productivity. If the machine is found in poor working condition, it will lead to bad quality of the product. Moon and Terziovski (2014), Mwaza and Mbohwa (2015), and Al Mannai et al. (2016) found the major contributions for the success of TPM in various manufacturing industries and examined the implications of strategic TPM in association with quality. When the quality of the product is concerned, the idea that comes to mind is the implementation of TPM. Here, a small-scale manufacturing unit is considered. A reduction of inventory to a minimum level is the main objective of the Kanban system. The successful deployment of the Kanban system reduces operational costs and, as a result, increases marketplace competitiveness [Pramod et al. (2006), Parihar et al. (2012), Praveen and Rudramurthy (2013), Reeda et al. (2000)].

Ravikant et al. (2011), Salaheldin and Eid (2007), and Wickramasinghe (2012) explained the implementation of TPM in the manufacturing sector as well as in service sector. They exposed that TPM adoption provides output to increase production and quality. Scannell et al. (2012), Schimke and Brenner (2014), and Savrul and Incekara (2015) presented the problems faced by industry while implementing TPM in their paper. The main motto of their paper was to examine critically the factors that influence TPM implementation and formulate comprehensive strategies to overcome these difficulties. Seth and Tripathi (2005), Sarouhas (2007), and Vasantharayalu and Surajit (2016) determined various factors that influence production improvement with better quality through the implementation of TPM and TQM in manufacturing and service industries. After implementing TPM, Ribeiro et al. (2019) have shown quite encouraging results like motivated employees and reduction in the number of accidents.

Section 3.1 contains the basic introduction of manufacturing and the implementation of TPM through literature review. Based on the review process, Section 3.2 explains how the manufacturing sector will get better quality and bring about improvement in its production by controlling the machine breakdown time through proper planning of TPM. Section 3.3 tells about the TPM tools and their implementation through identifying the problem to take necessary action to increase overall plant efficiency. Monitoring and analysis of implemented TPM methodology are well defined in Section 3.4 by providing a suitable case study. This section concludes with a comparison between before and after application of TPM in the manufacturing industry. Output of the work has been provided through the conclusion in Section 3.5 and suggests that industries will get more benefit if they implement and maintain TPM technique in their corporation.

3.2 QUALITY MANAGEMENT SYSTEM

A quality management system (QMS) is involved in the planning and execution process through a set of policies, processes, and procedures in an organization. With a better quality management system, an industry can meet its customer satisfaction.

3.2.1 Principles of Quality Management System

The seven principles of QMS are engaging people in various activities, focusing on customer needs and satisfaction, improving the leadership quality of team members, process approach, continuous improvement in the process, decision making through various analyses of data, and proper coordination and good relationship among employees (Figure 3.1).

3.2.2 Maintaining Productivity

The term derives from preventive maintenance, and it refers to maintenance that encompasses both corrective and preventive maintenance, and this is carried out in order to improve manufacturing efficiency. These days, the preventive maintenance is important to avoid failure or total collapse of the machinery or equipment. The maintenance can be scheduled to attend for correctness with a group of expert or technician in an early stage.

TPM is a method of keeping up with and improving the honesty of creation and best frameworks through the machines, gear, workers, and the aiding strategies. The goal is to have an overall equipment effectiveness (OEE) rating of 100% to represent best manufacturing.

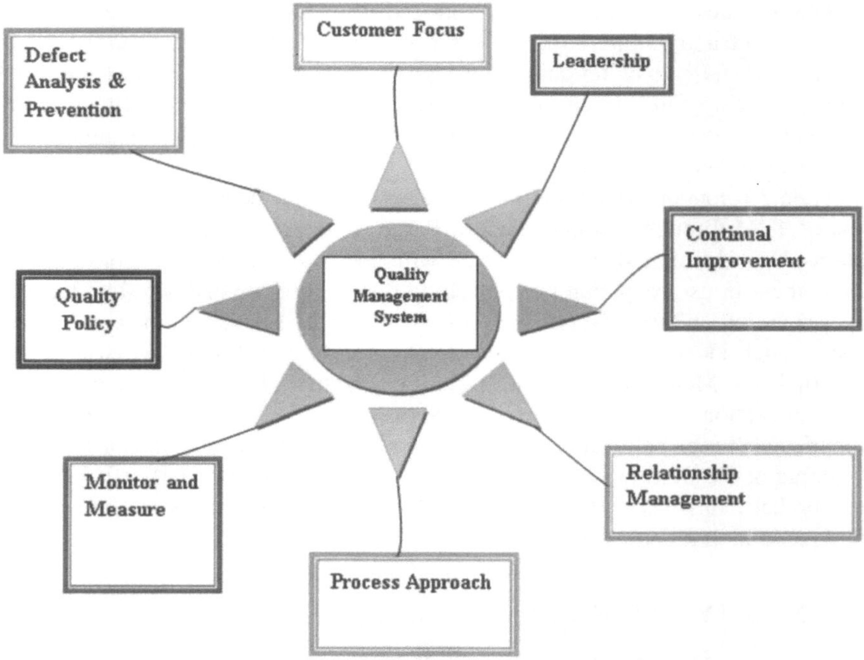

FIGURE 3.1 Principles of quality management system.

3.2.3 Pillars of TPM

Forming small, interdisciplinary teams to address fundamental areas such as preventive and independent protection, teaching staff who operate equipment, and the security and uniformity of work practices are all common ways to improve OEE through TPM. Participation of all departments is highly essential while implementing TPM for the efficient and effective use of production resources. Collaboration of these groups enhances manufacturing and reduces downtime by ensuring equipment dependability.

3.2.3.1 Focused Enhancement

Focused enhancement is the first pillar of TPM. It focuses on a planned and team-oriented approach to reduce the loss in a process. Here, a series of milestones are used to carry out the Plan, Do, Check, Act in a successful operation. This TPM pillar is significant because it increases efficiency, reduces errors, and improves the performance of equipment.

3.2.3.2 Maintenance that is Self-Contained

The second pillar of TPM is autonomized maintenance. It takes a purposeful approach to improve employees' skill levels in order to increase equipment efficiency by properly handling the device. By reducing performance failure and increasing device supply, the OEE can be increased.

3.2.3.3 Planned Maintenance

Arranged maintenance is the third of TPM's seven columns to achieve zero breakdowns. Proper machine maintenance is required to maintain the machine efficiency and productivity. TPM reduces the machine breakdown and helps to increase production with quality.

3.2.3.4 Training and Education

Training and education is considered the fourth pillar of TPM. It focuses on the training of employees, which leads to improvement in the skill set for successful implementation. This leads to the development of an eco-friendly system to support a manufacturing unit with high potential manpower utilization and industrial quality culture improvement.

3.2.3.5 Early Management

Because it's built far from other pillar groups, early management follows the first four pillars, including innovations to the future generation of produced goods and equipment plan.

3.2.3.6 Quality Maintenance

The main goal of TPM's sixth pillar, quality maintenance, is to keep faulty conditions to a minimum. This is made possible by improving communication between the 4M (manpower, machines, materials, and methods). Various inspections guide the prevention of flaws from the beginning of the manufacturing process.

3.2.3.7 Office TPM

Office TPM is the last pillar of TPM (Figure 3.2). It focuses on every aspect of the company/organization/industry that performs either administrative or maintenance activities. But, by reducing waste and damages established in manufacturing industries, office TPM implements the key TPM concepts. Zero accidents, zero health losses, and zero fires are the main goals of Safety Health Environment.

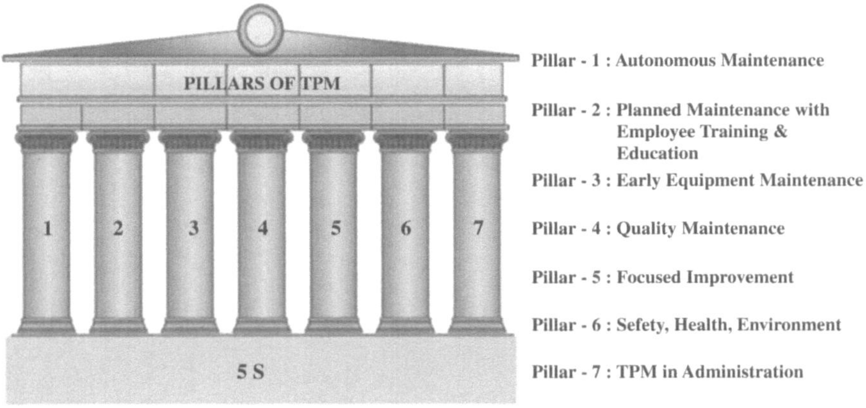

FIGURE 3.2 Seven pillars of TPM.

3.3 IMPLEMENTATION OF TPM

The implementation of TPM has a major objective: to increase the equipment effectiveness through proper maintenance of the machines. The increasing trend in production can only be achieved through active participation of operators. Various steps of implementation of TPM are represented in a flow chart.

Step 1: Declaration of TPM to establish a climate which upholds TPM.
Step 2: Launching of formal education programme to educate everyone.
Step 3: Production of hierarchical help design to advance and support the different TPM exercises.
Step 4: Establishment of basic TPM policies and holistic, measurable, attainable, and time-based goals.
Step 5: Approach towards a detailed master plan to identify the requirement of various resources.
Step 6: Implementation of TPM will begin at this stage.
Step 7: In this process, the efficiency of each piece of equipment will be enhanced.
Step 8: Creating a maintenance schedule for operators that is self-contained.
Step 9: Scheduling a preventative maintenance plan.
Step 10: Conducting a training programme to increase the operator's operating and maintenance abilities.
Step 11: Planned programme for early equipment management.
Step 12: Ongoing improvement.

3.3.1 Problem Identification

Closures, creation changes, measure disappointments, wastage in typical creation, unusual creation misfortunes, quality deformities, and part reprocessing were found as significant industry loss. During the manufacturing process, some of the losses listed below occur (Table 3.1):

TABLE 3.1
Production Process Failure vs Electric Failure

Production Process Failure	Electrical Failure
Blocking of cooling bed	Failure of HV line
Jamming of material	DTR coil car not drawn
Issue with gear box reduction (because of oil pollution, low oil level, teeth harm)	Mill drive tripped
Problem in section change	Source voltage fluctuates from 70 kV to 80 kV
Wearing of roller	POR side mill hood light faulty
Jamming of ejector	Drive tripped on RGC E-stop
Failure of sprocket and chain	
Failure of V belts	
Breakage of pulley	
Failure of motor	

The problems associated with the cold rolling mill are tabulated in Table 3.2.

3.4 MONITORING AND ANALYSIS OF THE METHODOLOGY THROUGH CASE STUDY

The means engaged with carrying out TPM in an association are recorded as follows:
Initial assessment of TPM level

a. TPM IEP (Introductory Education and Propaganda).
b. TPM committee formation.
c. Development of master plan to implement TPM.
d. Employees and stakeholders are given stage-by-stage training on all TPM pillars.
e. Process of preparing for implementation.
f. Developing a road plan for TPM implementation and establishing TPM policies and goals.

TABLE 3.2
Problems Associated with the Cold Rolling Mill

Sl No.	Problem	No. of Times Breakdown	Corrective Action	Preventive Maintenance
1	Gear box failure	7	Gear box repaired	Check oil level Abnormal sound Teeth match Bolt condition Steel condition
2	Motor failure	4	Motor repaired	Repeated checking
3	Cooling roll jam	9	Proper alignment	Cleaning and adjustment
4	Choke	4	Replacement	Cleaning
5	Pinions	2	Replacement	Oiling
6	Hose pipe	6	Cut the pipe from the leak portion and bypass the pipe	Inspection and leakage repair
7	Roller table	3		Proper lubrication
8	Stripper wiper	3	Replacement	Checking rough surface
9	Nozzle damaged	7	Replacement	Coolant water check
10	Drive problem coiler	10	Shaft oiling	Oiling
11	V belt	5	Replacement	Lubrication
12	ETR side gauge variation	4	Clean the gauge surface and adjust the C-frame	Cleaning
13	Roll stack lift down proxy malfunction	3	Proxy replacement	Repeated checking
14	DTR jog not happening	2	Selector switch contact faulty. So use spare contact	
15	Roll gap not getting closed	5	Replace its faulty rotary switch	

TABLE 3.3
Production Report before TPM Implementation

Date	Prod. in MT	Run. HRS.	Rejection in MT
1-10-2015	203.172	24	7
2-10-2015	144.623	20.56	7
3-10-2015	146.05	20.5	10
4-10-2015	209.794	24	5
5-10-2015	146.441	22	12
6-10-2015	138.285	19	10
7-10-2015	140.278	22.1	11
8-10-2015	159.372	23	12
9-10-2015	181.989	24	5
10-10-2015	103.071	14.56	13
11-10-2015	147.508	20.1	7
12-10-2015	31.027	7	5
13-10-2015	94.965	11.1	10
14-10-2015	123.506	18.55	8
15-10-2015	210.351	24	7
16-10-2015	123.276	18.45	7
17-10-2015	138.452	20.2	5
18-10-2015	143.031	22.45	8
19-10-2015	224.147	24	13
20-10-2015	17.511	4	7
21-10-2015	161.214	24	4
22-10-2015	209.379	24	8
23-10-2015	143.78	21.45	13
24-10-2015	189.593	24	10
25-10-2015	109.259	15.15	15
26-10-2015	121.55	17.1	8
27-10-2015	125.333	18.25	10
28-10-2015	121.6	16.25	12
29-10-2015	114.864	14.25	7
30-10-2015	130.176	18	6
31-10-2015	171.356	24	8
TOTAL	4424.953	600.02	270

It is the goal to have trouble-free machines and equipment that do not break down, as well as to produce high-quality components that provide absolute customer pleasure. Preventive support, breakdown upkeep, remedial maintenance, and support avoidance are instances of sustainability of machine. Planned maintenance is a proactive method to implement protection that employs trained to help operators in correctly preserving their system. The goal of strategic maintenance is to succeed and maintain machine accessibility, reduce repairing costs, increase machine reliability and maintainability, eliminate equipment failure and breakdown, and ensure that spare parts are always available. Tables 3.3 and 3.4 show the production report before and after TPM implementation, respectively.

TABLE 3.4
Production Report after TPM Implementation

Date	Prod. in MT	Run. HRS.	Rejection in MT
1-1-2016	166.099	23.25	3
2-1-2016	139.492	21.2	4
3-1-2016	134.936	20.55	7
4-1-2016	104.206	19.55	7
5-1-2016	181.256	23.55	4
6-1-2016	126.988	20.56	6
7-1-2016	131.542	23.2	2
8-1-2016	165.189	24	5
9-1-2016	153.906	24	0
10-1-2016	157.198	23.1	4
11-1-2016	157.694	24	0
12-1-2016	153.719	22.4	6
13-1-2016	162.339	24	0
14-1-2016	149.536	22.35	6
15-1-2016	118.706	20.2	8
16-1-2016	174.679	24	0
17-1-2016	199.729	24	5
18-1-2016	231.937	24	07
19-1-2016	140.266	21	9
20-1-2016	158.298	21.55	8
21-1-2016	160.415	24	0
22-1-2016	100.01	18	08
23-1-2016	104	19.1	6
24-1-2016	121.315	20	7
25-1-2016	193.13	24	7
26-1-2016	122.239	20.57	10
27-1-2016	178.454	23	5
28-1-2016	202.989	24	4
29-1-2016	147.358	21.35	8
30-1-2016	102.945	22	12
31-1-2016	171.321	22	12
TOTAL	4711.891	688.48	170

3.4.1 Methodology

We investigated the reasons of time waste and losses in the industry during its function. By computing OEE, some common equations were utilized to determine how this squandering of time and losses influence the industry. Coming up next are the factors and conditions that are utilized to compute OEE:

$$\text{Total shift time} : 24\,\text{hrs}$$

$$\text{Total Planned Production Time} = \text{Total shift time}(\text{TST}) - \text{Production Break}(\text{PB})$$

$$\text{Total Operating Time} = \text{Total Planned Production Time}(\text{TPPT}) - \text{Machine Down Time}(\text{MDT})$$

$$\text{Cycle Time (Time taken to produce one unit)}$$
$$= \text{Total Operating Time}(\text{TOT}) / \text{Total Units Manufactured}(\text{TUM})$$

$$\text{Valuable Operating Time per unit}$$
$$= \text{Total Planned Production Time}(\text{TPPT}) / \text{Total units Manufactured}(\text{TUM})$$

$$\text{Availability} = \text{Total Operating Time}(\text{TOT}) / \text{Total Planned Production Time}(\text{TPPT})$$

$$\text{Performance Rate} = \text{Cycle Time}(\text{CT}) / \text{Valuable Operating Time per unit}(\text{VOT / Unit})$$

$$\text{Quality Rate}$$
$$= \text{Total units Manufactured}(\text{TUM}) - \text{Total units Rejection}(\text{TUR}) / \text{Total units Manufactured}$$

$$\text{OEE} = \text{Availability} \times \text{Performance} \times \text{Quality Rate}$$

Tables 3.5 and 3.6 show calculation of OEE before and after implementation of TPM, respectively.

3.4.2 Comparative Study of Both Implementations

Figure 3.3(a) shows the correlation of TPM appropriation previously, then after the fact execution, OEE, Performance Rate, Quality Rate, and Availability. OEE,

TABLE 3.5
Calculation of OEE before Implementation of TPM

Factors/Months	Oct'2015	Nov'15	Dec'15
Total Shift Time	44640min	43200min	44640min
Machine Breakdown Time	8640min	7920min	6583.8min
Total Units Manufactured	4424.95MT	3681MT	4279.51MT
Total Units Rejected	270MT	245MT	230MT
Final Goods	4154.95	3436	4049.51
Total Planned Production Time	44640min	43200min	44640min
Total Operating Time	36000min	35280min	38056.2min
Cycle Time	8.14min	9.58min	8.89min
Valuable Operating Times Per Unit	10.09	11.74	10.43
Availability	80.64%	81.66%	85.25%
Performance Rate	80.67%	81.60%	85.23%
Quality Rate	93.90%	93.34%	94.62%
OEE	61.08%	69.74%	68.75%

TABLE 3.6
Calculation of OEE after Implementation of TPM

Factors/Months	Jan'16	Feb'16	March'16
Total Shift Time	44640 min	41760min	44640min
Machine Breakdown Time	3330min	2880min	2220min
Total Units Manufactured	4711.89MT	5203.11MT	5314.68MT
Total Units Rejected	170MT	130MT	92MT
Final Goods	4541.89	5073.11	5222.68
Total Planned Production Time	44640min	41760min	44640min
Total Operating Time	41310 min	38880 min	42420 min
Cycle time	8.76min	7.47min	7.98min
Valuable Operating Times Per Unit	9.47	8.02	8.40
Availability	92.60%	93.10%	95.02%
Performance Rate	92.50%	93.14%	95.00%
Quality Rate	96.39%	97.50%	98.27%
OEE	82.56%	84.54%	88.69%

Performance Rate, Quality Rate, and Availability were all lower before TPM implementation, as shown in the graph. Whereas these have significant value after TPM was adopted [Figure 3.3(b)]. After three months of implementation, OEE increased by about 20%.

3.4.3 Observation of Losses and Its Overcoming Through TPM

Table 3.7 gives a detailed description of losses.

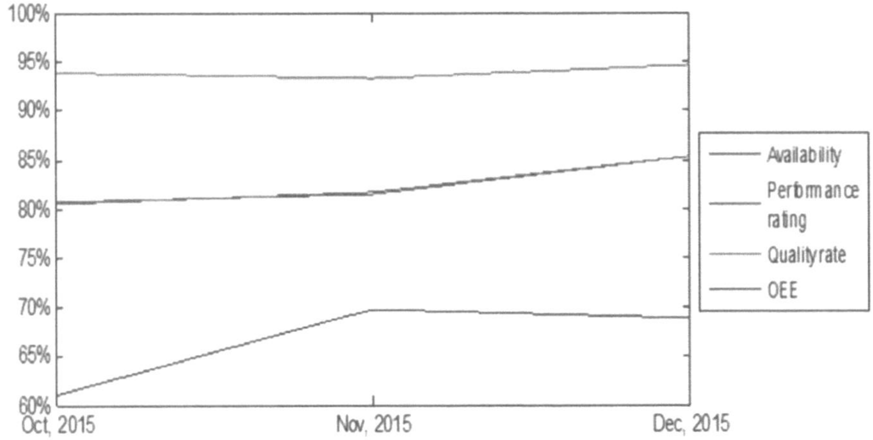

FIGURE 3.3 (a): Before TPM implementation.

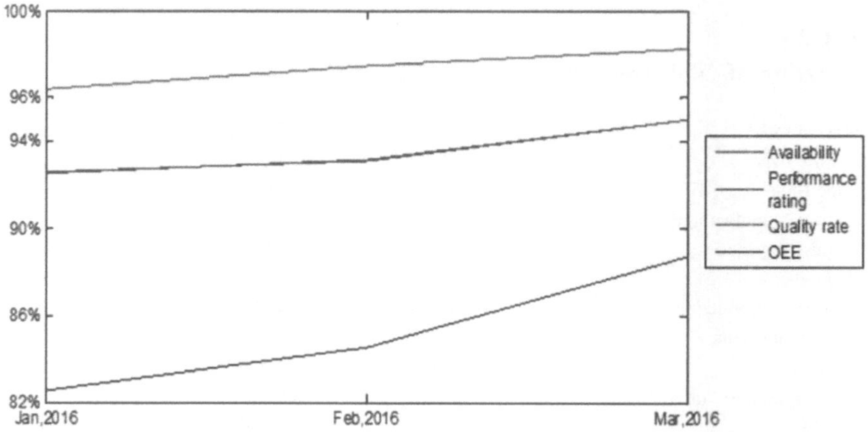

FIGURE 3.3 (b): After TPM implementation.

TABLE 3.7
Detailed Description of Losses

Sl. No	Losses Description	Unit	Oct-15	Nov-15	Dec-15	Jan-16	Feb-16	Mar-16
1	Calendar hours	mins	44640	43200	44640	44640	696	744
2	Machine shutdown	mins	0	0	0	0	0	0
3	Available hours	mins	44640	43200	44640	44640	41760	44640
4	Total stoppages	mins	8640	7920	6583.8	3330	2880	2220
5	Running Hrs.	mins	36000	35280	38056	41310	38880	42420
	EQUIPMENT FAILURE LOSS							
6	Mechanical breakdowns	mins	2200	2865	2700	1945	1657	1405
7	Electrical/Inst breakdowns	mins	1231	937	735	165	110	100
8	**TOTAL**	mins	**3431**	**3802**	**3435**	**2110**	**1767**	**1505**
9	Mechanical breakdowns	Nos	7	9.00	8.00	6.00	3.00	2.00
10	Electrical breakdowns	Nos	4	3.00	3.00	2.00	1.00	1.00
	TOTAL	Nos	11	12.00	11.00	8.00	4.00	3.00
	PROCESS FAILURE LOSS:							
11	Work roll change	mins	0	540	0.00	100.00	0.00	45.00
12	Back up roll changing	mins	1800	0.00	600.00	0.00	300.00	0.00
13	Stripper-wiper change	mins	230	180	250.00	55.00	80.00	60.00
14	Cooling nozzle checking	mins	260	280	210.00	100.00	100.00	80.00
15	Jamming of Equipment	mins	30	0.00	0.00	0.00	0.00	0.00
16	Cleaning Other Equipment	mins	0	240	100.00	0.00	0.00	0.00
17	Belt fails	mins	0	100	30.00	30.00	0.00	0.00
18	Wear of roller	mins	100	0.00	0.00	0.00	0.00	0.00
19	Cooling roll jam	mins	0	0.00	0.00	0.00	0.00	0.00
20	Gear box change	mins	45	50.00	30.00	12.00	0.00	0.00

(Continued)

TABLE 3.7
Detailed Description of Losses ***(Continued)***

Sl. No	Losses Description	Unit	Oct-15	Nov-15	Dec-15	Jan-16	Feb-16	Mar-16
21	Start up	mins	150	120	60.00	40.00	13.00	30.00
22	Motor Failure	mins	100	168	0.00	38.00	0.00	0.00
24	Misc	mins	274	300	100	60.00	50.00	60.00
	TOTAL	mins	**2989**	**2128**	**1380**	**435**	**543**	**275**
25	No of process failures	Nos	9	6.00	5.00	4.00	4.00	2.00
	SETUP LOSS:							
26	Quality change loss	mins	500	440	640	405	260	250
27	Caliper	mins	140	200	220.00	120.00	110.00	90.00
	TOTAL	mins	**640**	**640**	**860**	**525**	**370**	**340**
	MANAGEMENT LOSS							
28	Material shortage	mins	0	0.00	128	0.00	0.00	0.00
29	Power outage	mins	800	750	400	200	150	50.00
30	Operator shortage/poor planning	mins	0	0.00	0.00	0.00	0.00	0.00
31	Waiting for instructions	mins	0	30.00	0.00	0.00	0.00	0.00
32	Waiting for tools	mins	60	550.00	200.00	0.00	0.00	0.00
33	Problem at stock		120	170	180	60.00	50.00	50.00
	TOTAL	mins	**980**	**1350**	**908**	**260**	**200**	**100**
34	Actual production	MT	4424.95	3681	4279.51	4711.89	5203.11	5314.62
35	Rejections	Tons	270	245	230	170	130	92.00
Sl. No	**DESCRIPTION**	**UNIT**	**Oct-15**	**Nov-15**	**Dec-15**	**Jan-16**	**Feb-16**	**Mar-16**
1	Availability	%	80.6	81.7	85.3	92.6	93.1	95.0
2	Performance	%	80.7	81.1	85.2	92.5	93.1	95.0
3	Quality	%	93.9	93.3	94.6	96.4	97.5	98.3
4	OEE	%	61.1	69.7	68.8	82.6	84.5	88.7

Minor blemishes are the main driver of most of hardware disappointments, and they should be eliminated from all equipment.

i. Minor shortcomings in hardware will consistently discover better approaches to fizzle, and improvement exercises won't ever have the option to stay aware of the machine's disappointment rates.
ii. Almost all random equipment failure can be avoided with well-planned maintenance regimens. All TPM activities are built on the foundation of scheduled maintenance.
iii. Cross-departmental groups can propel hardware execution effortlessly than endeavors made by a solitary division working alone.
iv. The essence of ongoing machine improvement is constant learning.
v. Ones with great preventive upkeep plans produce more than machines that are possibly overhauled when they separate.
vi. Preventive maintenance plans take less time from technicians than repairing machines that have been neglected.

As per the case study executed in this chapter, it is observed that by implementing TPM:

a. The industry's working environment has improved, and the OEE has increased to 88.69% on average from December 2015 to March 2016.
b. The availability of the equipment was increased to 95% from December 2015 to March 2016.
c. There is a significant change in production rate by reducing the various losses in the industry.
d. The production was 5314.6 tons with a minimum rejection of material of 92 tons compared from October 2015 to March 2016.

3.5 CONCLUSION

This chapter described the concept of TPM and its implementation through a case study for better production. This increases the overall efficiency in production system also. The examination centers upon the critical commitments of TPM execution achievement factors like top administration authority and contribution, customary support rehearses and all-encompassing TPM execution drives towards influencing enhancements in assembling execution in an assembling industry.

With the present expanded seriousness and troublesome business sectors, TPM decides an organization's prosperity or disappointment. TPM has been a reliable programme for a long time, and organizations, especially those in the assembling business, can utilize it without risk. TPM is an adaptable process that evolves to meet the demands of more complex situations. When combined with other continuous improvement initiatives like Lean Manufacturing, Total Quality Management, or 6-Sigma, TPM becomes a powerful tool. Although more research and experience are needed, TPM has the potential to be a successful tool for improving factory automation performance. TPM Maintenance Prevention movement pushes dependability and execution improvement opportunity from the production line floor to hardware end process plan and development.

REFERENCES

Ahuja, I. P. S., & Khamba, J. S. (2008). An evaluation of TPM initiatives in Indian industry for enhanced manufacturing performance. International Journal of Quality & Reliability Management, Vol. 25 (2):147–172.

Al Mannai, B., Suliman, S., & Al Alawai, Y. (2016). TQM implementation effect on Bahrain industrial performance. International Journal of Industrial Engineering Research and Development, Vol. 8 (1):20–26.

Chana, F., Laub, H., Ipc, R., Chana, H., & Konga, S. (2005). Implementation of total productive maintenance: A case study. International Journal of Production Economics, Vol. 95 (1):71–94.

Eti, M. C., Ogaji, S. O. T., & Probert, S. D. (2004). Implementing total productive maintenance in Nigerian manufacturing industries. Applied Energy, Vol. 70 (4):385–401.

Flynn, B. B., Sakakibara S., & Schroeder R. (1995). Relationship between JIT and TQM: Practices and performance. Academy of Management Journal, Vol. 38 (5):1325–1360.

Green, K. W. Jr., Zelbst, P. J., Meacham, J. & Bhadauria, V. S. (2011). Green supply chain management practices: Impact on performance. Supply Chain Management: An International Journal, Vol. 17 (3):290–305.

Jeon, J., Kim, C., & Lee, H. (2011). Measuring efficiency of total productive maintenance (TPM): A three-stage data envelopment analysis (DEA) approach. Total Quality Management and Business Excellence, Vol. 22 (8):911–924.

Kutucuoglu, K. Y., Hamali, J., Irani, Z., & Sharp, J. M. (2001). A framework for managing maintenance using performance measurement systems. International Journal of Operations & Production Management, Vol. 21 (1/2):173–194.

Kumar, P., Wadood, A., Ahuja, I. P. S., Singh, T. P., & Sushi,l M. (2004). Total productive maintenance implementation in Indian manufacturing industry for sustained competitiveness. 34th International Conference on "Computers and Industrial Engineering", San Francisco, CA, November 14–16: 602–607.

Khanna, H. K., Laroiya, S. C., & Sharma, D. D. (2010). Quality management in Indian manufacturing organizations: Some observations and results from a pilot survey. Brazilian Journal of Operations and Production Management, Vol. 7 (2):141–162.

Kigsirisin, S., Pussawiro, S., & Noohawm, O. (2016). Approach for total productive maintenance evaluation in water productivity: A case study at Mahasawat water treatment plant. Procedia Engineering, Vol. 154:260–267.

Moon, S. W., & Terziovski, M. (2014). The impact of operations and maintenance practices on power plant performance. Journal of Manufacturing Technology Management, Vol. 25 (8):1148–1173.

Mwaza, B. G., & Mbohwa, C. (2015). Design a total productive maintenance model for effective implementation: Case study of a chemical manufacturing company. Procedia Manufacturing, Vol. 4:461–470.

Pramod, V. R., Devadasan, S. R., Muthu, S., Jagathyraj, V. P., & Moorthy, G. D. (2006). Integrating TPM and QFD for improving quality in maintenance engineering. Journal of Quality in Maintenance Engineering, Vol. 12 (2):150–171.

Parihar, S., Jain, S., & Bajpai, L. (2012). Calculation of OEE for an assembly process. International Journal of Research in Mechanical Engineering and Technology, Vol. 2 (2):25–29.

Praveen, K. R., & Rudramurthy (2013). Analysis of breakdowns and improvement of preventive maintenance on 1000 ton hydraulic press. International Journal of Emerging Technology and Advanced Engineering, Vol. 3 (8):636–645.

Reeda, R., Lemakb, D., & Meroc, N. (2000). Total quality management and sustainable competitive advantage. Journal of Quality Management, Vol. 5:5–26.

Ravikant, V. P., Shrikant, R. J., & Prasad, A. H. (2011). Implementing approach of total productive maintenance in Indian industries and theoretical aspects: An overview. International Journal of Advanced Engineering Sciences and Technologies, Vol. 6: 270–276.

Ribeiro, I. M., Godina, R., Pimentelab, C., Silva, F. J. G., & Matiasac, J. C. O. (2019). Implementing TPM supported by 5S to improve the availability of an automotive production line. Procedia Manufacturing, Vol. 38:1574–1581.

Seth, D., & Tripathi, D. (2005). Relationship between TQM and TPM implementation factors and business performance of manufacturing industry in Indian context. International Journal of Quality & Reliability Management, Vol. 22 (3):256–277.

Salaheldin, S. I., & Eid, R. (2007). The implementation of world class manufacturing techniques in Egyptian manufacturing firms: An empirical study. Industrial Management & Data Systems, Vol. 107 (4):551–566.

Sarouhas, T. P. (2007). Implementation of total productive maintenance in food industry: A case study. Journal of Quality in Maintenance Engineering, Vol. 13 (1):5–18.

Scannell, J. W., Blanckley, A., Boldon, H., & Warrington, B. (2012). Diagnosing the decline in pharmaceutical R&D efficiency. Nature Review of Drug Discovery, Vol. 11 (1):191–200.

Schimke, A., & Brenner, T. (2014). The role of R&D investments in highly R&D-based firms. Studies in Economics and Finance, Vol. 31 (1):3–45.

Savrul, M., & Incekara, A. (2015). The effect of R&D intensity on innovation performance: A country level evaluation. Procedia – Social and Behavioral Sciences, Vol. 210 (1): 388–396.

Singh, K., & Ahuja, I. S. (2020). Structural equation modelling of transfusion of TQM-TPM model for Indian manufacturing industries. International Journal of Management Practice, Vol. 13 (1): 47–73.

Singh, J., & Singh, H. (2020). Justification of TPM pillars for enhancing the performance of manufacturing industry of Northern India. International Journal of Productivity and Performance Management, Vol. 69 (1):109–113.

Vasantharayalu, & Surajit, P. (2016). An empirical study of total quality management (TQM) practices on operational performance of Indian manufacturing and service firms. International Journal of Management, Vol. 7 (6):192–202.

Wickramasinghe, V. (2012). Influence of total quality management on human resource management practices: An exploratory study. International Journal of Quality & Reliability Management, Vol. 29 (8):836–850.

4 Application of AI in Smart Cities

Mehtab Alam and Ihtiram Raza Khan

CONTENTS

DOI: 10.1201/9781003229018-4

4.1 INTRODUCTION

Artificial Intelligence (AI) is a field in computer science which deals with creating machine intelligence. In other terms, we can say that the intelligence exhibited by software or machines is termed as AI (Evans 1991). It is an important technology nowadays that supports our daily social life and economic activities. Smart cities use data sensors, actuators, smart devices and wireless communication to generate and gather big data. Handling big data generated by smart cities can be best utilized by implementing and using AI. AI can process and learn from this big data and bring about inferences which can, further, be used to make smart cities more efficient and resourceful. AI can be used to predict failures, usage patterns, demand ratio of resources as well as infrastructure (Islam and Manivannan 2017).

Deployment of AI in smart cities will be a daunting task. The challenge is that most of the AI's resources and power are limited to a few companies. Some of them being Microsoft, Google and Facebook. They all have their own AI ecosystems. We need to make AI as a service and a utility. We can make AI as a paid service on the basis of the volume of data processed or on per hour basis or on per project basis, in the future. Universities and educational institutions should fast track AI courses and put in efforts in the field to get a better workforce and AI systems. In the future, AI should transform into a utility, available to all. It will become as normal as cloud computing within the next 8–10 years.

This chapter is organized into four sections. Section 4.1 is the introduction to the chapter. In Section 4.2, we have discussed AI as a technology, its history, its working, a number of AI technologies, its applications in a few major industries worldwide and a few of its challenges. Then we compared AI with natural intelligence. In Section 4.3, we have introduced smart cities with a number of smart cities driving forces. Then we discussed a number of real-time implementations of smart city use cases in major industries worldwide, including healthcare, transportation, etc. In Section 4.4, we have tried to implement AI as a technology in smart cities. We discussed how AI can be used and provide benefits in a smart city. We then discussed AI in smart city management. Then we highlighted the transformative applications of AI in some of the smart city use cases and discussed some of its advantages and disadvantages. In Section 4.5, we have concluded our chapter. References are listed at the end.

4.2 ARTIFICIAL INTELLIGENCE

AI is a wide-ranging field of computer science which is concerned with making and building machines which are smart and capable of performing tasks and operations that typically require human intelligence (Wang, Li and Leung 2015). It is an interdisciplinary science which consists of multiple approaches. AI is evolving and can prove to be a potential game changer in computing algorithmic patterns and complex data structures, Ad-Hoc Networks and many other areas (Alam, Khan and Khan 2016, Swarm Intelligence in MANETs: A Survey). Advancement in the field of Machine Learning and deep learning are creating a paradigm shift in almost all the sectors of the technological industries. With use in applications like SIRI (Wikipedia, Siri n.d.), AlphaGo (Wikipedia, AlphaGo n.d.), Amazon Alexa (Wikipedia, Amazon n.d.), AI is being used by almost every industry and business and everyone nowadays (Alam and Khan 2019, Internet of Things as key enabler for Efficient Business Processes). AI can further be used in e-commerce prediction algorithms.

4.2.1 AI History

AI was first talked about in 1956 (Mijwil 2015). But it did not get any popularity at that time. But nowadays with big data, advanced and fast algorithms, immensely improved computing power and availability of storage space, AI is gaining popularity. Early research in AI included problem solving and symbolic methods. These works made way for more AI intensive tasks like automation and formal reasoning. The system included decision support system, smart search system which were able to mimic or augment some of the human abilities.

We can broadly consider the growth of current AI in three levels.

4.2.1.1 Artificial Neural Networks (ANN)

A neural network is an interconnected network of neurons. ANN is a computer system which is specifically designed to simulate and copy the way the human brain processes and analyses the data and information provided to it (G. P. Zhang 2000). It can be thought of as the foundation of AI. It helped scientists to solve problems which were proving impossible or very difficult to be solved by humans and other statistical methods. ANN has to be trained with training data before it can be used to perform tasks and functions. It had basic algorithms. Figure 4.1 depicts a simple neural network, having 1st layer as the input layer with 3 nodes, and intermediate/hidden layer with 5 nodes and the 3rd layer as the output layer with a single node.

4.2.1.2 Machine Learning

Machine learning is the concept in which the computer system or program is trained in such a way that it can learn and adapt to the new and unrecognized data without human involvement (Obulesu, Mahenrda and ThrilokReddy 2018). It has advanced working algorithms. Machine learning was the phase when AI started to get attention from scientists and people. In this stage, masses of people recognized AI as the future technology and started adopting and working on the technology extensively as shown in Figure 4.2.

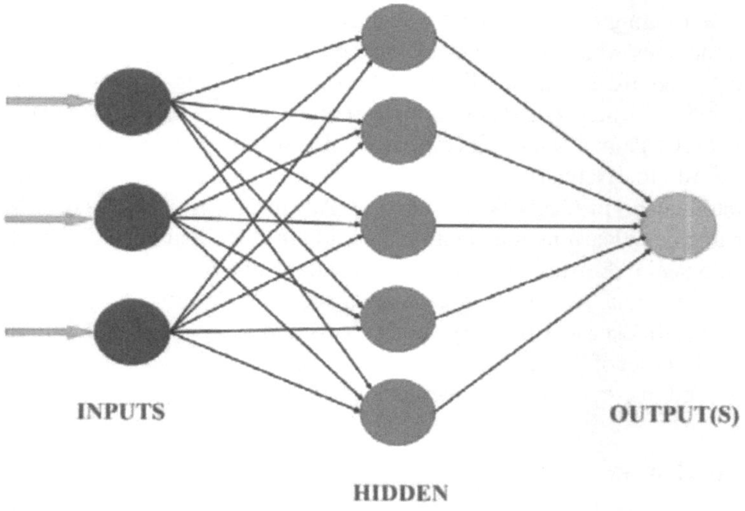

FIGURE 4.1 An ANN.

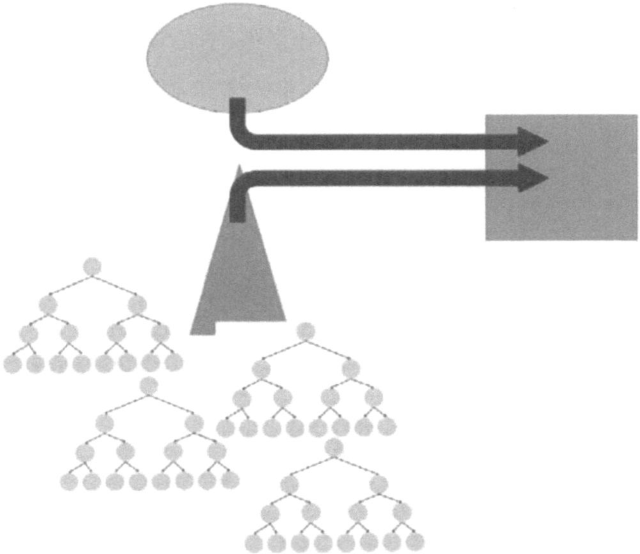

FIGURE 4.2 AI becoming popular in machine learning stage.

4.2.1.3 Deep Learning

Deep learning emulates the working of the human brain in processing data and creating patterns and finding relationships from within the data and uses these patterns and relationships for decision making as shown in Figure 4.3 (Shrestha and Mahmood 2019). It includes complex algorithms. Deep learning is the current stage in which AI is well established and is being used in almost every part of the society.

4.2.2 AI

AI works by combining big data with fast, robust and iterative processing with the help of intelligent and efficient algorithms. This allows the software programs to learn automatically from patterns, features or knowledge present in the data. This data is then used to determine solutions and/or to create a propensity model (Davenport et al. 2019). This model then starts to make predictions using the historical data as shown in Figure 4.4. Nowadays, AI can do a great deal than this but making predictions is one of the most common and first use of AI.

4.2.3 AI Technologies

AI is rapidly and radically changing almost every area of our daily lives. There is a notable competition between many technological giants and new start-ups to acquire these AI technologies before the others. AI market is in huge demand and is flourishing since it is being considered as a business investment (Alam and Khan 2020, Business 4.0-A New Revolution). According to a survey by Narrative Science, 38% of the enterprises are using AI since 2016. The figure increased to 61% in 2018

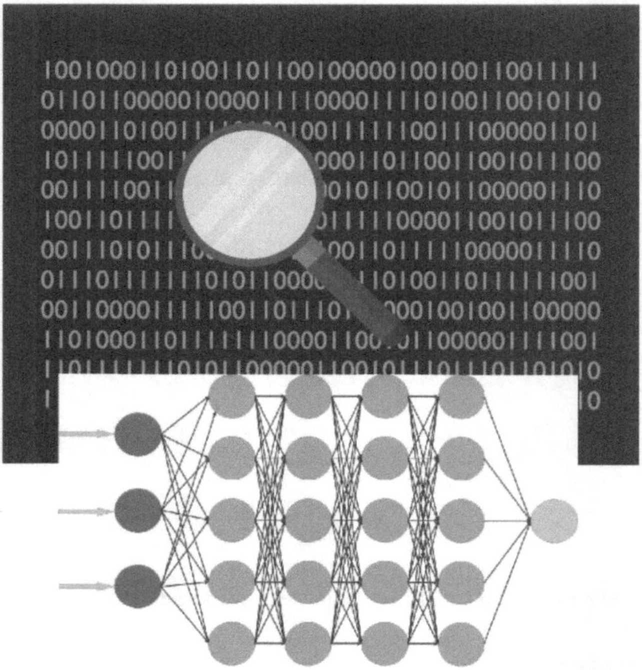

FIGURE 4.3 Deep learning in AI.

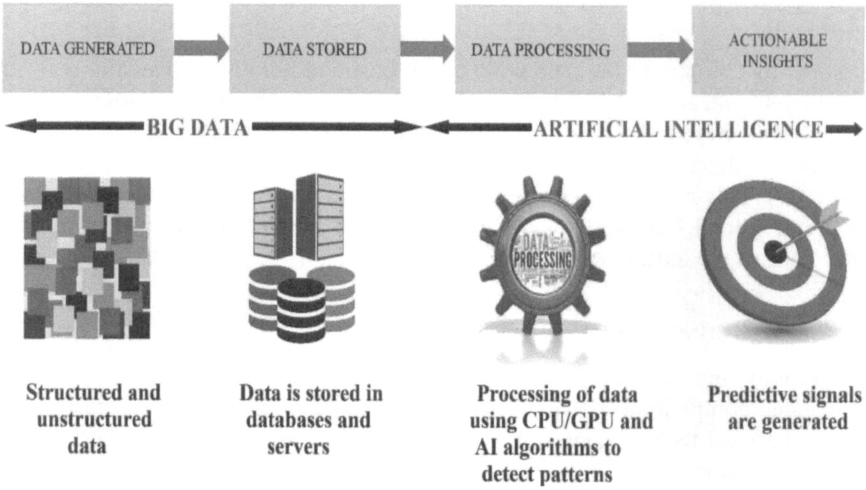

FIGURE 4.4 Working of AI using Big Data.

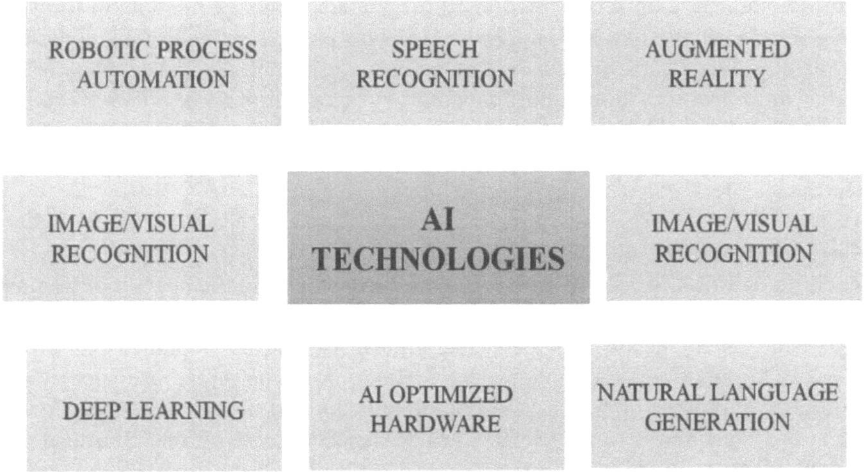

FIGURE 4.5 AI technologies.

(Science 2020). IDC estimates that the AI market is expected to grow to a total of $156.5 billion in 2020, which is more that 300% compared to 2016 and 12.3% when compared to 2017 (IDC 2020). At present there are a number of different AI technologies and tools. In this section we would discuss a few of them and these are depicted in Figure 4.5.

4.2.3.1 Natural Language Generation

As communication between humans can get tricky if two people don't know the same language, communication between machines can be a daunting and complex task. Processing information by the machines is very different when compared to the human brain. Natural Language Generation (NLG) (Mellish 1998) is a software process that automatically transforms big data into plain text in English language. It helps the system to transfer ideas and thoughts in a clear fashion. Neural network is undoubtedly one of the fastest growing technology being adopted by the enterprises.

4.2.3.2 Augmented Reality

Augmented reality is a technology that uses a device which superimposes a computer-generated image on to the users' view of the real world, in the meantime providing an augmented view of the surrounding (Khan, Goyal and Alam 2021). It gives the user a 3D view of the environment that can be interacted with, in a seemingly manner. It can be used in healthcare in transforming the healthcare tools and experiences (Riva and Wiederhold 2015). It is also used in virtual customer service by a number of enterprises (Poushneh and Vasquez-Parraga 2017).

4.2.3.3 AI-optimized Hardware

With immense improvement in CPU's (Arora et al. 2012) and GPU's (Zha and Sahni 2013) and processing power, systems are being configured and structured specifically to execute and carry on AI oriented tasks. Engineers are using big data to train the system and learn the derived patterns. Hardware processors nowadays consist of multiple cores which solely focus on low-precision arithmetic, novel dataflow architectures or in-memory computing capability. A general-purpose GPU (GPGPU) (Kim et al. 2012) and a separate graphics card can be coupled together to give massively parallel computing implementations, which can deliver up to 20 times the performance of a CPU.

4.2.3.4 Speech Recognition

Speech recognition is used in converting and transforming the speech of humans into a more useful and comprehensive format for the computer applications to process and use. The use of speech recognition is gaining popularity at a very fast pace. Applications and devices like Siri and Amazon Alexa use speech recognition technology to perform and complete tasks and actions. Hidden Markov models are considered to be useful tools for speech recognition (Bourlard and Morgan 1993).

4.2.3.5 Deep Learning

In deep learning, we try to duplicate the dense and very complex neural network of the human brain in order to process the data and bring out patterns which help in decision making. In this, the algorithms use ANN. GPUs are used to train such types of neural networks. Scientists have used multiple setups to train their ANNs. A Deep Belief Network (DBN) model of 1000 billion parameters was trained with the help of the Nvidia GTX280 graphics card having 240 cores (Raina, Madhavan and Ng 2009).

4.2.3.6 Robotic Process Automation

Robotic Process Automation uses algorithms and software programs to automate actions performed by humans, which in turn supports and makes efficient business processes. A software program or, sometimes, a physical robot is used instead of humans for doing the desired task such as typing, clicking or analysing data in a number of applications. At present, such technology is being used at places where it is not efficient or is dangerous for humans to complete the tasks. It is important to remember that AI is not meant to replace the human kind but to support and complement the skills and talent possessed by them (Gupta, Rani and Dixit 2019).

4.2.3.7 Text Analytics

Text analysis and Natural Language Processing (NLP) (Wilcock 2009) has the prime focus on the communication and interaction between human language and the language of the computers. It makes use of text analytics to analyse and examine the structure of the input sentence and further studies the interpretation and intention through various machine learning procedures. We can state that NLP is the method in which computers understand and analyse the given data, and derive meanings in an efficient, smart and useful way. The main use of this technology is in fraud detection and for security systems (Suryanarayana, Balaji and Rao 2018).

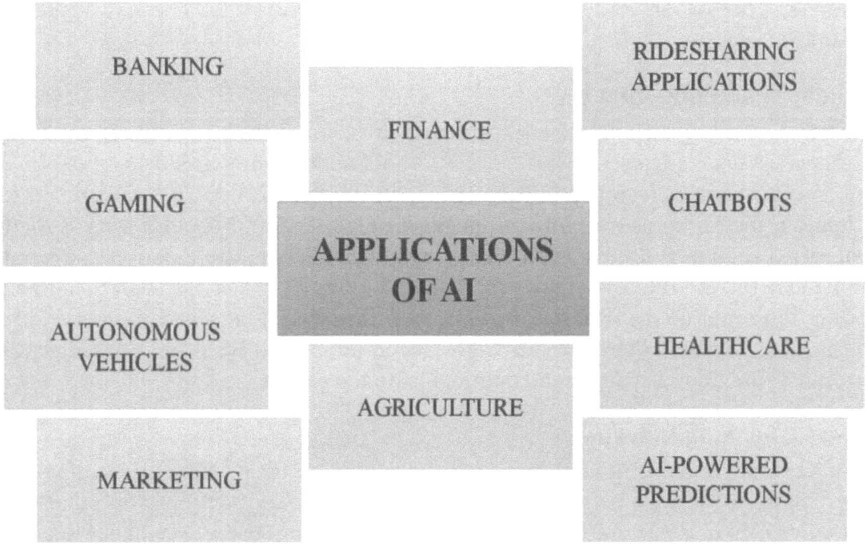

FIGURE 4.6 Applications of AI.

4.2.3.8 Image/Visual Recognition

Image recognition is the process of detecting or identifying a feature, place or face, etc., in an image or a video file. Currently it is being used in a number of tasks like reading and detecting licence plates of vehicles, detecting anomalies in medical imaging and studying people and eminent personalities (Narayanan 2019).

4.2.4 Applications of AI in Major Industries

AI is helping our planet in a number of ways. In this section, we see how AI is being used and implemented in various sectors of the society and these are depicted in Figure 4.6.

4.2.4.1 Google's AI-Powered Predictions

With everyone having smartphones, Google is using anonymous data about the current location from these smartphones to study and analyse the speed and movement of the traffic (Google n.d.). User can input and report traffic incidents like construction, accidents and other problems on the road which is used by AI algorithms to provide the best and fastest route between the user's current location and the destination.

4.2.4.2 Ridesharing Applications Like Uber

When someone books a ride, the AI algorithm captures the current location and the destination of the requested ride, then it searches for the nearest cab and calculates the cost of the ride (Schreieck et al. 2016). Cost calculations heavily depend on the demand, time, distance, road conditions and other physical factor. With all these as parameters, the AI engine calculates and predicts the cost of the ride and books the ride for the user.

4.2.4.3 Marketing Industry

Marketing is a way of advertising a product or business to attract more and more customers. AI is being used to advertise relevant and matching products on online as well as offline stores (Jarek and Mazurek 2019). In early 2000's, if we searched for an item on the online stores without its exact name, it would become very difficult to pin the product. But now, even if do a spelling mistake, the e-commerce would list the exact item with a few more recommendations. It is like the search engine is reading the mind of the user. Best example of this would be searching a movie or TV show on Netflix. They provide highly accurate predictive results based on the customers' previous reactions and comments on the shows.

4.2.4.4 Banking Industry

AI in banking is growing very fast. Banks are already using AI for providing customer support. It is also being used to detect fraudulent transactions and other such anomalies (Vijai 2019). AI solutions are nowadays being used to enhance security across retail sectors also. Banks are tracing card usage and endpoint access and the behaviours of the transactions and in the process have saved millions of dollars.

4.2.4.5 Finance Industry

Businesses are relying on data scientists and computers to predict and determine future patterns in the markets (Pau 1991). Trading also highly depends on predicting the future price with accuracy. Computer systems can process huge amount of data in little time. Algorithms are being used to make the machines learn and observe patterns in the data and give predictions about the future. Financial organizations are using AI to improve the performance of their stock trading platforms to boost their profits. Blockchain technology can be used for security purposes (Alam, Khan and Tanwer 2020, Blockchain Technology: A Critical Review and Its Proposed Use in E-Voting in India).

4.2.4.6 Agriculture Industry

With increase in population, food demand is also peaking. It is estimated that by 2050, 50% more food needs to be produced to meet the requirements. The only way to make this possible is to use the resources with utmost care. AI is helping farmers get maximum from their fields by using resources more efficiently and sustainably and improving crop yields (S. Y. Liu 2020). AI helps in deciding which crop is to be sown and when to start the harvest; it further helps in detecting the growth of insects and pests at an early stage and points out the pesticides needed. Blockchain technology with AI can be used for increasing agricultural output as well (Alam and Khan 2021, Blockchain for Indian Agriculture: A Revolution).

4.2.4.7 Healthcare Industry

Most of the medical care centres and organizations are relying on AI for saving lives and diagnosing diseases. AI is being used to detect anomalies and tumours in medical images and scans. AI digitalized device can help in finding cardiac diseases (Godber 2018). AI is being deployed in hospitals and nursing homes and patients' homes, which keeps track of how people are doing, monitoring their health conditions and capturing their vitals wherever needed. It is improving reliability, predictability and consistency with quality and patient safety. It is making doctors and nurses more effective, efficient, happier and confident in their task (Nguyen and Do 2019). AI can help in detecting and curbing COVID-19, which is very widespread nowadays (Alam, Parveen and Khan 2020, Role of Information Technology in Covid-19 Prevention).

4.2.4.8 Gaming Industry

AI has become an integral part of the online gaming domain. AI-based AlphaGo (Wang et al. 2016) software is considered one of the most advanced and significant accomplishment in AI today. The AlphaGo was trained over time by using large amount of data and supervision (Riedl and Zook 2013). Other games are chess, snooker and first-person shooter games. AI is extensively used in first-person shooter game FEAR. The game is designed in such a way that the bot opponents are trained throughout the game and they never repeat the same mistakes. They keep getting better and better and the game gets harder.

4.2.4.9 Autonomous Vehicles Industry

Autonomous and self-driving cars also use the AI technology. This is going to revolutionize the road transport system (Ma et al. 2020). The AI system collects the data via radars on the vehicles, cameras already installed on the road, cameras on the vehicles and GPS data, and this data is fed to the control system which makes decisions. Over time, the vehicles learn from the data and start making predictions and taking required actions. An example is Tesla's self-driving car having the autopilot feature.

4.2.4.10 Chatbots

Virtual assistants and conversational agents have become very popular today (Io and Lee 2017). Everyone is using Siri, Alexa or Cortana. They are being used to control household appliances and lights. It uses the AI technology to capture human sound, runs voice recognition algorithm and NLP to detect the said words and then searches for results and complete the tasks. With tremendous amount of data to learn from, there virtual assistants are improving day by day.

4.2.5 Challenges in AI

AI practitioners and researchers are claiming nowadays that machines can work in tandem with human beings in some areas; they can feel, think and have emotions just like humans. It is also being said that they are growing to be a competitor to the human brain. With all these advancements, there are still a lot of challenges in this field. One of the prime challenges is that these machines lack understanding and common sense, i.e., if we input a wrong value, in the learning phase, the machine will not be able to point it out and will act according to the wrong input provided.

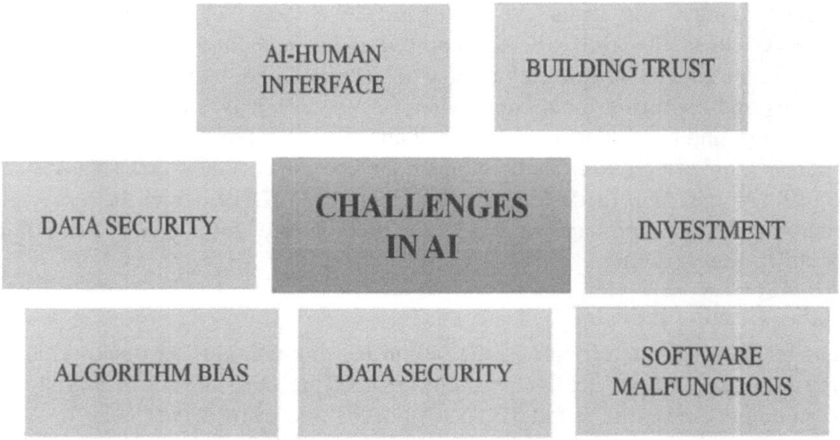

FIGURE 4.7 Challenges in AI.

We can say this to be "garbage in-garbage out," i.e., if we input garbage into the system, we will only get garbage out (Harkut and Kasat 2019). In this section, we discuss a few challenges in AI and these are depicted in Figure 4.7.

4.2.5.1 Building Trust

AI is all about science, technology, algorithms, machines, which even nowadays people are not aware about to that extent. This makes it difficult for the people to show trust in it and implement AI in their business or objectives.

4.2.5.2 AI-Human Interface

Since AI is relatively a new technology, working manpower is restricted. Data science and data analytics skilled manpower is required to get maximum output from AI. As AI is getting advanced day by day, businesses need skilled professionals who can match the requirements and leverage the technology. Professionals are required to be trained to be able to take out the maximum benefits of this technology.

4.2.5.3 Investment

AI is currently an expensive technology. Huge amount of computing power is required to execute machine learning models efficiently and effectively. CPUs paired with GPUs are used at very basic need scenarios. The cost of the hardware components can get very high, which in turn reduces the adoption rate for many businesses.

4.2.5.4 Software Malfunctions

Machines and algorithms are the sole controllers of AI and its applications. Mistakes are rare but when they occur, it is a very difficult and time-consuming task to pin point the error for rectification. Further, due to lack of experts in the field, businesses have little to no control over the system, which also increases the complications.

4.2.5.5 Non-Invincible

As with the case with any other technology, AI also has to deal with some limitations. It can replace some tasks but cannot replace all the tasks being performed in a business.

4.2.5.6 Data Security

One of the main inputs in the AI system is big data. The more the training data, the better will be the system performance. This big data consists of huge volumes of structured, unstructured data, classified data; sometimes it includes data which is personal and sensitive in nature. This means AI is very vulnerable to data breaches and sometimes identity theft.

4.2.5.7 Algorithm Bias

In AI system, the accuracy of decision making is highly dependent on how accurately it has been trained. If the training data is biased in any form, the system is bound to fail. Bias can be in the form of racial, gender, communal, regional or ethnic based.

4.2.6 AI vs Natural Intelligence

It is believed that comparing AI with human intelligence may not be fair. In cases where the process is happening with repetition and/or regularities, AI prevails, but in cases where qualities like human intuition is needed, natural intelligence (NI) wins (Binning, Baatz and Schmidt 2002; Zohuri and Rahmani 2020).

4.2.6.1 AI Winning over NI

Remarkable speed: One doctor can make a single diagnosis in about 10 minutes; AI system can make near to a million in the same time or even less.

Operate 24X7: Computer systems do not need rest or sleep after a day's work, they can work continuously for days, months or even years without stopping or halting. Self-driving vehicles can serve the users at any time of the day, for any number of hours.

Less biased: NI is highly biased on the basis of emotions and feelings. On the other hand, AI systems can be trained on big data consisting of millions and billions of examples, which help in making important decisions without any bias.

Accuracy: AI systems can perform tasks with high precision and accuracy if the training data is error free and robust. They can precisely predict the stock prices.

4.2.6.2 NI Winning over AI

Energy efficient: Computers consume a huge amount of electric energy to function. With modern CPUs and GPUs working in tandem, demand of energy increases exponentially.

Universal: AI systems are generally trained to perform a few specific tasks. The number is limited to 10's to 100's. But NI can help to perform and manage thousands of different and unique skills.

Multi-tasking: A regular worker in the field may be responsible for a large number of tasks, but the AI system can perform only a few tasks in tandem.

Complex movements: Most advanced and efficient robots also have limited mobility in them. But humans have high mobility and flexibility.

4.3 SMART CITIES

Cities are one of the most complex structures of the physical world. But the cities are made with living entities such as people, societies and communities that interact with non-living entities such as roads, buildings, and spaces, across a broad range of settings and context (Al-Hader et al. 2009). Smart cities are the cities which are smart. Smart cities generally means that a number of vivid and different sensors collect data and information to get insights from them and further use them to manage these objects efficiently and with sustainability in mind. If we consider them to be assets, then proper and efficient asset management will have a large number of benefits. Reducing the asset downtime and increasing the life expectancy of the asset can be a few examples. This asset management will provide us with better performance, reduced management costs, satisfied asset users and finally better sustainability. With 70% of the population migrating to cities in this decade, resource management and efficient utilizations are the biggest need of the hour. Sustainable development can only be achieved by making the cities smart (Alam, Khan and Tanweer 2020, IOT in Smart Cities: A survey 9 May).

4.3.1 What's Driving Smart Cities

Powerful forces from around the globe are converging together to make and develop smart cities throughout the globe. There are both positive and negative forces governments and leaders need to be aware of as they plan and conduct their strategic implementation of smart cities. Here we list some of the driving forces for smart city development and these are depicted in Figure 4.8.

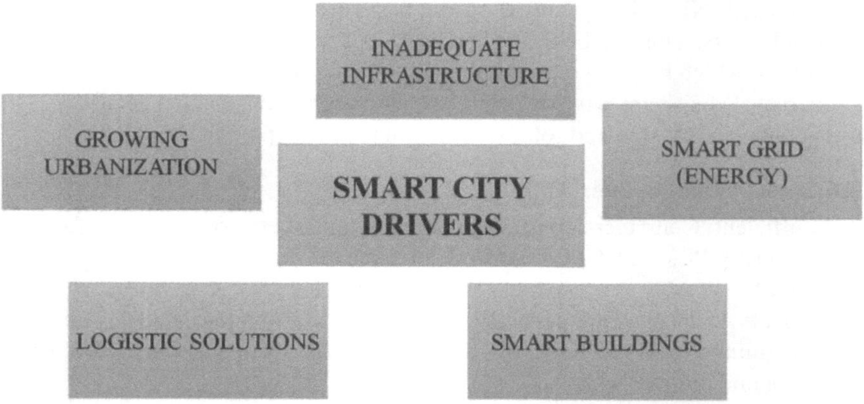

FIGURE 4.8 Smart city drivers.

4.3.1.1 Growing Urbanization

Cities deliver immense benefits to the people living in them. They create greater employment opportunities, access to better healthcare and access to better education. They also provide vivid access to entertainment, culture and arts. Due to these amenities, people are moving to the cities at an unprecedented rate (Tabane, Ngwira and Zuva 2016).

4.3.1.2 Inadequate Infrastructure

Urbanization is having a significant strain on the infrastructures of the cities. The current infrastructure is not enough for the population currently moving to the cities. Massive infrastructure upgrade is required for living, healthcare, parking, enterprises, industries, etc. (Neirotti et al. 2014).

4.3.1.3 Smart Grid (Energy)

As the population increases, so does the demand for power and energy. We need to provide electricity to new infrastructures, and to the different construction infrastructures. Intelligent management of energy resources is the need. Energy resources need to be utilized efficiently and responsibly (Rosario et al. 2017).

4.3.1.4 Smart Buildings

The buildings being constructed now are high rises and connected infrastructure, which help in accommodating maximum occupancy and with minimum space constraints. Using sensors and other smart devices help in minimizing energy consumption without compromising the comfort and safety of the users (Yarime 2017).

4.3.1.5 Logistic Solutions

Logistics is the lifeline of any city. Transportation of goods, materials and healthcare supplies to reach the end user on time is a requirement. Logistics chain should be optimized in such a way that is efficient and consumes minimum time between the origin and the consumer. Stocking, storage, transportation and distribution of the products will make the city sustainable (Alam and Khan 2020, Role of Sustainable Development in Smart Cities for a Better Life Style; Oberg and Graham 2016).

4.3.2 Real-time Implementation

Nowadays, there have already been a number of smart implementations in smart cities. In this section, we discuss a few of them and these are depicted in Figure 4.9.

4.3.2.1 Healthcare Industry

With the increase of urbanization, increases the population of the city, which in turn increases the demand for healthcare facilities and amenities. Smart healthcare combines the latest mobile devices and the smart technology with a person's health. Smart wearable devices like fitness bands, fitness trackers, health assessment devices and apps are being used. These devices monitor the health of the wearer and have the ability to provide solutions if needed at the right time. Data produced by these devices can be analysed by doctors and physicians, researchers

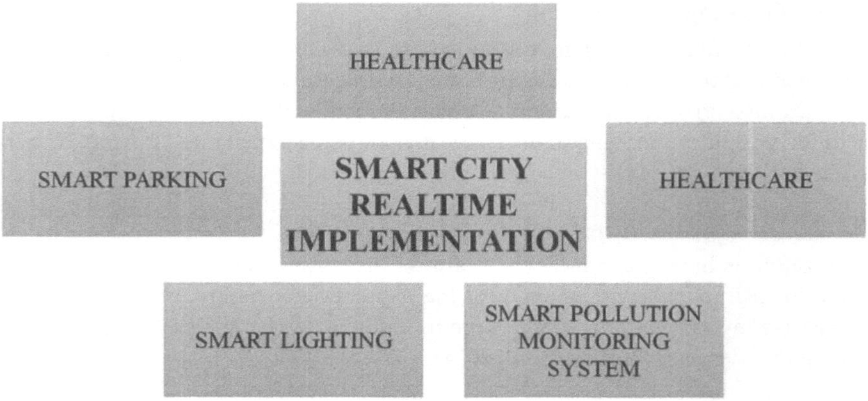

FIGURE 4.9 Smart city real-time implementations.

and healthcare professionals for better and enhanced personalized solutions and diagnosis (Hui 2016).

4.3.2.2 Smart Transport Industry

Smart transportation can help in making huge improvement in the way the passengers commute and travel in the dense urban cities. It can further help municipalities in saving costs and providing better quality services to the citizens, and in managing better safety and security of the citizens. Smart transportation uses technologies like sensors, actuators and internet to provide access to smarter, safer and faster travel. Smart transportation will help in reducing pollution as well (Rathore et al. 2015). Smart transportation will also help in fighting pandemics and diseases and help in smart tourism (Alam and Parveen 2021, COVID-19 and Tourism).

4.3.2.3 Smart Parking

With immense increase in the number of vehicles in the city, finding parking spots and utilizing parking areas to their maximum is very much required. With the correct infrastructure, sensors, security cameras and internet connectivity, cities will be able to mitigate the problems of parking in busy urban cities. Cities can share parking data on free and occupied spots in the area through a public portal or app, and users can access this data and move to the exact spot without delay. This will save time, fuel and make space usage efficient (Khanna and Anand 2016).

4.3.2.4 Smart Lighting Industry

The primary aim of implementing smart lighting in smart city paradigm was to conserve energy and making lighting devices efficient and versatile. A connected, smart lighting system is a part of a wireless, decentralized local network with intelligence. It is seamlessly connected to the internet and further to a data centre and various other management platform in the cloud. It consists of integrated devices, smart sensors and sometimes high-resolution cameras to respond to its surrounding conditions (Martirano 2011).

4.3.2.5 Smart Pollution Monitoring Systems

With people moving to the cities at unprecedented rates, and industrialization picking up the pace, pollution is also at its peak. Government needs to come up with ideas and technologies to curb the pollution. It is necessary to monitor the environmental conditions. Smart devices need to be installed and deployed to monitor air quality, water quality and soil quality across the city. As per the reading from these sensors, people and government can take corrective measures which will help in improving environmental health (Parmar, Lakhani and Chattopadhyay 2017).

4.4 AI IN SMART CITIES

Smart cities are now a reality. We have a number of smart cities around the world. But these cities still need development in one or the other sectors. Currently smart cities deploy Internet of things (IoT) (Alam, Khan and Tanweer 2020, IOT in Smart Cities: A survey 9 May), Internet of Drones (IoD) (Zhang et al. n.d.), Cloud computing (Bahrami 2015), Edge computing (Ngoko and Cerin 2017) and other technologies. AI, when implemented correctly, will play a major role in smart city development and upgradation.

AI will start to learn how people use their cities. It will start analysing the data and give predictions on the basis of previous experiences.

4.4.1 How Can AI Be Used in Smart Cities

The massive amount of big data being generated in cities every day is making AI very different to the AI of the past. Huge amount of data is the main driver for AI. When this big data is paired with efficient and robust algorithms, the capabilities of AI increase exponentially. The use of AI in smart cities is becoming interesting as the developers are integrating and creating systems that are capable of learning from the past experiences (Navarathna and Malagi 2018, Artificial Intelligence in Smart City Analysis). For example, in a smart city system where demand of energy tends to spike under certain circumstances, AI can learn where the spike is usually occurring and under what conditions and circumstances. Engineers and scientists can then make better use of the power grid. Other examples could be, by learning, AI can provide services to disabled and elderly people who might not be able to go for grocery shopping.

4.4.2 AI in Smart City Management

AI will soon play the central role in smart cities along with wireless networks and other communication medias. As discussed earlier, AI can be extensively implemented in vivid sectors from security purpose, stock market, to transportation and rescue management. AI can be the next step towards achieving sustainable development. This can be possible by using sensors, actuators, electronic devices and software driven systems in the form of AI, which read and understand the surroundings and necessary actions are taken to improve and optimize the system performance (Adio-Moses and Asaolu 2016). AI is a viable and convenient solution for design, construction, maintenance and time scheduling of transportation system. It can further be used for resolving complex transportation system problems at a faster and more efficient scale since AI can process huge amount of data in less time (Agarwal et al. 2015). Integration of AI into a system helps in providing real-time reports on different environmental parameters such as traffic accidents, predicting traffic conditions, predicting the sale of an item based on the outside weather condition, etc.

4.4.3 Transformative Applications of AI in Smart Cities

AI can play a crucial role in smart city development and urbanization. AI system, when implemented within the current smart city use cases, can help in growth and management with traffic monitoring and disposal management, advance security systems, to make the cities more and more secure and liveable. People will get access to and control of their homes and will have the peace of mind of security and safety, making their living and staying experience more comfortable and enjoyable. In this section, we take a look at some AI-enabled smart city use cases and these are depicted in Figure 4.10.

4.4.3.1 Advance Security Camera and Surveillance System

Cameras available nowadays are enabled with AI technology. These sensors and cameras can keep an eye on the surrounding environment and greatly increase the level of security in the cities. AI-enabled cameras can recognize people with their faces and help in tracking them if need be. High-resolution cameras can track the movement of vehicles and report them if they disobey the traffic and road safety rules. They can further be used to monitor crowd density at public places, which would help in maintaining social distancing wherever required. With big data available at hand, police and other officials can predict the kind of crimes and the intensity of crimes, observing all such activities in selected areas (Shidik et al. 2019).

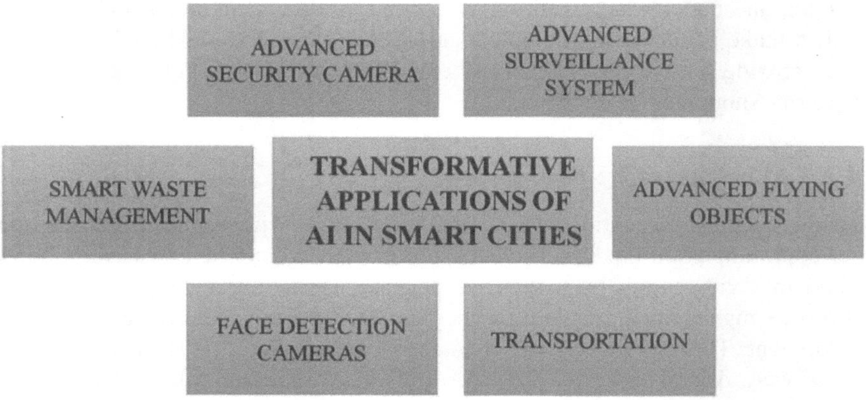

FIGURE 4.10 Applications of AI in smart cities.

4.4.3.2 Transportation Industry

Increasing traffic flows nowadays points to an important issue for the smart cities around the world. Management of the traffic requires high quality physical infrastructure combined with new ways of thinking and use of technologies. New and advanced technologies like AI and intelligent control need to be used to reduce traffic and route decision in order to curb the increased traffic congestion from the cities (Hawi, Okeyo and Kimwele 2015).

Application of AI for development of Intelligent Transport Systems is essential because it can solve problems like traffic congestion, overcrowding, environmental degradation, etc., in smart cities.

4.4.3.3 Autonomous Flying Objects for Overhead Vision

Drones enabled with AI technology can be used to keep an eye and constantly monitor the internal city and other concerning areas. They consist of cameras which help in providing real-time visuals of various places where humans cannot reach quickly and easily, thereby preventing the crime or unauthorized access. They help the authorities and security departments to take immediate and timely actions. AI-enabled drones can be used to track humans as well as animals for their safety and wellbeing. They can help in monitoring the movement of the traffic and lay out two-dimensional aerial view which can be utilized for better urbanization of the cities (Lee, Gyu La and Kim 2018).

4.4.3.4 Face Detection Cameras for Public Safety

Face recognition technology can exploit the AI technology and can be used to make smart cities smarter. It is capable of detecting divergent people with their facial features, which discloses their personal identities as and when needed. Images captured by security cameras and flying drones can be used to recognize the various facial features and then compare them to a central database with the information. This helps in tracing the identity of the individual and authenticate people entering or leaving a city, society, restricted area or even a historical monument (Anghelescu, Serbanescu and Ionita 2013).

4.4.3.5 Smart Waste Management and Disposal System

As urbanization increases, the population also increases. This increase in population leads to production of increased amount of waste. Proper management and disposal of waste and garbage is a very important aspect of a smart city. It helps in keeping the environment clean and in maintaining the hygiene of the society (Anuj, Alam and Khan 2021). AI technology can detect the nature of the trash lying on the roads and pavements. They can also detect the type of waste for grouping purpose. Installation of sensors, enabled with AI, on dustbins can make the process of garbage collection more hygienic and efficient. Authorities responsible for cleaning can receive notifications in the office as and when the waste bins are almost full. This will reduce operational cost by eliminating unnecessary pickups. AI-enabled system will also provide dynamically updated garbage collection routes and schedules which will help in optimization of waste management (Pardini et al. 2020).

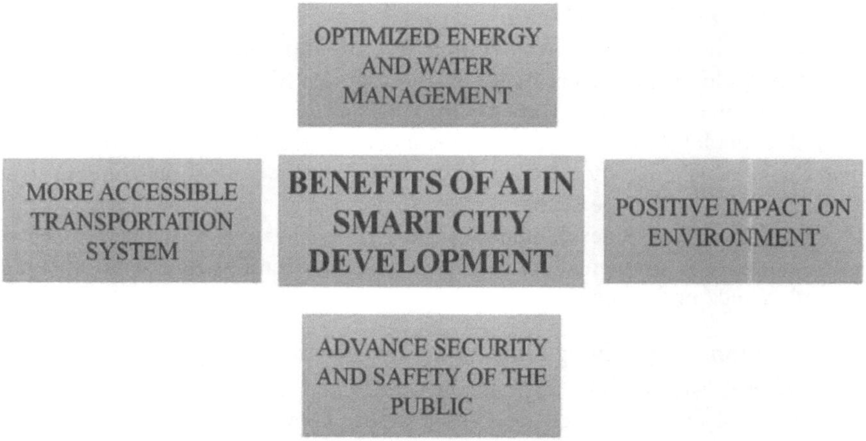

FIGURE 4.11 Benefits of AI in smart city development.

4.4.4 Benefits of AI in Smart City Development

Integration of AI into smart cities has a large number of benefits for the humans, animals and the environment. It will help us make an eco-friendly environment which will lead to sustainable development. Implementation of AI in smart cities will come with various types of advantages for the society. Some of them are discussed below and are depicted in Figure 4.11 (Khan et al. 2018; Navarathna and Malagi 2018, Artificial Intelligence in Smart City Analysis).

4.4.4.1 Good Impact on the Environment

One of the most essential benefit of smart city and AI integration is the fact that we will be able to control the CO_2 emission. Since the increase in the CO_2 emission is one of the biggest concerns throughout the globe, any measure to curb it is seen as a sustainable task. Smart cities and AI with energy efficient, eco-friendly, better traffic and waste management and smart lights and appliances will help us reduce pollution, which will help in making the environment less polluted and will conserve energy.

4.4.4.2 Optimized Energy and Water Management

Power generating grids and water management system are the major producers of energy. Implementation of AI and making these infrastructures will make them more efficient and smarter. It will help in reducing pollution by maximizing output. It will also help in getting clean drinking water and keeping our environment clean.

4.4.4.3 Greater Accessibility to the Transportation System

Transportation is the lifeline of any city. AI-powered smart transportation system will help in reducing fuel consumption and waste emission by providing the shortest and fastest route available. Smart use cases help in goods transportation in required time. Smart public transportation will help users save time by checking the timing and location of the transport in real-time and planning their travel accordingly.

4.4.4.4 Advanced Safety and Security of the Public

Security and safety of the people is the main concern for any city. AI-powered smart cities are equipped with high-end smart cameras which help in 24×7 surveillance of areas under the scanner. Facial recognition upgrade to the cameras helps recognize people who disobey the law or enter restricted areas. AI-powered cameras are equipped with motion and smoke detection technologies which further help in keeping the city safe.

4.4.5 AI Smart City Challenges

Implementing and achieving AI technology in smart cities is not an easy task. There are a large number of challenges in making cities smart, let alone implementation of AI in them (Alam et al. 2020). In this section, we will discuss few challenges for AI implementation and these are depicted in Figure 4.12 (Khan et al. 2018) (Shrivastava, Bisht and Narayan 2017).

4.4.5.1 Infrastructure and Cost

Cameras, sensors, actuators, drones, etc. used in making the cities smarter currently cost a fortune. These devices are priced high and the installation cost only increases it to new highs. Further, implementation of AI systems again needs huge investments. AI systems require high-end computer systems, databases and energy to function, everything adding to the overall cost.

4.4.5.2 Security and Privacy Concern

Another concern in AI-powered smart cities is the security and privacy concern for the local people. With cameras everywhere and drones hovering above them all the time, people may get offended and may get concerned for their privacy and their personal space. The recordings and data being saved on the server may get compromised and stolen by cyber-attacks.

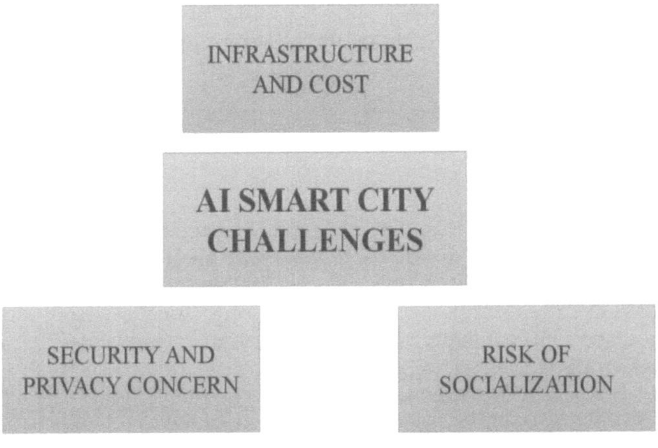

FIGURE 4.12 AI smart city challenges.

4.4.5.3 Risk of Socialization

In the process of developing such cities, one must give priority to inclusive urbanization, which deals with increasing vulnerability of the poor and the slum population. It should be made sure that no population should be excluded from the big data collected for the AI systems. Including all age groups, gender, class and income group of people from the society is very important.

4.4.6 Who Will Get Access to AI?

With the current trend for AI adoption and utilization, sooner or later AI is going to become a utility. It is going to become as normal and typical as cloud computing. Cloud computing has become a standard and almost every business and companies have accepted it. But it took nothing less than 10 years to penetrate to the levels it is currently being used in. Similarly, AI needs time to be accepted and implemented by the businesses, governments and societies. Currently, if we look at AI, it is Facebook, Google, Apple and Microsoft. Only a few leading companies have been working on AI technology but a huge number of start-ups are coming up with great and innovative ideas (Liu et al. 2018).

4.5 CONCLUSION

Smart city is the future of our cities. Smart city concept is being adopted and implemented by most countries worldwide. With the increase in the number of smart cities throughout the world, AI and its implementation in making the cities secure and comfortable for the citizens has to be the next step. AI will help the smart city engineers, organizers and caretakers to check and maintain these cities and their use cases. This work recognizes the transformative ability of AI when it comes to the smart city context. It can prove to be a paradigm shifting force that can revolutionize mobility in an unprecedented way. AI systems should operate and function with a responsible, sustainable and user-centred architectural framework that understands and satisfies the human user.

The promise of an autonomous, connected, shared and digitized transport service provision is not enough if it does not facilitate efforts that lead to improved environmental conservation, resource efficiency, productivity gains, social inclusion, integration, health and wellbeing.

REFERENCES

Adio-Moses, D, and O S Asaolu. 2016. "Intelligence for Sustainable Development of Intelligent Buildings." *Proceedings of the 9th CIDB Postgraduate Conference*. University of Cape Town, South Africa.

Agarwal, P K, J Gurjar, A K Agarwal, and R Birla. 2015. "Application of artificial intelligence for development of intelligent transport system in smart cities." *Journal of Traffic and Transportation Engineering* 1 (1): 20–30.

Alam, M, and I R Khan. 2019. "Internet of Things as key enabler for efficient business processes." *SSRN Electronic Journal*. Doi:10.2139/ssrn.3806408

Alam, M, and I R Khan. 2020. "Business 4.0-A New Revolution." In *Information Technology for Management*. Delhi: KD Publications. Doi:10.6084/m9.figshare.14369636

Alam, M, and I R Khan. 2020. "Role of Sustainable Development in Smart Cities for a Better Life Style." In *Environmental Sustainability and Human Health*. Delhi: SRF Research Journal & Book Publication House. Doi:10.6084/m9.figshare.14369633

Alam, M, and I R Khan. 2021. "Blockchain for Indian agriculture: A revolution." *Information Technology in Industry* 9 (2): 513–518. Doi:10.17762/itii.v9i2.378

Alam, M, and R Parveen. 2021. "COVID-19 and tourism." *International Journal of Advanced Research* 9 (4): 788–804. Doi:10.21474/IJAR01/12766

Alam, M, A H Khan, and I R Khan. 2016. "Swarm intelligence in MANETs: A survey." *International Journal of Manufacturing Technology and Management* 5 (5): 141–150. Doi:10.6084/m9.figshare.14309384

Alam, M, I R Khan, S A Siddique, R Wiquar, and H Anwar. 2020. "IoT and AI as Key Enabler of Growth of Smart Cities." *2nd International Conference on ICT for Digital, Smart and Sustainable Development*. New Delhi. Doi:10.4108/eai.27-2-2020.2303467

Alam, M, I R Khan, and S Tanweer. 2020. "IOT in smart cities: A survey." *Juni Khyat* 10 (5): 89–101. Doi:10.6084/m9.figshare.14329718

Alam, M, I R Khan, and S Tanwer. 2020. "Blockchain technology: A critical review and its proposed use in E-voting in India." *SSRN Electronic Journal*. Doi:10.2139/ssrn.3570320

Alam, M, R Parveen, and I R Khan. 2020. "Role of information technology in Covid-19 prevention." *International Journal of Business Education and Management Studies* 5 (1): 65–75. Doi:10.6084/m9.figshare.14369627

Al-Hader, M, A Rodzi, A R Dharif, and N Ahmad. 25–27 Nov 2009. "SOA of Smart City Geospatial Management." Computer Modeling and Simulation, EMS '09. Third UKSim European Symposium.

Amazon n.d.. https://en.wikipedia.org/wiki/Amazon_Alexa.

Anghelescu, P, I Serbanescu, and S Ionita. 2013. "Surveillance system using IP camera and face-detection algorithm." *Proceedings of the International Conference on Electronics, Computers and AI*. Pitesti, Romania.

Anuj, R, M Alam, and I R Khan. 2021. "Smart Garbage Monitoring System using IoT." *2nd International Conference on Emerging Trends in Mathematical Sciences & Computing (IEMSC-21)*. Kolkata, West Bengal. Doi:10.2139/ssrn.3902056

Arora, M, S Nath, S Mazumdar, S B Baden, and D M Tullsen. 2012. "Redefining the role of the CPU in the era of CPU-GPU integration." *IEEE Micro* 32 (6): 4–16. Doi:10.1109/MM.2012.57

Bahrami, M. 2015. "Cloud Computing for Emerging Mobile Cloud Apps." *3rd IEEE International Conference on Mobile Cloud Computing, Services, and Engineering*. Doi:10.1109/mobilecloud.2015.40

Binning, G, M Baatz, and J A Schmidt. 2002. "Artificial versus natural intelligence." *Europhysics News* 33 (2): 1–7.

Bourlard, H A, and N Morgan. 1993. *Connectionist Speech Recognition*. Boston, MA: Springer US.

Davenport, T, A Guha, D Grewal, and T Bressgott. 2019. "How artificial intelligence will change the future of marketing." *Journal of the Academy of Marketing Science* 48 (1): 24–42.

Evans, G. 1991. "Solving home automation problems using artificial intelligence techniques." *IEEE Transactions on Consumer Electronics* 37 (3): 395–400. Doi:10.1109/30.85542

Godber, E. 2018. "Uses of Artificial Intelligence in Health." *International Conference on Artificial Intelligence Applications and Innovations (IC-AIAI)*. Nicosia, Cyprus. 31–36. Doi:10.1109/IC-AIAI.2018.8674444

Google. n.d. Google AI. Google. https://ai.google/

Gupta, S, S Rani, and A Dixit. 2019. "Recent Trends in Automation-A study of RPA Development Tools." *3rd International Conference on Recent Developments in Control, Automation & Power Engineering (RDCAPE)*. Noida, India. Doi:10.1109/RDCAPE47089.2019.8979084

Harkut, D G, and K Kasat. 2019. "Introductory Chapter: Artificial Intelligence - Challenges and Applications." In D. G. Harkut (Ed.), *Artificial Intelligence - Scope and Limitations*. IntechOpen. Doi: 10.5772/INTECHOPEN.84624

Hawi, R, G Okeyo, and M Kimwele. 2015. "Techniques for smart traffic control: An in-depth review." *International Journal of Computer Applications Technology and Research* 4 (7): 566–573.

Hui, K. 2016. "A secure IoT-based healthcare system with body sensor networks." *IEEE Access* 4: 10288–10299. Doi:10.1109/access.2016.2638038

IDC. 2020. IDC Forecasts Strong 12.3% Growth for AI Market in 2020 Amidst Challenging Circumstances. IDC. 4 August. https://www.idc.com/getdoc.jsp?containerId=prUS46757920

Io, H N, and C B Lee. 2017. "Chatbots and Conversational Agents: A Bibliometric Analysis." *IEEE International Conference on Industrial Engineering and Engineering Management (IEEM)*. Singapore. Doi:10.1109/IEEM.2017.8289883

Islam, T, and D Manivannan. 2017. "Predicting Application Failure in Cloud: A Machine Learning Approach." *IEEE International Conference on Cognitive Computing (ICCC)*. Honolulu, USA. Doi:10.1109/IEEE.ICCC.2017.11

Jarek, K, and G Mazurek. 2019. "Marketing and artificial intelligence." *Central European Business Review* 8 (2): 46–55. Doi:10.18267/j.cebr.213

Khan, I R, A Goyal, and M Alam. 2021. "Augmented Reality Application for Newspapers." *International Conference on Engineering, Applied Sciences and Management*. Buraimi, Sultanate of Oman. 252–258.

Khan, S, D Paul, P Momtahan, and M Aloqaily. 2018. "Artificial Intelligence Framework for Smart City Microgrids: State of the Art, Challenges, and Opportunities." *Third International Conference on Fog and Mobile Edge Computing (FMEC)*. Barcelona, Spain.

Khanna, A, and R Anand. 2016. "IoT Based Smart Parking System." *International Conference on Internet of Things and Applications (IOTA)*. Pune. 266–270. Doi:10.1109/IOTA.2016.7562735

Kim, H, R Vuduc, S Baghsorkhii, J Choi, and W Hwu. 2012. "Performance analysis and tuning for general purpose graphics processing units (GPGPU)." *Synthesis Lectures on Computer Architecture* 7 (2): 1–96. Doi:10.2200/S00451ED1V01Y201209CAC020

Lee, D, W G La, and H Kim. 2018. "Drone Detection and Identification System using Artificial Intelligence." *International Conference on Information and Communication Technology Convergence (ICTC)*. Jeju, South Korea.

Liu, J, X Kong, F Xia, X Bai, L Wang, Q Qing, and I Lee. 2018. "Artificial intelligence in the 21st century." *IEEE Access* 6: 34403–34421.

Liu, S Y 2020. "Artificial intelligence (AI) in agriculture." *IT Professional* 22 (3): 14–15.

Ma, Y, Z Wang, H Yang, and L Yang. 2020. "Artificial intelligence applications in the development of autonomous vehicles: A survey." *IEEE/CAA Journal of Automatica Sinica* 7 (2): 315–329. Doi:10.1109/JAS.2020.1003021

Martirano, L. 2011. "A Smart Lighting Control to Save Energy." *Proceedings of the 6th IEEE International Conference on Intelligent Data Acquisition and Advanced Computing Systems*. Prague. 132–138.

Mellish, C. 1998. "Natural language generation." *IEEE Colloquium Speech and Language Engineering - State of the Art*, (Ref. No. 1998/499), 5: 1–5. Doi: 10.1049/ic:19980959

Mijwil, M M. 2015. "History of Artificial Intelligence." *Computer Science, College of Science* 1(6): 1–6.

Narayanan, S. 2019. "AI in image analytics." *International Journal of Computational Intelligence Research* 15 (2): 81–95.

Navarathna, P J, and V P Malagi. 2018. "Artificial Intelligence in Smart City Analysis." *International Conference on Smart Systems and Inventive Technology (ICSSIT)*. Tirunelveli, India. Doi:10.1109/ICSSIT.2018.8748476

Neirotti, P, A De Marco, A C Cagliano, G Mangano, and F Scorrano. 2014. "Current trends in smart city initiatives: Some stylised facts." *Cities* 38: 25–36.

Ngoko, Y, and C Cerin. 2017. "An Edge Computing Platform for the Detection of Acoustic Events." *IEEE International Conference on Edge Computing (EDGE)*. Doi:10.1109/ieee.edge.2017.44

Nguyen, T L, and T T H Do. 2019. "Artificial Intelligence in Healthcare: A New Technology Benefit for Both Patients and Doctors." *Portland International Conference on Management of Engineering and Technology (PICMET)*. Portland, OR, USA. 1–15. Doi:10.23919/PICMET.2019.8893884

Oberg, C, and G Graham. 2016. "How smart cities will change supply chain management: A technical viewpoint." *Production Planning & Control* 27 (6): 529–538.

Obulesu, O, M Mahenrda, and M ThrilokReddy. 2018. "Machine Learning Techniques and Tools: A Survey." *International Conference on Inventive Research in Computing Applications (ICIRCA)*. Coimbatore. 605–611. Doi:10.1109/ICIRCA.2018.8597302

Pardini, K, J P C Rodrigues, O Diallo, A K Das, V H C de Albuquerque, and S A Kozlov. 2020. "A smart waste management solution geared towards citizens." *Sensors* 20 (8): 2380.

Parmar, G, S Lakhani, and M K Chattopadhyay. 2017. "An IoT Based Low Cost Air Pollution Monitoring System." *International Conference on Recent Innovations in Signal processing and Embedded Systems (RISE)*. Bhopal. 524–528.

Pau, L F. 1991. "Artificial intelligence and financial services." *IEEE Transactions on Knowledge and Data Engineering* 3 (2): 137–148. Doi:10.1109/69.87994

Poushneh, A, and A Z Vasquez-Parraga. 2017. "Discernible impact of augmented reality on retail customer's experience, satisfaction and willingness to buy." *Journal of Retailing and Consumer Services* 34: 229–234. Doi:10.1016/j.jretconser.2016.10.005

Raina, R, A Madhavan, and A Y Ng. 2009. "Large-scale Deep Unsupervised Learning using Graphics Processors." *26th International Conference on Machine Learning*. Montreal, Canada.

Rathore, M M, A Ahmad, A Paul, and G Jeon. 2015. "Efficient Graph-Oriented Smart Transportation Using Internet of Things Generated Big Data." *11th International Conference on Signal-Image Technology & Internet-Based Systems (SITIS)*. Bangkok. 512–519. Doi:10.1109/SITIS.2015.121

Riedl, M O, and A Zook. 2013. "AI for Game Production." *IEEE Conference on Computational Intelligence in Games (CIG)*. Niagara Falls. 1–8. Doi:10.1109/CIG.2013.6633663

Riva, G, and B K Wiederhold. 2015. "The new dawn of virtual reality in health care: Medical simulation and experiential interface." *Studies in Health Technology and Informatics* 219: 3–6. Doi:10.3233/978-1-61499-595-1-3

Rosario, M, S C Mukhopadhyay, Z Liu, D Slomovitz, and S R Samantaray. 2017. "Advances on sensing technologies for smart cities and power grids: A review." *IEEE Sensors Journal* 17: 7596–7610.

Schreieck, M, H Safetli, S A Siddiqui, C Pflugler, M Wiesche, and H Krcmar. 2016. "A matching algorithm for dynamic ridesharing." *Transportation Research Procedia* 19: 272–285.

Science, Narrative. 2020. *Outlook on Artificial Intelligence in the Enterprise*. Narrative Science.

Shidik, G F, E Noersasongko, A Nugraha, P N Andono, J Jumanto, and E J Kusuma. 2019. "A systematic review of intelligence video surveillance: Trends, techniques, frameworks, and datasets." *IEEE Access* 7: 170457–170473.

Shrestha, A, and A Mahmood. 2019. "Review of deep learning algorithms and architectures." *IEEE Access* 7: 53040–53065. Doi:10.1109/ACCESS.2019.2912200

Shrivastava, S, A Bisht, and N Narayan. 2017. "Safety and Security In Smart Cities Using Artificial Intelligence — A Review." *7th International Conference on Cloud Computing, Data Science & Engineering - Confluence*. Noida, India.

Siri. Apple n.d.. https://en.wikipedia.org/wiki/Siri.

Suryanarayana, S V, G N Balaji, and G V Rao. 2018. "Machine learning approaches for credit card fraud detection." *International Journal of Engineering & Technology* 7 (2): 917. Doi:10.14419/ijet.v7i2.9356

Tabane, E, S M Ngwira, and T Zuva. 2016. "Survey of Smart City Initiatives Towards Urbanization." *International Conference on Advances in Computing and Communication Engineering (ICACCE).* Durban.

Vijai, C. 2019. "Artificial intelligence in Indian banking sector: Challenges and opportunities." *International Journal of Advanced Research* 7 (4): 1582–1587. Doi:10.21474/IJAR01/8987

Wang, F-Y, J J Zhang, X Zheng, X Wang, Y Yuan, X Dai, J Zhang, and L Yang. 2016. "Where does AlphaGo go: From church-turing thesis to AlphaGo thesis and beyond." *IEEE/CAA Journal of Automatica Sinica* 3 (2): 113–120. Doi:10.1109/JAS.2016.7471613

Wang, X, X Li, and V C M Leung. 2015. "Artificial intelligence-based techniques for emerging heterogeneous network: State of the arts, opportunities, and challenges." *IEEE Access* 3: 1379–1391. Doi:10.1109/ACCESS.2015.2467174

Wikipedia. n.d. AlphaGo. Deepming Technologies. https://en.wikipedia.org/wiki/AlphaGo.

Wilcock, G. 2009. *Introduction to Linguistic Annotation and Text Analytics.* San Rafael, California: Morgan & Claypool.

Yarime, M. 2017. "Facilitating data-intensive approaches to innovation for sustainability: Opportunities and challenges in building smart cities." *Sustainability Science* 12 (6): 881–885.

Zha, X, and S Sahni. 2013. "GPU-to-GPU and host-to-host multipattern string matching on a GPU." *IEEE Transactions on Computers* 62 (6): 1156–1169. Doi:10.1109/TC.2012.61

Zhang, G P. 2000. "Neural networks for classification: A survey." *IEEE Transactions on Systems, Man, and Cybernetics, Part C (Applications and Reviews)* 30 (4): 451–462. Doi:10.1109/5326.897072

Zhang, P, C Wang, Z Qin, and H Cao. n.d. "A multidomain virtual network embedding algorithm based on multiobjective optimization for Internet of Drones architecture in Industry 4.0." *Software: Practice and Experience.* Doi:10.1002/spe.2815

Zohuri, B, and F M Rahmani. 2020. "Artificial intelligence versus human intelligence: A new technological race." *Acta Scientific Pharmaceutical Sciences* 4 (5): 50–58.

5 Effect of Sustainable Energy Sources for Load Frequency Control (LFC) of Single-Area Wind Power Systems

D. Boopathi, K. Jagatheesan, B. Anand, V. Kumarakrishnan, and Sourav Samanta

CONTENTS

5.1 INTRODUCTION

Demand of the electric power always upturns because of industrial development and urbanization. In order to meet the load, demand generation has to be increased. While increasing power generation, complications also increase; complications like power quality (standard limit voltage, frequency) and quantity. Mostly frequency

DOI: 10.1201/9781003229018-5

deviation is a big issue; to control the frequency deviation, load frequency control (LFC) scheme has to be adopted in power generating units. For LFC scheme, many research peoples are utilized different controllers as secondary controller and also, various computation techniques to optimize controller gain parameters [1, 2]. Some of the particular researchers' works are considered in this literature review to find the problem for the proposed research work.

Grasshopper optimization algorithm-based integral controller developed for LFC issues of three-area unequal sources power network, and result compared with PSO and GA techniques by [3]. Distributed mode predictive controller (DMPC) has been demonstrated for LFC of wind form by [4]. Moth swarm algorithm-based PID controller is adopted by [5] for real-time single-area Egyptian power network with thermal, wind form and hydro sources to control the oscillation of frequency at emergency loading time. Author in [6] clearly explained the effect of flywheel-energy-storage-system (FESS) and fuel cell (FC) with wind turbine generator (WTG) while examining the LFC scheme with integral controller. A micro grid examined by integral controller for LFC issue in [7] and effects of each power sources such as wind, PV, ESS and synchronous generator are analyzed.

Battery energy storage system (BESS) in a micro grid has performed well to control the frequency oscillation with wind generator [8]; LFC problem of a hybrid power system (thermal, aqua electrolyzer [AE], WTG, FC, diesel engine generator [DEG] and BESS) is solved by using PSO-tuned linear matrix in equality-based controller in [9]. Sunflower optimization (SFO) techniques optimized by fuzzy PDF-PI controller is implemented by [10] for AGC issue of three-area grid-connected power network; performance improvement of the proposed controller is found out by comparing with GA, PSO technique-based simulation results. An ant colony optimization (ACO)-optimized PID controller designed for single-area nuclear power network LFC is demonstrated in [11]; PSO-PID controller is demonstrated in [12] for LFC problem of a multi-source single-area power network with thermal, gas, hydro, nuclear sources with solar energy.

Classical controllers such as I, PI, PID controllers area designed for solving AGC problem of an unequal three-area power generating network [13], for AGC problem of equal six area thermal power network, ACO-optimized PID controller was developed by considering different cost function in [14]; result shows that integral-time-absolute-error (ITAE) cost function gives better performance than other cost functions. Chaotic firefly algorithm (CFA)-PID controller was developed for interconnected thermal power network AGC problem in [15], ant bee colony-PID regulator was employed for AGC of three-area reheated thermal power network in [16]. ACO-optimized PID controller was designed for AGC of grid-connected power network consisting of non-linearity [17], and PSO technique was also implemented for various applications like medical field for dengue fever classification in [18, 19]. Author in [20] solved the LFC crises of a micro grid power network by adopting the PSO technique-based PID controller, and the results are compared with conventional method. Bacterial foraging optimization technique-based controller was implemented for AGC of two-area power network in [21].

A dynamic gain-tuned control method is involved in the wind energy system in [22]. A two-area wind power plant with energy storage system is investigated for AGC problem by using the classical controller in [23]. In [24], the impact of time-to-time variations of wind energy system on the power network operation is analyzed. The impact of wind energy on the power system with fast variations and the uncertainty has been studied in [25]. Firefly algorithm-optimized PID controller was involved in [26] for the AGC issue of a two-area interconnected thermal power plant; in [27], online wavelet filter is included with the thermal power plant. A Quantum-inspired evolutionary algorithm-based PID controller is designed in [28] for the AGC problem of interconnected power grid. From the above literature review, we can clearly see that a secondary controller with optimization technique is much needed for solving AGC/LFC problem of power system. The above literature review helps to solve the AGC of the proposed power system.

5.1.1 Structure of the Article

This chapter is constructed as follows: In Section 5.1, introduction of the research work and the literature review for finding the research gap is analyzed in detail. Section 5.2 illustrates the proposed power system design and mathematical equations of the model. The controller design with cost function is discussed in Section 5.3. In Section 5.4, details of the proposed optimization technique (ACO) along with tuned gain parameters of the controller and the performance are discussed. Simulation result analyses with comparisons are explained in detail in Section 5.5. Finally, conclusion is given in Section 5.6.

5.1.2 Highlights of the Research Work

- Zero carbon (wind) power generating system is considered as a leading power source in this research work.
- The proposed optimization technique supremacy is verified by comparing response with conventional and PSO-tuned controller response.
- The impact of the energy storage units (FC and BESS) is in detail analyzed in wind power system during emergency load situation.

5.2 SYSTEM MODELLING

A single-area multi-source power system including wind energy, FC and BESS is developed to investigate LFC. Wind energy system is a nonlinear power production unit; it needs a regulator to control the power output. Due to this property, the proposed power system produces many oscillations during time-varying load. To control the oscillation, primary controller like speed regulator is adopted in the system, but that controller is not enough to control the oscillation during the sudden loading time. So this proposed power system has secondary controller such as PID controller. The Simulink diagram for the proposed power network is shown in Figure 5.1. The nominal parameters in the mathematical model are given in annexure 1.

5.3 CONTROLLER STRUCTURE AND COST FUNCTION

The PID regulator is executed to implement the LFC scheme of the proposed network because this controller is efficient for the industrial application and cost effective. It incorporates proportional controller, integral controller and derivative controller.

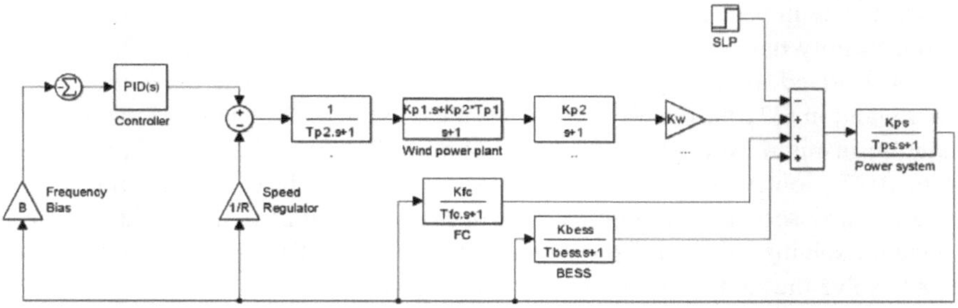

FIGURE 5.1 Simulink diagram of single-area power network (wind, FC and BESS).

Equation 5.1 shows the mathematical expression of the controller. A simple structure of the PID regulator is shown in Figure 5.2.

$$G_C(s) = k_P + \frac{K_I}{s} + \frac{K_D}{s} \tag{5.1}$$

Where

K_p = P controller gain value
K_i = I controller gain value
K_d = D controller gain value

ITAE is an objective function applied for getting the optimal controller gain parameters of the proposed controller for its better performance. The ITAE cost function provides minimum overshoots and oscillations during sudden loading conditions. The mathematical expression of ITAE objective function is given in Equation 5.2. In the equation, e (t) represents error signal of power system.

$$\text{JITAE} = \int t.|ACE|\,dt \tag{5.2}$$

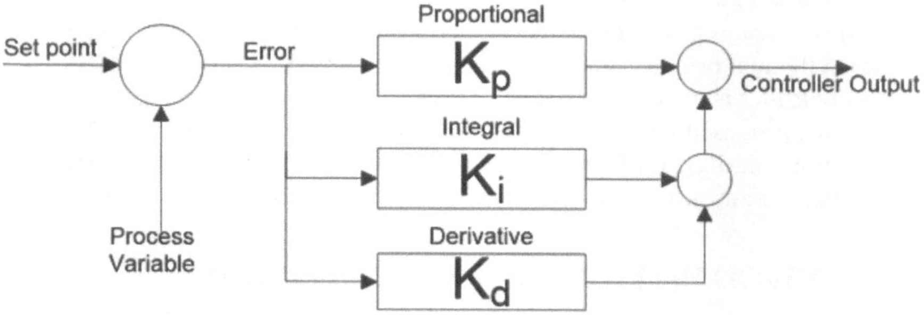

FIGURE 5.2 Simple structure of PID controller.

TABLE 5.1
Conventional (Simple Method) Controller Gain Parameter

Gain Parameter/Controller	K_P	K_I	K_D
I	-	5	-
PI	7.5	5	-
PID	7.5	5	1

5.4 TUNING OF CONTROLLER GAIN VALUES

5.4.1 Conventional Tuning Method

The conventional tuning method is a modest method for acquiring the finest gain values of controller. To tune the controller gain values, firstly tune the integral gain KI value by keeping KP and KD as zero. Afterwards, the proportional gain value KP is tuned by keeping KI constant and KD as zero. Later to find derivative gain value KD, keep KP and KI constant [11]. The optimized controller gain parameters are reported in Table 5.1 and performance index curve for PID controller is shown in Figure 5.3.

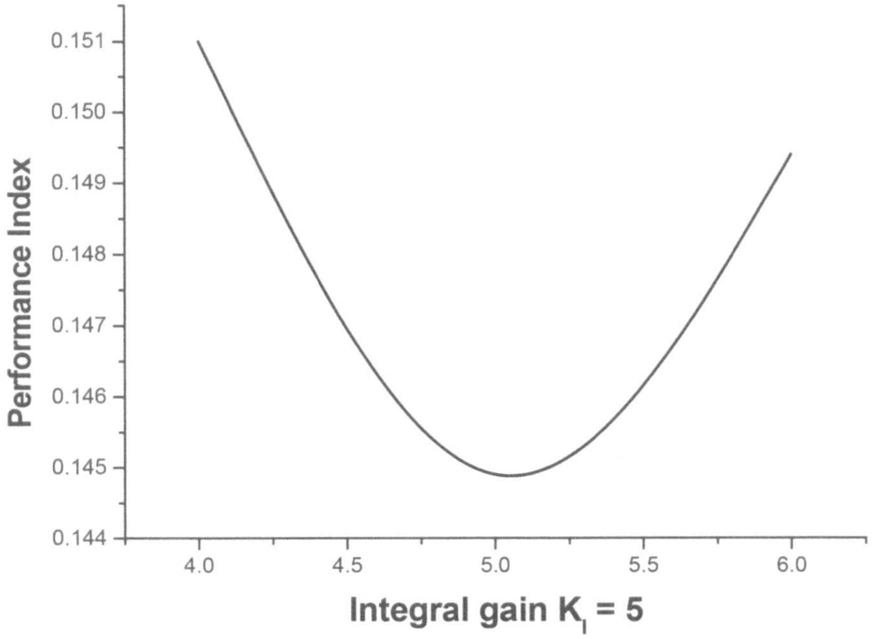

FIGURE 5.3 Performance index curve.

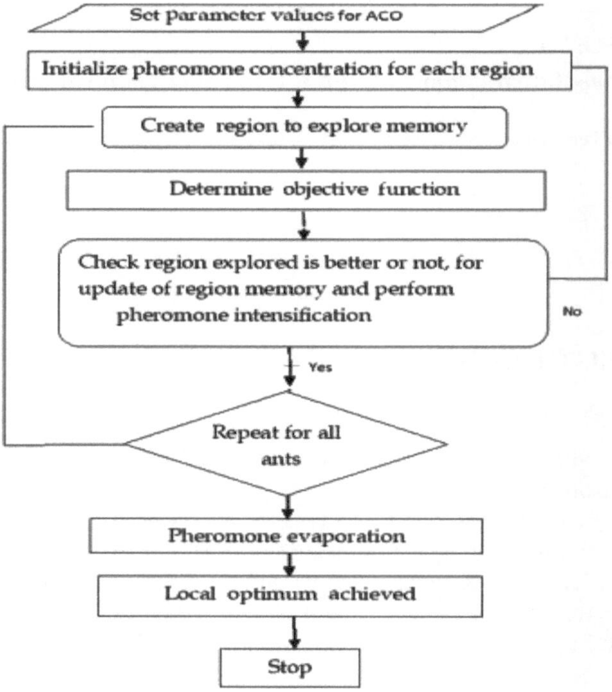

FIGURE 5.4 Flowchart of ACO technique.

5.4.2 ACO Method

Ant colony optimization is a probabilistic technique, developed for searching the shortest path in the graph based on the behavior of ants searching a path between their colony and source of prey. In swarm intelligence approaches, this algorithm is a member of the ant colony algorithm family, and it has certain metaheuristic optimizations. Marco Dorigo proposed it in his doctoral thesis in 1992 [29, 30]. The functional flowchart of the ACO algorithm is shown in Figure 5.4. This algorithm explains the behavior of the ants with the design of an artificial ant. In nature, the real ants search for the food center with different paths based on the pheromone concentration in the paths; ants find the shortest path to the food center from the nest.

The optimum gain values of PID controller by using ACO technique and ITAE objective function are acquired and reported in Table 5.2.

TABLE 5.2
ACO-based PID Controller Gain Values

Gain Value/PID Controller	K_P	K_I	K_D
Without FC+BESS unit	9.8	8.8	0.5
With FC+BESS unit	10	7.4	0.1

5.5 SIMULATION AND RESULT DISCUSSION

The proposed power system Simulink model is studied in MATLAB 2014a version, simulated for 60 seconds with one percentage step-load-perturbation (1% SLP). The results of the proposed power system which is tuned by optimization techniques are discussed here as three different cases. In case 1, performance of the conventional tuned I, PI and PID controllers are studied. In case 2, conventional, PSO and ACO-based PID controller are compared. In case 3, the impact of the energy storage units in the power system is analyzed.

5.5.1 Case 1: Performance Comparison of Conventional Tuned I, PI, and PID Controller

The conventional method-based I, PI and PID controller's performance were compared to ensure the dominance of the PID regulator over other conventional controllers. The conventional controller time domain parameters such as relaxing time, overshoot and undershoots are reported in Table 5.3. The conventional controller frequency response is shown in Figure 5.5.

From Figure 5.5 and Table 5.3, it is clear that conventional method-based PID controller performs better than other controllers.

5.5.2 Case 2: Performance Comparison of Conventional, PSO, and ACO Tuned PID Controller

In this case, the performance of conventional, PSO and ACO tuned PID controller is compared to find the superiority of tuning method for controlling the frequency deviation of system during emergency loading conditions. The graphical comparison is plotted in Figure 5.6 and the numerical values (time domain parameters) are reported in Table 5.4.

The superiority of the proposed ACO-based PID controller performance is evaluated by comparing simulation results of conventional method (simple method) and PSO-based PID controller. From the comparison of graph in Figure 5.6 and numerical values in Table 5.4, it is clearly identified that ACO-based PID controller performs better than conventional and PSO methods in terms of quick settling from the oscillation of frequency during unexpected loading conditions.

TABLE 5.3
Time Domain Parameters of Conventional Controller

Time Domain Parameters/ Controller	TS (S)	OS (Hz)	US (Hz)
I	45	0.025	0.098
PI	25.5	1×10^{-3}	0.077
PID	24.5	0.9×10^{-3}	0.0756

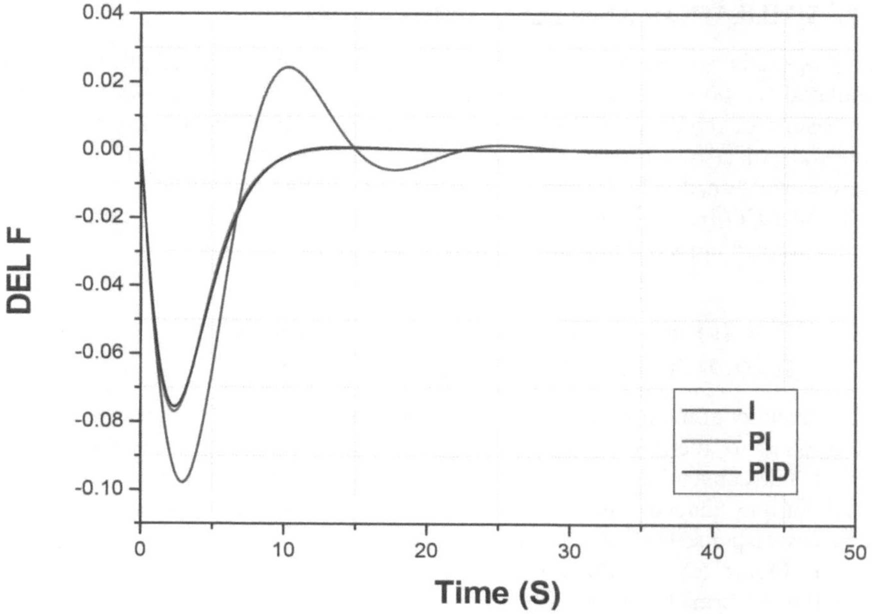

FIGURE 5.5 Frequency response of conventional controllers.

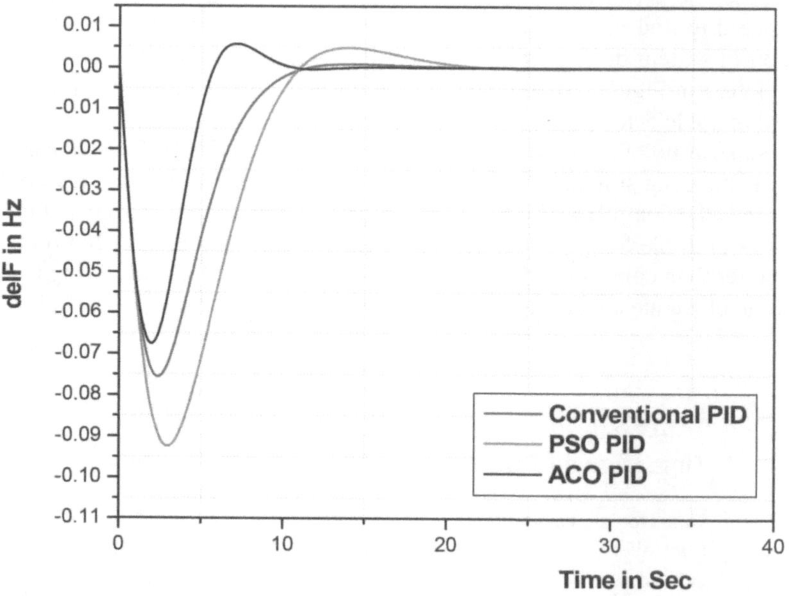

FIGURE 5.6 Frequency response of conventional, PSO and ACO PID controller.

TABLE 5.4
Time Domain Parameters of Conventional and PSO PID Controller

Time Domain Parameters/ Controller	TS (S)	OS (Hz)	US (Hz)
Conventional–PID	24.5	0.9×10^{-3}	0.0756
PSO–PID	19	6×10^{-3}	0.0685
ACO-PID	17	5.8×10^{-3}	0.0675

5.5.3 Case 3: Performance Comparison of ACO Tuned PID Controller with/without FC and BESS

Effects of the FC and BESS with wind power plant are analyzed by the proposed ACO-PID controller. Adding FC and BESS with wind form takes more time to settle the oscillation than wind form alone, because FC and BESS are nonlinear power generating sources. The optimized parameters are reported in Table 5.5. Graphical comparison of the wind power system with FC and BESS is shown in Figure 5.7.

5.6 CONCLUSION

The studied system elucidates the LFC for single-area wind power system including FC and BESS unit. The ACO-based PID controller is studied for proposed power system. At first, the classical controllers' results were analyzed and it was found that PID controller is superior to other classical controllers. Then the dominance of ACO-PID regulator is demonstrated by comparing the performance of classical PID and PSO-PID regulator; it is shown that the proposed ACO-PID controller is superior. Further, ACO tuned PID regulator is deliberated for wind power plant with/without FC and BESS unit.

5.7 ANNEXURE 1

$$
\begin{aligned}
K_{\mathrm{p1}} &= 1.25, K_{\mathrm{p2}} = 1.4, T_{\mathrm{p1}} = 0.6\mathrm{S}, T_{\mathrm{p2}} = 0.041\mathrm{S}, K_{\mathrm{BESS}} \\
&= -1/300, K_{\mathrm{FC}} = 1/100, T_{\mathrm{BESS}} = 0.1\mathrm{S}, T_{\mathrm{FC}} = 4\mathrm{S}, K_{\mathrm{PS}} \\
&= 120, T_{\mathrm{PS}} = 20\mathrm{S}, K_{\mathrm{W}} = 0.125, R = 2.4, B = 0.045.
\end{aligned}
$$

TABLE 5.5
Time Domain Parameters of Wind Power System with and without FC+BESS Unit

Time Domain Parameters/Sources	TS (S)	OS (Hz)	US (Hz)
Without FC+BESS unit	19	6×10^{-3}	0.0685
With FC+BESS unit	28	0.025	0.083

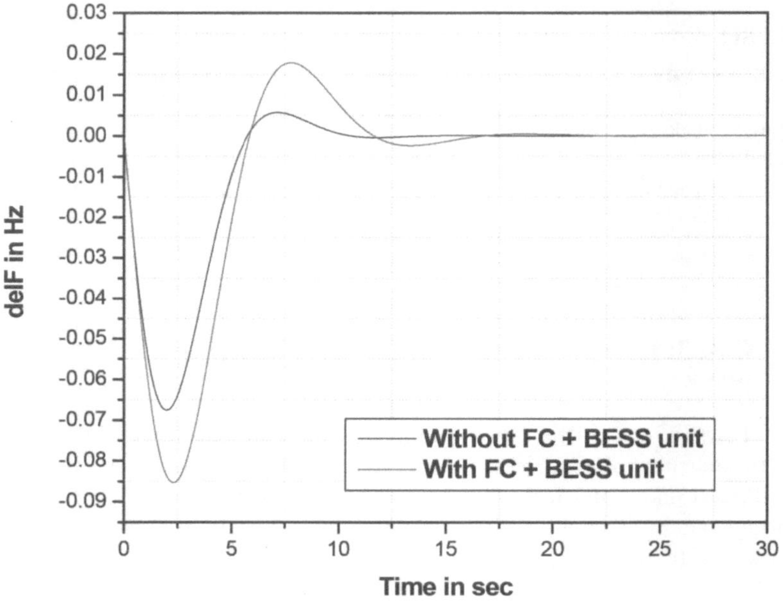

FIGURE 5.7 Frequency response of wind power system with and without FC+BESS unit.

REFERENCES

1. Elgerd, Olle Ingemar. Electric energy systems theory: An introduction. Tata McGraw-Hill, 1982.
2. Kothari, D. P., and I. J. Nagarath. Power System Engineering. Tata McGraw-Hill Publishing Company Limited, New Delhi, 2015.
3. Nosratabadi, Seyyed Mostafa, Mosayeb Bornapour, and Mohammad Abbasi Gharaei. "Grasshopper optimization algorithm for optimal load frequency control considering predictive functional modified PID controller in restructured multi-resource multi-area power system with redox flow battery units." Control Engineering Practice 89 (2019): 204–227.
4. Liu, Xiangjie, Yi Zhang, and Kwang Y. Lee. "Coordinated distributed MPC for load frequency control of power system with wind farms." IEEE Transactions on Industrial Electronics 64, no. 6 (2016): 5140–5150.
5. Magdy, Gaber, Emad A. Mohamed, G. Shabib, Adel A. Elbaset, and Yasunori Mitani. "SMES based a new PID controller for frequency stability of a real hybrid power system considering high wind power penetration." IET Renewable Power Generation 12, no. 11 (2018): 1304–1313.
6. Abazari, Ahmadreza, Hassan Monsef, and Bin Wu. "Coordination strategies of distributed energy resources including FESS, DEG, FC and WTG in load frequency control (LFC) scheme of hybrid isolated micro-grid." International Journal of Electrical Power & Energy Systems 109 (2019): 535–547.
7. Chowdhury, A. Hasib, and Md Asaduz-Zaman. "Load frequency control of multi-microgrid using energy storage system." In 8th International Conference on Electrical and Computer Engineering, pp. 548–551. IEEE, 2014.

8. Such, Matthew Clayton, and Glenn Y. Masada. "BESS control on an microgrid with significant wind generation." In 2017 IEEE Power & Energy Society General Meeting, pp. 1–5. IEEE, 2017.
9. Pandey, Shashi Kant, Soumya R. Mohanty, Nand Kishor, and João PS Catalão. "Frequency regulation in hybrid power systems using particle swarm optimization and linear matrix inequalities based robust controller design." International Journal of Electrical Power & Energy Systems 63 (2014): 887–900.
10. Nayak, Pratap Chandra, Sonalika Mishra, Ramesh Chandra Prusty, and Sidhartha Panda. "Performance analysis of hydrogen aqua equaliser fuel-cell on AGC of Wind-hydro-thermal power systems with sunflower algorithm optimised fuzzy-PDFPI controller." International Journal of Ambient Energy 2020 (2020): 1–14.
11. Dhanasekaran, Boopathi, Saravanan Siddhan, and Jagatheesan Kaliannan. "Ant colony optimization technique tuned controller for frequency regulation of single area nuclear power generating system." Microprocessors and Microsystems 73 (2020): 102953.
12. Kumarakrishnan, V., G. Vijayakumar, D. Boopathi, K. Jagatheesan, S. Saravanan, and B. Anand. "Optimized PSO technique based PID controller for load frequency control of single area power system." Solid State Technology 63, no. 5 (2020): 7979–7990.
13. Jagatheesan, Kaliannan, and B. Anand. "AGC of multi-area hydro-thermal power systems with GRC non-linearity and classical controller." Journal of Global Information Management (JGIM) 26, no. 3 (2018): 11–24.
14. Jagatheesan, Kaliannan, B. Anand, Nilanjan Dey, M. Omar, and Valentina E. Balas. "AGC of multi-area interconnected power systems by considering different cost functions and Ant Colony Optimization technique based PID controller." Intelligent Decision Technologies 11, no. 1 (2017): 29–38.
15. Naidu, K., H. Mokhlis, A. H. A. Bakar, and V. Terzija. "Comparative performance analysis of firefly algorithm for load frequency control in automatic generation control of interconnected reheat thermal power system." 3rd IET International Conference on Clean Energy and Technology (CEAT) 2014, pp. 24–26, 2014.
16. Mokhlis, H., K. Naidu, A. H. A. Bakar, and V. Terzija. "Performance investigation of ABC algorithm for automatic generation control of a three area interconnected power system." 5th Brunei International Conference on Engineering and Technology (BICET 2014) pp. 1–8, 2014.
17. Nguyen, Gia Nhu, K. Jagatheesan, Amira S. Ashour, B. Anand, and Nilanjan Dey. "Ant colony optimization based load frequency control of multi-area interconnected thermal power system with governor dead-band nonlinearity." In X.-S. Yang, A. K. Nagar, and A. Joshi (Eds.), Smart Trends in Systems, Security and Sustainability, pp. 157–167. Springer, Singapore, 2018.
18. Chatterjee, Sankhadeep, Sirshendu Hore, Nilanjan Dey, Sayan Chakraborty, and Amira S. Ashour. "Dengue fever classification using gene expression data: A PSO based artificial neural network approach." In Proceedings of the 5th International Conference on Frontiers in Intelligent Computing: Theory and Applications, pp. 331–341. Springer, Singapore, 2017.
19. Azab, Mohamed. "Global maximum power point tracking for partially shaded PV arrays using particle swarm optimisation." International Journal of Renewable Energy Technology 1, no. 2 (2009): 211–235.
20. Boopathi, D., S. Saravanan, K. Jagatheesan, and B. Anand. "Performance estimation of frequency regulation for a micro-grid power system using PSO-PID controller." International Journal of Applied Evolutionary Computation (IJAEC) 12, no. 2 (2021): 36–49.
21. Gupta, S. K., Yogendra Arya, Shivank Shukla, and Pankaj Chawla. "Two-area AGC in interconnected system under the restructured power system using BFO controller." In 2014 6th IEEE Power India International Conference (PIICON), pp. 1–6. IEEE, 2014.

22. Xu, Yao, Fangxing Li, Zhiqiang Jin, and Maryam Hassani Variani. "Dynamic gain-tuning control (DGTC) approach for AGC with effects of wind power." IEEE Transactions on Power Systems 31, no. 5 (2015): 3339–3348.
23. Liang, Liang, Jin Zhong, and Zaibin Jiao. "Frequency regulation for a power system with wind power and battery energy storage." In 2012 IEEE International Conference on Power System Technology (POWERCON), pp. 1–6. IEEE, 2012.
24. Banakar, Hadi, Changling Luo, and Boon Teck Ooi. "Impacts of wind power minute-to-minute variations on power system operation." IEEE Transactions on Power Systems 23, no. 1 (2008): 150–160.
25. Guo, Yufeng, Qi Wang, Dongrui Zhang, Jie Wan, Daren Yu, and Jilai Yu. "Anticipatory AGC control strategy based on wind power variogram characteristic." IET Renewable Power Generation 14, no. 7 (2020): 1124–1133.
26. Naidu, K., H. Mokhlis, and A. H. A. Bakar. "Application of Firefly Algorithm (FA) based optimization in load frequency control for interconnected reheat thermal power system." In 2013 IEEE Jordan Conference on Applied Electrical Engineering and Computing Technologies (AEECT), pp. 1–5. IEEE, 2013.
27. Naidu, K., Hazlie Mokhlis, Ab Halim Abu Bakar, V. Terzija, and H. A. Illias. "Application of firefly algorithm with online wavelet filter in automatic generation control of an interconnected reheat thermal power system." International Journal of Electrical Power & Energy Systems 63 (2014): 401–413.
28. Jagatheesan, K., S. Samanta, A. Choudhury, N. Dey, B. Anand, and A. S. Ashour. Quantum inspired evolutionary algorithm in load frequency control of multi-area interconnected thermal power system with non-linearity. In A. E. Hassanien, M. Elhoseny, and J. Kacprzyk (Eds.), Quantum Computing: An Environment for Intelligent Large Scale Real Application (pp. 389–417). Springer, Cham, 2018.
29. Colorni, Alberto, Marco Dorigo, and Vittorio Maniezzo. "Distributed optimization by ant colonies." In Proceedings of the First European Conference on Artificial Life, vol. 142, pp. 134–142. 1991.
30. Colorni, Alberto, Marco Dorigo, and Vittorio Maniezzo. "An investigation of some properties of an 'Ant Algorithm.'" In Conference: Parallel Problem Solving from Nature 2, PPSN-II. Brussels, Belgium, 1992.

6 AI Technology in Networks-on-Chip

Biswajit R. Bhowmik

CONTENTS

DOI: 10.1201/9781003229018-6

6.1 INTRODUCTION

With the rapid advancements of technology, a multiprocessor system-on-chip (MPSoC) produces gaps and complexity in communication that affects the high-performance requirements (latency, bandwidth, and power consumption, etc.), scalability, and robustness of the system on seamless integration of IP cores [Bhowmik (2021); Bhowmik et al. (2018)]. Altering the bus-based communication in systems-on-chip (SoCs), a network-on-chip (NoC) has solved communication and productivity problems in MPSoCs. An NoC provides efficient on-chip communication and improves both performance and scalability of the on-chip systems. An NoC is an embedded switching network that interconnects IP cores in SoCs and provides parallel communication, fault tolerance, reliability, and eliminates the communication bottlenecks [Bhowmik, Biswas and Deka (2015); Bhowmik et al. (2016a); Lee et al. (2006)].

The on-chip network, commonly known as NoC on a die as an alternate prevalent interconnection infrastructure, has been continuously occupying the space of the SoCs. In other words, the SoC-based systems are continually transformed to NoC-based multiprocessor and multicore designs. These designs would prospectively facilitate high core utilization and the need for high performance. An NoC comprises high-speed channels (interconnects) routers and IP cores that collectively construct a structure (topology) spanning the chip [Bhowmik et al. (2019b); Wolf et al. (2008)]. The on-chip architecture represents a new communication paradigm for a large-scale system of integrated circuits. It includes several advantages such as scalability, predictability, and higher bandwidth to support the basic requirements of concurrent communications. NoC-based communication systems are studied both architecturally and theoretically. The studies on NoCs include design methodology, exploration of topology, guarantee for quality-of-service (QoS), resource scheduling and management, verification, validation, test, etc. [Bhowmik et al. (2017, 2019a)]. In all these cases, a system designer must evaluate the NoCs. The most common approach for analyzing the studies on NoCs is performance analysis. To date, performance evaluation of NoCs is largely based on simulation employing a set of NoC simulators, e.g., Noxim, BookSim, Nirgam, etc. [Catania et al. (2016); Chen and Wang (2018); Joseph and Pionteck (2014); Pérez et al. (2020)]. However, the use of simulators has several limitations. For example, an NoC simulator (1) is slow to provide the results while NoC size increases; (2) provides little insight into the actual network performance changes in different design parameters, etc. An NoC simulator practically may become impossible to use for accurately, quickly evaluating NoCs via simulation, especially for optimization purposes. Therefore, it has become essential to opt for an alternate approach to overcome the issues raised due to NoC simulations.

The rapid development of the artificial intelligence (AI) industry currently leads the next generation of computation. Consequently, AI technology in every aspect becomes an essential milestone in surpassing the human level [Bhowmik et al. (2021a)]. Recent research shows that AI technology, including machine learning (ML), deep learning (DL), etc., is often preferred to improve data transformation and hardware efficiency in on-chip communication networks. Such networks are called AI chips, where an NoC is one feasible solution to SoCs. This chapter seeks to critically situate the AI technology in evaluating the NoC performance at different perspectives: theoretical, technological, analytical, etc. Major contributions of the chapter elucidate some of the salient problems, interests, and issues for its organization around NoC-based AI chips, including the following broad themes: (1) Topology and Design Space Exploration for NoCs using Deep Learning; (2) Exploiting Machine Learning for Tradeoffs and Design Considerations in NoCs; (3) Customization of NoCs for AI Chips; (4) Efficient Training of Deep Convolution Networks on Heterogeneous Manycore Systems; (5) Performance Evaluation of NoCs via Full System Simulation; (6) Analytical Analysis of Network-on-Chip Performance; (7) Learning-Based Quality Management in NoC-based System Communications; (8) Machine Learning to Predict Performance Metrics of On-Chip Networks; (9) Machine Learning for Power, Energy, and Thermal Management in NoCs.

The overall thrust of this chapter is to provide "Core Perspective" and "Further Perspective" of the application of AI technology in predicting several NoC performance metrics. This chapter examines how a custom model in DL, ML exploits the NoC-based AI chips on a diversity of perspectives on the accuracy, quality, speed up, prediction error, etc., toward high-speed communication and voluminous computation by MPSoCs. As long as the network is efficiently trained with a large set of collected data, the model would improve performance. Thus, one can see how this AI technology dramatically improves the speed and accuracy of predicting various performance metrics.

The rest of the chapter is organized as follows. Section 6.2 describes the transformation of SoCs to NoCs. Basics of AI technology are described in Section 6.3. Exploration of NoC using DL is discussed in Section 6.4. Exploitation and customization of NoCs are seen in Sections 6.5 and 6.6, respectively. Section 6.7 evaluates NoCs by simulations. Learning-based performance prediction is described in Section 6.8. Section 6.9 discusses quality management schemes for NoCs. Section 6.10 provides analytical analysis of NoC performance. Sections 6.11 and 6.12 give future directions and concludes this chapter, respectively.

6.2 TRANSFORMATION OF SYSTEMS-ON-CHIP (SoCs) TO NETWORK-ON-CHIP (NoCs)

In the era of nanoelectronics, SoC design brings opportunities and challenges. The demand of integrating a large number of transistors on a single chip requires large numbers of processing units and semiconductor intellectual property blocks. Degradation of performance and synchronization problems is caused because of wire delays—the problem of synchronization between IPs increases when clock frequency increases and the feature size decreases. In addition, a large amount of overall power is consumed by communication between IPs. The performance of SoCs is evaluated by the interconnect efficiency and accommodation of the communication requirement of IPs [Bhowmik (2019); Lee et al. (2006)]. A multiprocessor system-on-chip, i.e., MPSoC, is an architecture consisting of numerous multi-core SoCs. It contains multiple processing elements (PEs), memory hierarchy, input/output components, etc., connected by on-chip interconnects. Applications of an MPSoC are in various domains like telecommunication architectures, network security, multimedia, etc. Its communication infrastructures help build a high-performance system by reducing power consumption via specialized architectures and PEs [Bhowmik (2019); Wolf et al. (2008)].

A large number of PEs can be integrated into a single MPSoC [Bhowmik (2021); Sant'Ana et al. (2019)]. In computer architecture, MPSoCs represent a crucial and distinct category since the architectures satisfy the needs of embedded applications [Wolf et al. (2008)]. Though it may add to the advantage of executing complex applications, it may also bring hurdles for the on-chip interconnect design. Interconnects are required to provide higher bandwidths and lower latencies for better service. According to the requirement of highly scalable and robust multiprocessor systems, SoCs are expected to satisfy these requirements. With the increase in the number of cores, SoC architectures fail to meet the performance requirements due to shared bus-based communications [Bhowmik (2021); Bhowmik et al. (2020a); Wu et al. (2015)]. Further, MPSoCs produce gaps and complexity in communication that affect the on-chip system's scalability and robustness.

Alternatively, NoC can cope with the requirements and acts as an optimal solution in which heterogeneous NoCs are embedded on a single chip multiprocessor (CMP). Figure 6.1 shows a typical heterogeneous CMP architecture where an NoC provides parallelism, reusability, scalability, and platform infrastructure for modular on-chip communication [Amano (2013); Bhowmik (2021); Jiang et al. (2013)]. It means an NoC is a solution that brings improvements to the traditional bus-based architecture [Bhowmik et al. (2019b); Morgan et al. (2020)]. An NoC has thus become an efficient technology to provide energy and power-efficient system with reliable and cost-efficient applications. NoC designs use a network of packet switches having links and routers using the layered protocol. Hence, NoCs are the attempt for improvements

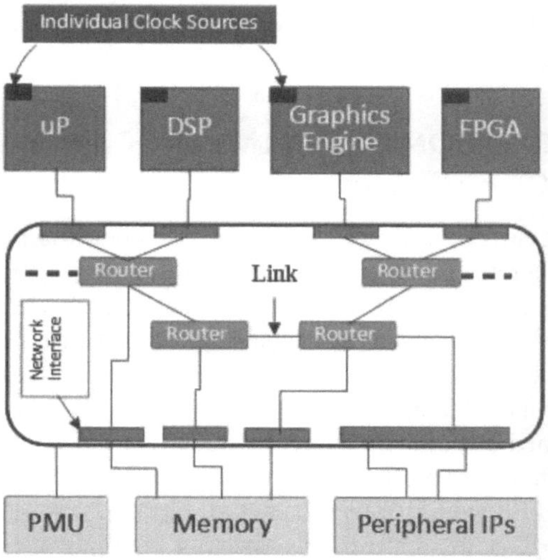

FIGURE 6.1 Heterogeneous CMP architecture.

over the traditional bus-based architectures [Bhowmik (2020); Bhowmik, Biswas and Deka (2015); Bhowmik et al. (2020b)].

6.3 BASICS OF AI TECHNOLOGY

NoC is the choice of chip designers due to its efficient on-chip communications. NoC-based communication architectures are the new era for enabling high-performance communication and computation [Bhowmik et al. (2016b)]. An NoC design is validated through its evaluation. One primary method of evaluating an NoC is measuring its performance metrics via simulations. However, a simulation method is time-consuming, and the evaluation time increases with the NoC size. Several AI, ML, and DL-based techniques [Bhowmik et al. (2021a); Chen and Louri (2020); Chi et al. (2019); Gao and Zhou (2019)] are used as alternatives. This section gives a brief about these technologies.

6.3.1 Artificial Intelligence

Nowadays, everybody has heard about AI. Tentatively, the concept of AI was promptly coined during the early 80s of the previous century. AI has been progressing rapidly since then. Initially, AI was portrayed as a robot with human-like characteristics, but currently, it can encompass anything. A few popular definitions [Pagallo (2020)] of this technology are provided below.

Definition 1. "*The exciting new effort to make computers think ... machines with minds, in the full literal sense* — Haugeland, 1985".

Definition 2. "*The study of mental faculties through the use of computational models* — Charniak and McDermott, 1985".

Definition 3. "*A field of study that seeks to explain and emulate intelligent behavior in terms of computational processes* — Schalkoff, 1990".

Definition 4. "*The study of how to make computers do things at which, at the moment, people are better* — Rich and Knight, 1991".

The word "intelligence" means the capacity to learn, particularly "the ability to act like humans". Therefore, the primary objective of AI is to build and understand intelligent entities that are also known as agents. AI is considered a mixture of science and engineering for making intelligent machines, primarily through smart computer programs. A rational agent view of the AI is shown in Figure 6.2. It summarizes that the current state of AI systems in practice is spread everywhere, including speech processing, computer vision, learning, planning, and reasoning. This technology is currently viewed from two perspectives: narrow and general intelligences. Artificial narrow intelligence (ANI) systems are AI systems that complete one specific task after the design and training phases. For example, performance evaluation, topology exploration, etc., in NoCs. Such systems are programmed in such a way that the systems become intelligent at performing specific tasks. An ANI system is also known as a weak AI (WAI) system. Artificial general intelligence (AGI), also known as strong AI (SAI) systems, is designed to perform on par with other humans. Besides, the third type of AI system

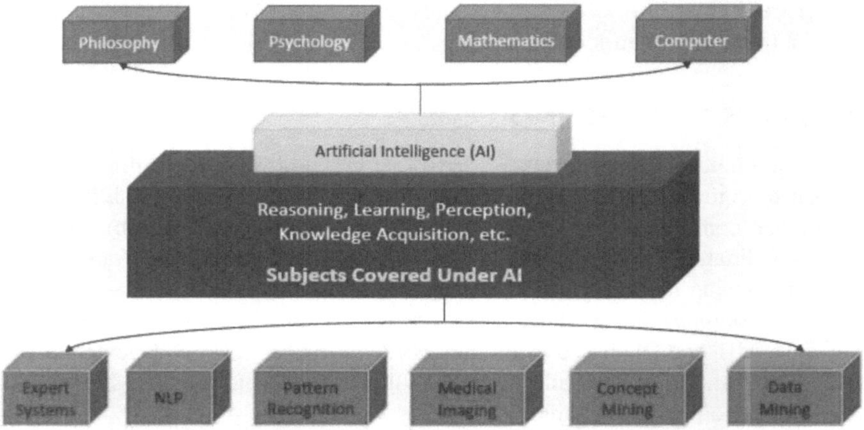

FIGURE 6.2 A rational agent view of the AI.

is rapidly evolving. It is known as the artificial superintelligence (ASI) system, an AI machine that becomes self-aware and surpasses human intelligence and ability.

6.3.2 Machine Learning

The AI is segmented into multiple subareas. ML is naturally a subset of AI. In other words, ML has treated a way to achieve the goals of AI techniques. The AI technology progresses as much as the ML capabilities continue to evolve.

The ML provides different statistical methods and algorithms that enable computers to learn. This learning takes place automatically from their previous experiences and training datasets. Subsequently, their behavior is changed accordingly. Different techniques and algorithms in ML that make the computer learn include k-means clustering, support vector machine (SVM), random forests, decision trees, etc. Like AI, ML is also applied to various areas like hardware and software systems, healthcare systems, etc. For example, if the training datasets generated by a simulation method for NoCs are sufficient by size, their performance can easily be predicted. Recent applications include e-commerce and the OTT platform. E-commerce companies like IndiaMart, Snapdeal, Amazon, Flipkart, etc. and OTT platforms like Amazon Prime Video, Netflix, Hotstar, etc. use ML algorithms to recommend their products and movies user based on other users' past viewing data. Such a recommendation is continuously getting improved by learning from past experiences. A learning algorithm works well if the input data, e.g., knowledge, experience, is reasonably good enough.

6.3.3 Deep Learning

ML models do not perform very well if a dataset size and complexity increase in multi-directions. In this case, the models need some human intervention and guidance, which results in a more efficient learning algorithm. This is the place where DL finds its hotspot. DL is another subarea of AI and is thought of as the evolution

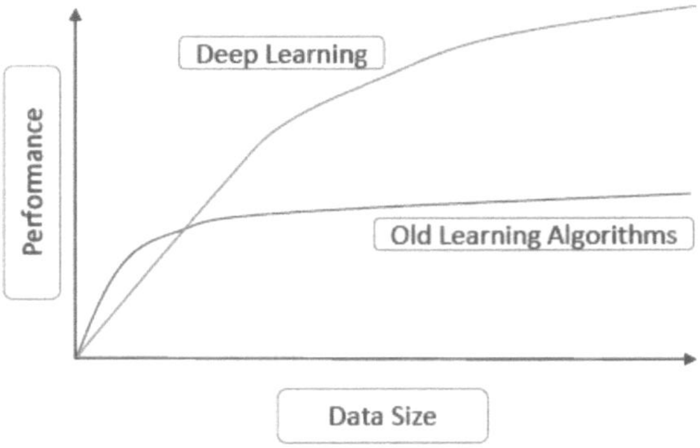

FIGURE 6.3 Performance behavior of deep learning algorithm with the scaling of data.

of ML. Like the previous two technologies, this new technology can be applied to many and in the same areas. However, the difference is that the new technology is used to solve more complex problems. Figure 6.3 shows the performance behavior of deep learning algorithm over the old learning algorithms at the scaling of data. The complexity of the problems is realized in terms of various data characteristics, e.g., volume/size, diversity, less structure, etc. DL uses different learning algorithms to solve real-world problems. A few unique algorithms are Artificial Neural Networks (ANN), Convolutional Neural Networks (CNN), and Recurrent Neural Networks (RNN), which mimic how the human brain works. Thus, to summarize, a lot of AI systems are powered by algorithms that belong to both ML and DL technology. Therefore, DL is on top of ML which is built on top of AI. In other words, AI is first tried to achieve through ML and then via DL. Therefore, these technologies are not the same but are interrelated. Figure 6.4 shows their relationship.

6.4 DL-BASED NoC EXPLORATION

DL techniques have achieved great success in diverse application domains. These techniques are a class of ML methods employed for the applications. For example, CNN are treated as the particular class of DL architectures which are

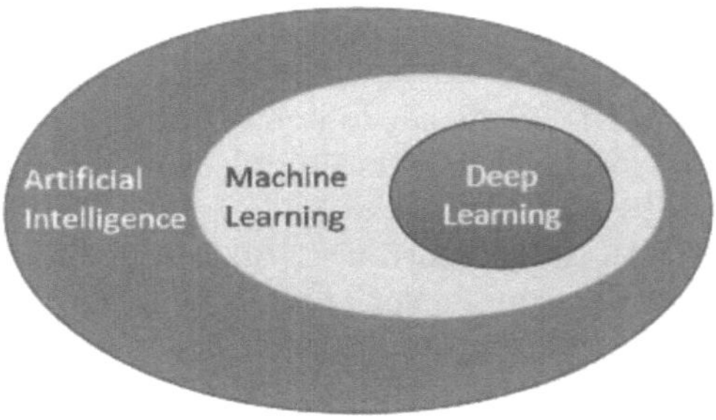

FIGURE 6.4 Relationship of artificial intelligence with machine learning and deep learning.

employed for a heterogeneous single-chip multiprocessor (HSCMP) [Choi et al. (2018)]. Although NoCs can overcome many issues raised in SoCs, the NoCs cannot provide scalable, low-latency, and energy-efficient communication infrastructures for on-chip communications. DL often solves these problems, which are called DL problems and essential to solving [Ogras and Marculescu (2006)]. Note these problems duly arise on traditional wired NoCs. However, such problems do not occur in the case of wireless NoCs (WiNoCs). Instead, WiNoCs can achieve energy efficiency and low latency during communications on massive manycore chips. This is because of the mm-wave-based communication via the message-forwarding main, powerful, and central antenna [Deb et al. (2012); Wettin et al. (2014)].

NoC-enabled homogeneous CMP architectures generally target neuroscience applications. For example, a customized NoC architecture is incorporated on a CMP design to implement spiking neural networks [Painkras et al. (2012)]. Similarly, reconfigurable neural networks are implemented with multicast-aware mesh NoC architectures [Firuzan et al. (2015)]. In all the cases, the neural networks are more efficient on discrete graphical processing unit (GPU) systems due to the highly data-parallelizable nature of the networks. Many works have also focused on discrete GPU platforms for improving system performance. In this case, the performance can be improved by enhancing NoC architectures employed on the platforms. For example, processes in a GPU-based system are executed independently on each GPU core of the system. Therefore, low inter-GPU communication is possible, resulting in improved performance. However, if there is a traffic imbalance observed in the communication channels of the NoCs, then asymmetric traffic pattern by GPUs may be exhibited for the applications running on. This phenomenon worsens the performance bottlenecks in the GPU-based system. The issue can be overcome by the involvement of a trained CNN on the GPU-enabled systems, e.g., NoC-enabled heterogeneous CPU-GPU CMP platforms that involve heavy communication over the traditional discrete-GPU system [Hestness et al. (2015)].

NoCs in heterogeneous systems expectedly handle various traffic patterns at varying constraints, e.g., quality of service (QoS), reliability, and fault-tolerance [Bhowmik, Biswas and Deka (2016); Bhowmik et al. (2017)]. The central processing units (CPUs), in this case, are sensitive to the memory access time. So, data exchanges should take place at low latency for the communications by these CPUs. Subsequently, the bandwidth requirements by NoCs differ. For discrete GPU-based systems, modern NoCs attempt to maximize overall bandwidth, whereas the centralized GPU demands high bandwidth. Consequently, the NoC designs must be altered for such heterogeneous systems that embed multiple CPUs and GPUs [Lee et al. (2013)]. Many DL techniques, including deep convolutional neural networks (DCNN) [Chen and Louri (2020); Choi et al. (2018); Gao and Zhou (2019)], help alter the modern NoC designs by exploring the design spaces and network topologies. Such NoC designs embed deep learning processing units (DPUs) for frequent arithmetic units while DL computations need repetitive operations [Chi et al. (2019)]. Figure 6.5 shows a typical DPU-based heterogeneous NoC system.

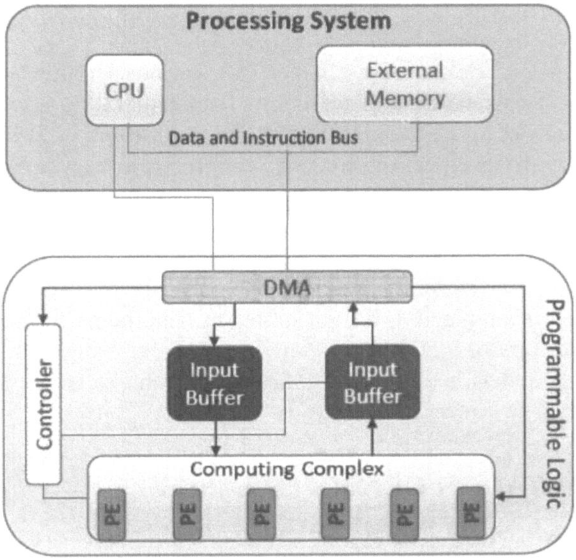

FIGURE 6.5 DPU-based heterogeneous NoC architecture.

6.5 EXPLOITING MACHINE LEARNING FOR TRADEOFFS

ML is the most popular branch of AI. It may be because of low error or increased speed that the ML framework provides [Das et al. (2015)]. ML is a mechanism for the analysis of data that improves automatically via experience and by the use of data [Bhowmik et al. (2021a)]. AI is a branch that builds systems that can learn from data (known as training data) and make decisions (predictions) based on the learning without being explicitly programmed to do so. In many commercial fields, AI has been used widely because of its efficiency and accuracy. Subsequently, ML is used in various real-time applications, such as computer vision, speech recognition, prediction tasks, spam detection, email filtering, etc. Several learning algorithms have proved their significant contributions in these applications.

Many people are showing interest in applying ML in NoC over the past years. ML can be used in on-chip communications to train a network as fast as possible at a low cost and low power budget, leading to evaluating performance parameters based on the learning. AI technology, e.g., ML, can find the optimal solution to a very high dimension without developing the analytical model. ML-based models are designed to fulfill the requirement in a complex and dynamic network having physical limitations due to noise and nonlinear distortions. In the case of NoCs, different algorithms of AI are used to analyze the parameters of NoC. Various aspects of NoC, such as fault tolerance, task scheduling, and performance prediction in power area, latency, and throughput, are done using ML algorithms and a neural network-based framework. Also, the NoC architecture analysis in congestion prediction and temperature prediction can be done efficiently using ML techniques. These techniques allow advancements in NoCs such as expandability, dynamic reconfiguration on demand, etc. Also, it has mapping more than one logical unit onto a single physical team, which improves the design of NoC.

Furthermore, ML models are used in NoCs for various activities like space management, power analysis, task mapping, routing [Lee et al. (2006)], switching, design strategies, etc. [Yin et al. (2020)]. Different ML algorithms used in literature are neural network models, regression models, reinforcement learning, support vector regression (SVR), etc., for measuring other performance parameters of NoCs at a reasonable accuracy rate and speedup. It becomes possible by the ML models that learn from the trained data and provide results. Thus, ML helps NoC build a high-performance NoC-based system to predict the temperature of NoC in real-time, indicate the NoC specification and its design strategies, and the performance evaluation of variable parameters of NoCs [Foroutan et al. (2013); Wang et al. (2019a)]. The advantage of ML models is that they are fast and can make predictions on unseen data with a tradeoff whenever necessary.

6.6 CUSTOMIZATION OF NoCs FOR AI CHIPS

NoC is emerging as a new technology for interconnecting cores in a chip. It is crucial to estimate power, performance, and area of NoC at an early stage. Automation in NoC is driven by using AI technologies such as ML, DL, regression, etc. Advanced analytics are used to estimate the performance parameters of NoC like latency, power, and area through ML [Bhowmik et al. (2021b); Hazarika et al. (2021); Ping et al. (2013)]. AI uses a computer to solve a complex problem by learning and self-correction method. AI is used in different applications. AI technology has three major steps: first, to learn from data, second is to implement decision-making, and third is to incorporate knowledge to handle uncertainty [Mishra and Srivastava (2014)]. In the field of NoCs, AI is considered a promising solution to the problem in different aspects of NoCs, as mentioned earlier [Bhowmik et al. (2021b); Hazarika et al. (2021)].

ML is an efficient tool in AI technology that automates data analytics by learning from a trained dataset. ML can be applied for designing an efficient NoC [Jain et al. (2021); Yin et al. (2020)]. Researchers use ML to create NoC before actual hardware implementation with accuracy and the speedup of the NoC designs. Also, ML provides ways to evaluate the performance of NoC with good accuracy and less cost. The optimal design of NoC should associate the building blocks-routers, links, network interfaces, and all other PEs in such a way that involves less power consumption, less time, less energy. The design of NoC should affect reliable, feasible, and cost-efficient implementations. For example, design space exploration and customization of NoC architecture are done using ML with a motive to optimize placements of link components (both vertical and planar), thus providing high performance and energy-efficient systems [Das et al. (2016)]. Also, ML can train a network with various parts of NoCs and handle different aspects of NoCs like space management, power analysis, traffic management, latency analysis, area management, etc. Further, other design parameters like latency, throughput, power, area, energy, etc. can be evaluated by ML models like prediction trees, regression algorithms, neural network models, reinforcement models, SVR, etc.

NoC is a powerful system that allows on-chip communication. Using ML, NoCs are automated to handle the uncertainty in the data. It results in a robust model providing a better solution in different parameters of NoC, such as dynamic energy-saving, congestion, and temperature prediction, which makes the system yield better results with improved accuracy. NoC architectures are powerful enough to provide robust on-chip communications. Before hardware implementation, chip designers use simulators to predict the performance parameters to achieve efficient design implementation. Researchers have provided ML for the customization of NoC network parameters, which directly influence the performance metrics- network latency, flit latency, hop count, etc. [Dong et al. (2009)].

Many NoC-based on-chip networks do high-performance computations and communications. Consequently, such networks are customized as AI chips using convolutional and deep neural networks to accelerate the extensive analysis and processing of application data for communications. The main objectives of designing such NoC-based AI chips are to achieve comparatively better energy efficiency and high volume data computational capacity over the traditional NoCs without the inclusion of neural network implementation [Gao and Zhou (2019)]. Figure 6.6 represents a high-level view of AI chip optimization by NoC designs. A limited number of prior works are available that customizes and optimizes NoCs for AI chips. For instance, Chen et al. (2019) presented a hierarchical mesh NoC using a deep neural network. The network allocates routers for data transmission. Note that both input data as packets and their weights are transferred from a global buffer to the PEs in NoCs. In addition to the mesh NoCs, Chen et al. (2014) have used torus topology to replace ring-shaped NoC architectures. However, AI techniques such as CNNs and DNNs used in the design of chips have ignored both router optimization and data reuse over the interchip communication. Plus, the neural networks have achieved a speedup up to 451× over a GPU while the energy consumption is reduced by 150× on an AI chip system with 64 cores. Few researchers have considered hybrid NoC customizations for designing a powerful AI chip system. For example, Neo-NoC is such a system in which Liu et al. (2018) presented a modern neuromorphic acceleration system. In the system, NoCs are used as accelerators to promote the computation ability of the system. However, both latency and throughput remain a distinct disadvantage. The limitations can be overcome by customizing NoC routers. The customizations

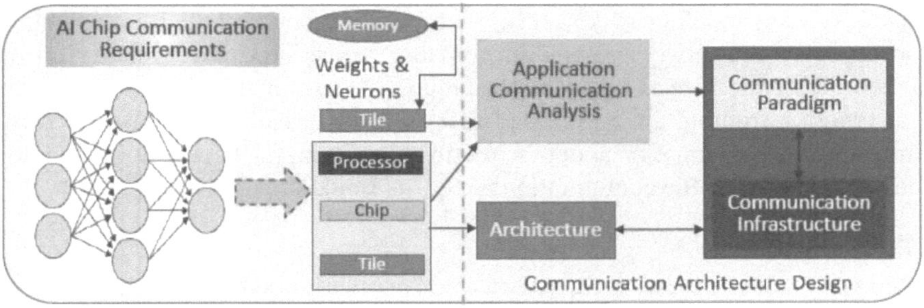

FIGURE 6.6 Optimization of AI chips via NoC designs.

primarily involve the high utilization of buffers to increase the throughput and lower the latency [Ramanujam et al. (2010); Thiem and Oyanagi (2011)].

6.7 NoC EVALUATION BY SYSTEM SIMULATION

A simulator is a software tool that can be used as a blueprint to model designs of devices (or components) to predict performance metrics and outputs on a given input. The simulator performs simulation. Simulation is the method of planning a model of an ideal physical system (architecture), executing the designed model on a computer, and investigating the completed output. It allows researchers to examine the performance and efficiency parameters by exploring the design space of the novel micro-architecture. Thus, a simulation decides the design on NoC before resorting to its actual implementation and provides a detailed output with good debugging options.

6.7.1 Need for NoC Simulation

The design space and architecture of NoC are vast and costly. So, it is crucial that before actual implementation, one should use a simulator to explore the design decisions for NoCs. The NoC simulators are cheap and flexible. A good NoC simulator should explore the design space fast, evaluate the quality of designs considering the parameters: power, performance, reliability, cost, etc., at different traffic patterns. Simulators can be classified in terms of programming abstraction.

- High-level simulators: High-level simulators are written in Java or any high-level language, e.g., ATLAS and GpNoCsim.
- Low-level simulators: Low-level simulators are written in SystemC language, e.g., Noxim, Nirgam, and Nostrum.

6.7.2 Various NoC Simulators

Researchers use various NoC simulators. Some of the famous and widely used simulators are listed below.

6.7.2.1 BookSim

BookSim [Pérez et al. (2020)] is a cycle-accurate type NoC simulator implemented using C++. Many topologies like a torus, flattened butterfly, and the BookSim simulator supports mesh networks. BookSim contains various alternatives for configuring the network architecture and gives several routing algorithms. It possesses modularity as topology, routing algorithm, and user choice that can be set for the router's architecture. All information about the routing algorithm, traffic, topology, and flow control are present in the configuration file of the BookSim simulator.

6.7.2.2 Noxim

Noxim simulator is another cycle-accurate and popularly used NoC simulator by the researchers. The computing team of the University of Catania has developed this simulator. This simulator is developed in the SystemC language. Noxim simulator generally works on 2D mesh topology at different simulation setups such as routing strategies, different traffic patterns, and distributions, buffer depth, and wormhole router design and switching, packet injection rate, and network size. Standard output statistic parameters are energy, communication delay, and throughput. Examples of some commonly used routing algorithms supported by this NoC simulator include XY routing, odd-even (OE) routing, lookup table-based routing, negative-first, and north-last routing, dyad and fully adaptive routing, and so on [Catania et al. (2016)].

6.7.2.3 SICOSYS

SICOSYS [Joseph and Pionteck (2014)] is a cycle-accurate simulator. SICOSYS implementation is in C++. It is very convenient to use and can run on all UNIX operating systems with a C++ compiler. SICOSYS simulator can be used to model the extended category of message routers accurately. The SICOSYS simulator results are very similar to hardware simulators, and the cost of computation is low in the SICOSYS simulator.

6.7.2.4 TOPAZ

TOPAZ is a simulator of general-purpose for interconnection networks. It is used to design a wide variety of routers with different combinations of speed and precision. TOPAZ is a simulator that originated from the SICOSYS simulator. TOPAZ implementation is in C++ language. It is very convenient to use and can run on all UNIX operating systems with a C++ compiler. Using POSIX threads, TOPAZ can perform parallel execution.

6.7.2.5 NIRGAM

NIRGAM is a discrete event, cycle-accurate type simulator [Chen and Wang (2018)]. Its implementation is in SystemC. It acts as a model to test and explore with an ideal NoC design in routing algorithms, switching mechanisms, and applications on various topologies. Topologies named 2-D mesh, 2-D torus are supported by the NIRGAM simulator. It supports deterministic XY, Source Routing, and Adaptive OE routing algorithms. NIRGRAM supports NoC parameters that can be configured- Frequency of Clock, Size of Flit, Topology (m x n), and Virtual Channels. It creates graphs in Matlab by generating and running Gnuplot graphs and Matlab scripts. Each simulation keeps a record of performance parameters for each channel. Some standard parameters include average packet latency packet, average flit latency, and average throughput.

6.7.2.6 Nostrum

Nostrum is a layered NoC simulator that is cycle-accurate too. The simulator is also known as Nostrum NoC Simulation Environment (NNSE). It is written in SystemC and python. This NoC simulator was developed at the Royal Institute of Technology (KTH), Stockholm, by a team called Nostrum [Millberg et al. (2004)]. The simulator supports both 2D mesh and torus NoCs. The supported switching techniques are wormhole, and store and forward switching mechanisms. Using the deflection order routing and XY routing algorithms, one can maintain and design a network according to flow control, routing algorithm, and topology—the simulator maps applications to nodes of the network. Users can change buffering options and arbitration policies and generate traffic patterns in the network to analyze different metrics, including network throughput and latency. The main bottleneck of the NNSE simulator is that only finite customization is possible. For the best effort and guaranteed communication, the NoCs are then required to be reconfigured.

6.7.2.7 ATLAS

ATLAS is a Java-based NoC simulator developed by the Hardware Design Support Group. The simulator is famously known as the generation and evaluation framework for NoCs [Mello et al. (2011)]. It has a user interface. This tool generates the result in the form of tables and graphs. Using this, one can evaluate the latency, bit rate in each router, and throughput.

6.7.2.8 GEM5

GEM5 tool is written in C++ by a set of designers [Lowe-Power et al. (2020)]. This tool is used to simulate components as cores and buses. It is created from two projects, M5 and GEMS. M5 allows high configuration for simulation of the different process models. GEMS provides a flexible memory system. It is a cycle-accurate simulator, and it has two modes of operation: full system (FS) and syscall emulation (SE). FS model simulates a complete system to run the complete OS. SE mode discards the need for device modeling and simulating OS by emulating system calls. This tool allows the modification of core type, core number, memory type and size, cache size, associativity, disk image, and kernel.

6.7.3 Merits and Demerits of Simulators

NoC evaluations avail different advantages of employing a simulator. As the name suggests, simulators help in the cost-effective simulation of various hardware structures. There is no need to use a physical system—creating ways to characterize non-existing systems and components. Simulators further can help increase the efficiency of the proposed features or system because executing it allows access to relatively more extensive performance statistics. Lastly, simulators save the toil of re-booting designs (as in real hardware) in debugging, as most of them provide a controlled opportunity to reverse code in case of error detection.

Besides many advantages, simulators have certain drawbacks such as:

1. When simulations run at a lower abstraction level, the results are precise at cycle level, but it takes too much time because high computation is done to simulate the components [Das et al. (2015); Rafie et al. (2014)]. Higher abstraction levels are used to solve this problem but the results obtained are of lower accuracy using this solution.
2. As the topology size of NoC increases, simulation can take days to complete [Jiang et al. (2013); Kumar and Talawar (2019)].
3. Output parameters are not the same with every simulator. Orion is a powerful model set for NoCs. Power consumption and area are estimated using this simulator [Kahng et al. (2012)]. Noxim simulator supports 2D mesh

topology with synthetic traffic patterns. Energy, communication delay, and throughput are the output statistic parameters [Catania et al. (2016)]. BookSim simulator mainly provides the latency analysis of NoC architecture. However, BookSim can be modified to achieve the power and area parameters, including the latency [Jiang et al. (2013)]. Thus, there is a need for standardization in simulators.

6.8 MACHINE LEARNING-BASED PERFORMANCE PREDICTION

Conventional performance measurement approaches rely on simulations at lower abstraction levels to run the architectures for accurate results. But running at lower abstraction levels takes a long time. On the other hand, running simulations at higher abstraction levels is time efficient but could not deliver accurate results [Silva et al. (2019)]. So traditional approaches can be avoided using alternate technology like ML-based methods. This section reviews previous works done on many areas of NoCs. This approach evaluates NoC concerning the following issues: traffic management network, power analysis, latency analysis, reliability analysis, space management, area analysis, etc. Table 6.1 shows the summary of the previous works done. It presents the related performance of the earlier works concerning the AI algorithms used, parameters analyzed, and observations.

Rafie et al. (2014) proposed an ML-based mapping algorithm. The proposed method uses the parameters- contention factor, robustness index, and communication cost for mapping. The process provides the best mappings with various metrics assessed by ML models: decision tree classifier (DTC), SVM, and multiple clustering. Experimental results showed that this algorithm could find the best mappings from various solutions with a 15× better performance than the standard mapping generator algorithm.

Silva et al. (2019) proposed ML-based techniques to improve the design phase of NoC architecture. Four real applications- audio/video, read/write, block transfer, and signaling, are analyzed, and the RedScarf tool is used to collect data. Authors have examined different ml techniques and picked Random Forest for audio/video with an accuracy of 90% for arbitration protocols and 85% for routing protocols and achieved 99% for application inference. They assessed different regression classifiers available in Weka software. They concluded the best classifier for read/write operation is a random tree, for block transfer is M5P, for signaling is random forest. Further, it is concluded that the best classifier is a tree-based classifier that can evaluate latency with an accuracy of 99%.

A Support Vector Regression-based model for prediction of traffic flow latency is discussed by Qian et al. (2015). By adopting an analytical queuing model, the source queue and channel queue waiting times are first evaluated. The model uses two equivalent queues. The queueing theory estimations are incorporated as features in the learning process to increase the accuracy of prediction. Then, the authors proposed SVR-based learning framework to collect training data. The framework analyzes different traffic scenarios and predicts the traffic flow latency. Results showed a prediction error of 12% in-network saturation load and a speedup of about 100× in comparison to accurate cycle simulations.

TABLE 6.1
Relative Performances by Previous Works on NoCs Using Machine Learning

Previous Works	Algorithm	Parameters	Observations
Ping et al. (2013)	Prediction tree	Latency	Accuracy = 82%
Rafie et al. (2014)	DTC, SVM	Performance	Performance = 15x better
Qian et al. (2015)	SVR	Latency, Speedup	Error = 12% and Speedup = 100×
Narayana (2016)	ANN	Temperature	Temperature = 2.7–2.9°C root mean square with respect to HotSpot
Rao et al. (2018)	Random forest, neural network, linear regression	Performance, Area	Designs NoC with a better performance rate by 40–54%, decrease in area by 6–15%
Fettes et al. (2018)	Reinforcement learning, supervised learning	Energy	RL achieved 15.4% energy savings, SL achieved 20.3% energy savings
Silva et al. (2019)	Tree-based classifier	Latency, Accuracy	Accuracy of latency prediction = 99%
Chen and Liao (2019)	ANN	Temperature	Drops by 39.4–54.2% maximum error, 40.6–51.5% average error
Wächter et al. (2019)	Linear Regression	Power, Temperature, Energy	Energy savings by 20%, prediction rate faster by 35%
Zheng and Louri (2019)	Reinforcement Learning, ANN	Energy, Power, Area	Rise in system performance by 7%, improves power consumption by 26%, reduction in area overhead by 67%
Chen and Liao (2020)	ANN	Temperature, Performance	Rise in performance by 9.16–38.37%, decreases average error by 37.2–62.3%
Zhao et al. (2020)	CNN	Power, Latency	Power consumption reduces by 50%, latency reduces by 4×
Hu et al. (2020)	RNN, Queueing Theory	Latency, Speedup	Speedup of RNN = 5.85–7.38×, speedup of QT = 6.78–8.21×

Hu et al. (2020) use graph rewriting for adjustment of application-specific NoC where the Design Space Exploration (DSE) is as a Markov Decision Process (MDP). It uses Monte Carlo Tree Search (MCTS), an reinforcement learning technique as a search heuristic, and it shows better efficiency compared to Simulated Annealing (SA) and Genetic Annealing (GA). The authors proposed an RNN-based model for latency estimation. It is faster compared to cycle-accurate SystemC simulations and offers similar speedup as queuing theory (QT). Compared to SystemC simulation (SC), RNN provides a speedup of 5.85× to 7.38×, and QT provides a speedup of 6.78× to 8.21×.

Fettes et al. (2018) proposed a model using reinforcement learning and supervised learning approaches for Learning-enabled Energy-Aware Dynamic (LEAD) frequency/voltage scaling of multicore architectures. Three different versions of Supervised Learning are presented based on the change in throughput/energy, utilization of buffer, and change in buffer utilization. Simulation results are generated using Splash-2 and PARSEC benchmarks on a mesh architecture of size 4×4. When supervised learning is used, LEAD attained 15.4% average dynamic energy savings for a 0.8% loss in throughput with no latency impact. And by using reinforcement learning, LEAD shows an increase of 20.3% in average dynamic energy savings at the cost of 1.7% rise in latency and 1.5% drop in throughput.

Chen and Liao (2019) proposed a method for temperature prediction of thermal aware NoC using an online learning model. The model can adapt the hyperplane of the temperature behavior of NoC during runtime. The model uses ANN to learn data and predict local nodes' temperature in the NoC system. The model is evaluated for three different traffic patterns- random, transpose, hotspot. This model brings a drop by 39.4–54.2% maximum error and 40.6–51.5% average error.

Wächter et al. (2019) proposed a linear regression-based model to predict the temperature of heterogeneous multicore systems. The method uses the information of the parameters- power consumption and the two previous successive sensing temperature. Experimental results showed energy savings by 20%, prediction rate faster by 35%. The model reduces the mean absolute error from 3.25 to 1.15°C for the evaluated applications scenarios.

Chen and Liao (2020) proposed an adaptive model for temperature prediction based on ML. The model is designed to combine the Least Mean Square filter theory and ANN. They have evaluated system performance under three synthetic traffic patterns- Uniform Random, Hotspot, Transpose. The model brings a rise in performance by 9.16–38.37% and decreases average error by 37.2–62.3%. The model also gets a drop-in area overhead by 18.59–22.11%.

Zheng and Louri (2019) proposed a model based on reinforcement learning for energy-efficient NoC design. To reduce power consumption (both static and dynamic), the proposed approach combines power gating and Dynamic Voltage Frequency Scaling. Experimental results showed a rise in system performance by 7% and improved power consumption by 26% compared to power gating and Dynamic Voltage Frequency Scaling without reinforcement learning. The authors also proposed an offline model based on ANNs to reduce RL's hardware costs. Results showed a reduction in area overhead by 67%.

An ML-based approach for NoC design is presented by Rao et al. (2018). ML-based designs are better in comparison to traditional manual methods on various quality metrics. Different standard pre-trained models- random forest, neural network, and linear regression- are exercised to build NoC design with optimized quality metrics. The models are implemented on some specific SoC features that ultimately affect the NoC design. Experimental results showed that MLNoC designs NoC with a better performance rate of 40–54% and decreased area overhead by 6–15%.

Ping et al. (2013) built a high accuracy model based on a prediction tree for estimation of latency. The method minimizes the requirements of mesh. Representation of the network overlay consisting of communication nodes as a tree structure is called a prediction tree. The latency of unmeasured network links is predicted based on the selected measured network links. The authors have defined three novel heuristics supported by removing imprecise prediction steps and selecting the optimal target node. Experimental results showed an accuracy of 82% with 50% measurement over a 120-node network.

Zhao et al. (2020) designed the strategy for allocation and mapping using CNN. The computation operations are allocated to PEs and then mapped to nodes in the network on a chip using a genetic algorithm for CNN. This method brings better performance results concerning parameters power and latency. The results demonstrated a 50% reduction in power consumption and reduced network latency by 4×.

Narayana (2016) proposed an ANN-based temperature prediction engine scheme. This model uses ANN to predict the thermal profile of the cores and NoC elements of the chip. The predictive dynamic thermal management (DTM) scheme then uses this thermal profile that combines core and network-level DTM techniques. Results showed that the system could exchange data between cores in a multicore environment in an energy-efficient manner and provide a medium capable of broadcasting to share thermal control messages to trigger and manage the DTM schemes efficiently. The proposed ANN model showed 2.7–2.9°C root mean square concerning HotSpot.

6.9 AI-BASED QUALITY MANAGEMENT IN NoCs

With the emergence of big data applications, communication loads in NoCs become heavy, resulting in costlier on-chip communications in multicore and manycore network systems. One of the primary reasons for cost increment is the amount of time taken to transfer the data, i.e., packet latency and the amount of energy consumed for delivering these packets over the interconnects. Indeed, an efficient scheme needs to be employed for data-incentive applications. Otherwise, one must computerize more with the quality for better performance requirements in NoC-based systems [Chen and Louri (2020)]. Consequently, AI-based algorithms as quality management techniques are developing to control and ensure accuracy on the latency and power consumption on the transmitted data. These techniques must predict the quality loss for the results and tolerable quality loss (TQL) by the ongoing applications on the NoC systems to ensure accuracy [Betzel et al. (2018); Stevens et al. (2018)]. These techniques directly move toward the approximation communication in the NoC-based systems that indicate reinforcement learning (RL) to overcome the aforementioned challenges. Thus, the employed reinforcement learning primarily focuses on two things- maximizing the data approximation and result accuracy assurance. However, the objectives can be fulfilled by integrating a suitable ML algorithm that makes the appropriate approximation on the on-chip data. Multiple approximation communication approaches are studied in the literature that enhances NoC performance.

Betzel et al. (2018) have done a thorough survey on the approximation communication in NoC-based systems. In the survey, different techniques based on three key factors- compression, value prediction, and relaxed synchronization- have addressed the design's communication bottleneck issues. Approximations through compression techniques are made either before or after the traffic injection in the NoCs. A few compression techniques [Boyapati et al. (2017); Stevens et al. (2018)] are discussed to reduce the resulting error in terms of the size of resilience data error before passing traffic in NoCs. Instead of compression, a few methods [Wang et al. (2017); Xiao et al. (2019b)] are based on the network congestion control mechanisms that target to reduce the amount of lost or dropped data packets before transmission in the NoCs. Xiao et al. (2019b) have presented an approximation communication technique for on-chip networks. The method is based on the threshold error from the dropped data by an application. The approximation methods [Boyapati et al. (2017); Stevens et al. (2018); Wang et al. (2017); Xiao et al. (2019b)] discussed above include a quality management framework that is software-oriented where a programmer can assign the threshold error to remove significant approximation error during computations in the on-chip systems.

Besides the approximation-oriented communication techniques for improving quality in NoC-based systems, many researchers invest in a quality control technique to ensure a tolerable data error for an on-chip application. Zhang and Xu (2020) presented a quality management model called "*ApproxIt*", targeted to approximate the computing application. The method is a runtime quality calibration scheme for controlling the quality of the applications. Another quality management scheme called "*Approxilyzer*" is modeled by Venkatagiri et al. (2016). This solution enables quantifying the impact of single-bit errors on the on-chip data. Authors think that if the resulting error is controllable by managing error on traffic injection, then the accuracy in NoC performance can be improved. An essential requirement for such quality management frameworks includes specifications of the approximate variables by the programmers involved in the frameworks. Because result error is controlled through code analysis, these variables may limit the approximate communication techniques employed for maintaining NoC performance with an acceptable quality loss.

It is well-known that ML applications have grown significantly in computer architecture and embedded systems, e.g., NoC-enabled heterogeneous multicore communications systems. ML algorithms used in such applications generally target to manage various kinds of on-chip resources, e.g., active IP cores or PEs, frequency level, etc. [Xiao et al. (2019a)]. To achieve enhanced NoC performance, these resources are adjusted by using Reinforcement and imitation learning. Authors in [Xiao et al. (2019a)]. have preferred Reinforcement learning based on the task mapping. Further enhancement of the system performance is possible if the learning algorithms can dynamically explore other NoC resources, e.g., communication infrastructures- links, topology, different design trade-offs. For example, Savva et al. (2012) have presented an ANN for controlling interswitch communication links between adjacent NoC routers. In this way, Wang et al. (2019b) have introduced another Reinforcement learning-based holistic model called "*IntelliNoC*" for controlling error mitigation policy dynamically in NoCs.

6.10 ANALYTICAL ANALYSIS OF NoC PERFORMANCE

Performance prediction is one of the essential aspects of NoCs. NoC evaluation is done via assessing different performance parameters to check if the specific instance of NoC designs satisfies the system's needs. Performance evaluation in NoCs has been simulation-based for the last two decades. However, this traditional simulation is extremely slow, mainly when the NoC-based systems target the NoC size increases and big-data applications. Additionally, a simulation method provides a little insight into how various performance metrics are affected by NoC design parameters. Therefore, the simulation-based NoC evaluation is often neglected for optimization purposes [Ogras et al. (2010)]. Recently, AI technology like ML and DL, as discussed above, are continuously applied in evaluating NoC performance, including addressing the simulators' runtime.

NoC's performance is generally evaluated based on the following metrics: latency, power consumption, area, hop count, throughput, etc. [Foroutan et al. (2013)]. These metrics are modeled mathematically for NoC routers so that resource utilization and proper measurement can be done analytically during NoC performance. In the early stage of NoC designs, performance evaluation is done to measure the end-to-end latency, router latency, etc. [Hou et al. (2019)]. Later the design space for NoCs is explored to find their optimum configurations. The optimality is based on the different candidates rather than only architectural parameters such as topology, network links, and dynamic configuration method (switching strategy and routing algorithm), which directly include evaluation of other performance metrics, e.g., throughput with latency.

A packet experiences the delay at each router of an NoC with excellent granularity. It determines the average latency of packets in the network. Latency is the time taken by a packet to reach a destination from a source. It is divided into network and packet latency. The **average network latency (ANL)** measures the time taken by a packet to get to its destination across the network. The ANL is defined in Equation 6.1. Generally, this latency measures the packet's time to get its destination and back to the source again, hence known as a round trip delay (RTD). Expression for average network latency where L_k is the network latency of flit k, P is the number of transmitted flits. The **average packet latency (APL)** ratio is the average latency in time slots to the minimum packet delay in the time slot.

$$ANL = \frac{\Sigma_{k=1}^{P} L_k}{P} \tag{6.1}$$

Throughput of a network is the rate of successfully transmitted packets to the destination nodes. Equation 6.2 defines the network throughput $\propto$. Here, T is the total execution time, S is the number of packets, P_k is the k-th packet, and R denotes the number of routing nodes. It is observed that the packet delivery rate (PDR) is the same as the packet injection rate (PIR) when the traffic load is low. On the contrary, the network throughput starts saturating while the traffic load increases.

$$\propto = \frac{\Sigma_{k=1}^{S} P_k}{T \times R} \tag{6.2}$$

Hop Count is the total number of links that data packets traverse from the source core to the destination core. It is a rough measurement parameter of the distance between the source host and destination host [Phing et al. (2017)]. The following Equation 6.3 defines the average hop count (AHC) traveled by packets in a network. Where P_h refers to the hop count of the kth packet, N is the total number of packets that arrive at the destination node.

$$AHC = \frac{\Sigma_{k=1}^{N} P_h}{N} \tag{6.3}$$

Power is an essential concern in chip design. In NoC, power dissipation is caused by the transmission of flits in the network. There are different power parameters like switch power, channel power, etc. The switching power controls the flow of energy in NoCs. In NoC, Area constitutes the space occupied by switches, routers, interconnects, links, etc.

6.11 FUTURE DIRECTIONS

An NoC is a general-purpose on-chip communication solution. It has been introduced to cope with the growing demand for high-speed communication requirements of today's multiprocessor systems-on-chip, which run intensive computing applications. NoC performance parameters such as bandwidth, throughput, latency, power, etc. are the dominant consideration in the applications. To date, these parameters are evaluated by the conventional simulation method despite a slow procedure and employing recent powerful AI technology. Initially, an NoC is introduced as a two-dimensional (2D) on-chip communication network. One disadvantage of adopting a conventional 2D NoC is that it has limited floor-planning choices limiting performance enhancements for data-incentive networks. Oppositely, an NoC is an enabling solution that integrates large numbers of embedded processors (IP cores) in a single die. Thus, the continuing demand for high-performance requirements in low-latency, low-power interconnects necessitates a new on-chip communication infrastructure beyond traditional 2D NoCs. Among different possible alternatives, the designers and researchers envision later three additional important types of NoC designs: three-dimensional (3D) NoC, wireless NoC (WiNoC), and photonic NoC (PNoC or PhNoC) as one of the future generation on-chip communication networks, respectively.

6.11.1 Research on 3D NoCs

With the increasing size, a 2D NoC may not be suitable for efficient communication due to long routing distance in the NoC. By stacking active silicon layers, such integration mitigates the problem of interconnection wire delay by proposing a 3D NoC. Further, the 3D ICs can simultaneously achieve better functionality, performance, and packaging density than 2D ICs. Thus, 3D NoC architectures with numerous benefits can be considered in place of 2D NoCs to secure an unprecedented performance gain. Although a 3D NoC is one of the upcoming NoC architectures and offers several advantages over a 2D NoC, one of the major bottlenecks for 3D NoC-based communication systems is post-manufacturing testing of different components such as cores, routers, and channels for various manufacturing and transient faults because of the increased density and adopting the NoC promising alternative. While the cores perform computation, defect-free routers and media in a routing path are needed for reliable data exchange between sender and receiver. Otherwise, faulty components in a 3D NoC will cause different system-level failures that may significantly impact its performance. Related research direction in the 3D NoCs can be directed as follows: (1) improving reliability and yield in 3D NoCs through the testing of faults in the basic components, e.g., communication channels; (2) latent defect tolerance: developing efficient and effective online fault detection techniques in future 3D-NoCs to observe and find the root cause of performance degradation at the router level; (3) optimal application mapping for energy-efficient with high throughput capable 3D NoCs; (4) designing a fault-tolerant routing algorithm for QoS and resiliency to faults; (5) fault tolerance for emerging 3D-NoC architectures; (6) resiliency against the failure or the fabrication defects of TSVs; (7) formal verification of existing routing algorithms of 3D NoCs; (8) study of the relation between TSVs failure and 3D NoC's reliability; and (9) solving arising deadlock and livelock problems in 3D NoCs, etc.

6.11.2 Research on WiNoCs

Recent research on silicon integrated antenna has established that such antenna can operate in the millimeter (mm)-wave range from 10–100 gigahertz (GHz) and be accepted as a viable technology for intra-chip as well as inter-chip communication [Kim et al. (2006); Lin et al. (2007)]. In recent observations, excellent emission and absorption properties also lead to antenna-like behavior in carbon nanotubes (CNTs), operating at optical frequencies [Kempa et al. (2007)]. Nowadays, some efforts are being put to address this problem by introducing a new NoC architecture with ultra-low-latency and low-power express wireless channels. As a result, WiNoCs are coming in next-generation NoC-based communications [Deb et al. (2012); Kim et al. (2016)]. In a WiNoC, every node is supported with a transceiver that exchanges message signal with an antenna placed in the architecture. Such implementation reduces the multihop communication between highly separated nodes to a single hop due to the wireless channels. Some fundamental problems that require immediate studies are: (1) topology exploration for high-performance WiNoCs on integrating basic elements: transceivers, antennas, and waveguides; (2) designing architectural and circuit level schemes to improve the reliability for future pure WiNoCs; (3) addressing reliability and fault tolerance issues with a new test methodology for hybrid WiNoCs; (4) enhancement of the broadcast communication with WiNoCs on developing an optimal MAC mechanism; (5) multi-objective task mapping approaches for WiNoCs; (6) reduction of the power consumption through power profiling of the transceivers, dynamic power management strategies; (7) study different issues, such as error detection, error correction of wireless networks in perspectives of WiNoCs; (8) performance evaluation of WiNoCs at the combination of an optical layer with a wireless layer; (9) end-to-End QoS assurance in WiNoCs; and (10) inference management techniques for high performance enabled WiNoCs, etc.

6.11.3 Research on PhNoC

The design and performance of modern 2D NoCs and upcoming 3D NoCs are bounded by the power dissipated on a single die. With recent advances in silicon nanophotonics, PhNoC architectures are being explored to replace traditional 2D and 3D NoCs and enable higher bandwidth and lower power dissipation in future CMP communications [Chittamuru et al. (2017)]. Riding on the achievements of photonic technologies, a new type of NoC, which is in progress to be called optical network-on-chip (ONoC) and widely known as a PhNoC, is becoming a next-generation on-chip network. The PhNoC consists of two planes: electronic and optical. The electronic plane is used to perform network management and distributed control functions, short message exchange, etc. On the contrary, high-bandwidth multi-wavelength dedicated for bulk message transmission is accomplished by the optical plane [Kundu and Chattopadhyay (2014); Shacham et al. (2007)]. Some cutting-edge research on PhNoCs may include: (1) finding a new integration technology that integrates all necessary optical components to build full optical on-chip NoCs; (2) the reliability establishment in terms of the testing of faults in the channels, routers, cores in the electronic plane of the PhNoCs; (3) investigation of fault tolerance methods for a PhNoC; (4) reduction of crosstalk and other issues like static power consumption, insertion losses in the PhNoCs; (5) devising channel allocation schemes for homogeneous and heterogeneous PhNoCs; and (6) application-driven task mapping for the PhNoCs, etc.

6.12 CONCLUSION

This chapter has considered the applications of AI technology in NoC performance evaluation. The chapter has shown how NoCs are customized to make AI chips. Additionally, it has discussed shifting conventional simulation methods by AI technology to maintain and control the QoS, performance accuracy, etc. Further, different mathematical models have been sought for performance analysis. Finally, a set of research dimensions might extend the literature in the combined field of NoCs and AI technology.

ACKNOWLEDGEMENTS

The author would like to express many thanks to his Post-Graduate Students, especially Ms. Pallabi Hazarika, Ms. Prachi Kale, and Mr. Sajal Jain, for preparing this chapter. The author also expresses his sincere gratitude to the anonymous referees for their corrections and comments on this chapter.

REFERENCES

Amano, H. (2013). Tutorial: Introduction to interconnection networks from system area network to network on chips. In 2013 First International Symposium on Computing and Networking, pp. 15–16.

Betzel, F., K. Khatamifard, H. Suresh, D. J. Lilja, J. Sartori, and U. Karpuzcu (2018). Approximate communication: Techniques for reducing communication bottlenecks in large-scale parallel systems. ACM Computing Surveys 51(1), pp. 1–32.

Bhowmik, B. (2019). A power-aware fault detection scheme for 2d mesh-based network-on-chip interconnects. Journal of Low Power Electronics 15(2), pp. 256–272.

Bhowmik, B. (2020). Maximal connectivity test with channel-open faults in on-chip communication networks. Journal of Electronic Testing 36(3), pp. 385–408.

Bhowmik, B. (2021). Dugdugi: An optimal fault addressing scheme for octagon-like on-chip communication networks. IEEE Transactions on Very Large Scale Integration (VLSI) Systems 29(5), pp. 1009–1021.

Bhowmik, B., S. Biswas, and J. K. Deka (2015). A packet address driven test strategy for stuck-at faults in networks-on-chip interconnects. In IEEE 23rd Mediterranean Conference on Control and Automation (MED), pp. 176–183.

Bhowmik, B., S. Biswas, and J. K. Deka (2016). A reliability-aware topology-agnostic test scheme for detecting, and diagnosing interconnect shorts in on-chip networks. In IEEE 18th International Conference on High Performance Computing and Communications (HPCC), pp. 530–537.

Bhowmik, B., S. Biswas, J. K. Deka, and B. B. Bhattacharya (2018). Reliability-aware test methodology for detecting short-channel faults in on-chip networks. IEEE Transactions on Very Large Scale Integration (VLSI) Systems 26(6), pp. 1026–1039.

Bhowmik, B., J. K. Deka, and S. Biswas (2016a). Towards a scalable test solution for the analysis of interconnect shorts in on-chip networks. In IEEE 24th International Symposium on Modeling, Analysis and Simulation of Computer and Telecommunication Systems (MASCOTS), pp. 394–399.

Bhowmik, B., J. K. Deka, and S. Biswas (2016b). When clustering shows optimality towards analyzing stuck-at faults in channels of on-chip networks. In IEEE 18th International Conference on High Performance Computing and Communications (HPCC), pp. 868–875.

Bhowmik, B., J. K. Deka, and S. Biswas (2017). Charka: A reliability-aware test scheme for diagnosis of channel shorts beyond mesh NoCs. In IEEE/ACM 20th Design, Automation, and Test in Europe (DATE) Conference, pp. 214–219.

Bhowmik, B., J. K. Deka, and S. Biswas (2020a). Improving reliability in spidergon network on chip-microprocessors. In 2020 IEEE 63rd International Midwest Symposium on Circuits and Systems (MWSCAS), pp. 474–477.

Bhowmik, B., J. K. Deka, and S. Biswas (2020b). Reliability monitoring in a smart NoC component. In 2020 27th IEEE International Conference on Electronics, Circuits and Systems (ICECS), pp. 1–4.

Bhowmik, B., J. K. Deka, S. Biswas, and B. B. Bhattacharya (2019a). A low-cost test solution for reliable communication in networks-on-chip. Journal of Electronic Testing 35(2), pp. 215–243.

Bhowmik, B., J. K. Deka, S. Biswas, and B. B. Bhattacharya (2019b). Performance-aware test scheduling for diagnosing coexistent channel faults in topology-agnostic networks-on-chip. ACM Transactions on Design Automation of Electronic Systems 24(2), pp. 1–29.

Bhowmik, B., S. A. Varna, A. Kumar, and R. Kumar (2021a). Reducing false prediction on covid-19 detection using deep learning. In 2021 IEEE International Midwest Symposium on Circuits and Systems (MWSCAS), pp. 404–407.

Bhowmik, B., P. Hazarika, P. Kale, and S. Jain (2021b). AI technology for NoC performance evaluation. IEEE Transactions on Circuits and Systems II (TCAS II) *(Special Issue on IEEE APCCAS 2021)* 68(62), pp. 3483–3487.

Boyapati, R., J. Huang, P. Majumder, K. H. Yum, and E. J. Kim (2017). APPROX-NoC: A data approximation framework for network-on-chip architectures. SIGARCH Computer Architecture News 45(2), pp. 666–677.

Catania, V., A. Mineo, S. Monteleone, M. Palesi, and D. Patti (2016). Cycle-accurate network on chip simulation with noxim. ACM Transactions on Modeling and Computer Simulation 27(1), pp. 1–25.

Chen, K.-C. and T.-Y. Wang (2018). Nn-noxim: High-level cycle-accurate NoC-based neural networks simulator. In 2018 11th International Workshop on Network on Chip Architectures (NoCArc), pp. 1–5.

Chen, K.-C. J. and Y.-H. Liao (2019). Online machine learning-based temperature prediction for thermal-aware NoC system. In 2019 International SoC Design Conference (ISOCC), pp. 65–66.

Chen, K.-C. J. and Y.-H. Liao (2020). Adaptive machine learning-based temperature prediction scheme for thermal-aware NoC system. In 2020 IEEE International Symposium on Circuits and Systems (ISCAS), pp. 1–4.

Chen, Y. and A. Louri (2020). Learning-based quality management for approximate communication in network-on-chips. IEEE Transactions on Computer-Aided Design of Integrated Circuits and Systems 39(11), pp. 3724–3735.

Chen, Y., T. Luo, S. Liu, S. Zhang, L. He, J. Wang, L. Li, T. Chen, Z. Xu, N. Sun, and O. Temam (2014). Dadiannao: A machine-learning supercomputer. In Proceedings of the 47th Annual IEEE/ACM International Symposium on Microarchitecture, MICRO-47, USA, pp. 609–622.

Chen, Y.-H., T.-J. Yang, J. Emer, and V. Sze (2019). Eyeriss v2: A flexible accelerator for emerging deep neural networks on mobile devices. IEEE Journal on Emerging and Selected Topics in Circuits and Systems 9(2), pp. 292–308.

Chi, Y., W. Cui, L. Qiao, X. Long, and L. Wang (2019). Hardware architecture design of the deep learning-based machine vision chip. In 2019 IEEE 3rd Information Technology, Networking, Electronic and Automation Control Conference (ITNEC), pp. 1110–1113.

Chittamuru, S. V. R., S. Desai, and S. Pasricha (2017). SWIFTNoC: A reconfigurable silicon-photonic network with multicast-enabled channel sharing for multicore architectures. Journal on Emerging Technologies in Computing Systems 13(4), pp. 1–27.

Choi, W., K. Duraisamy, R. G. Kim, J. R. Doppa, P. P. Pande, D. Marculescu, and R. Marculescu (2018). On-chip communication network for efficient training of deep convolutional networks on heterogeneous manycore systems. IEEE Transactions on Computers 67(5), pp. 672–686.

Das, S., J. R. Doppa, D. H. Kim, P. P. Pande, and K. Chakrabarty (2015). Optimizing 3d NoC design for energy efficiency: A machine learning approach. In 2015 IEEE/ACM International Conference on Computer-Aided Design (ICCAD), pp. 705–712.

Das, S., J. R. Doppa, P. P. Pande, and K. Chakrabarty (2016). Design-space exploration and optimization of an energy-efficient and reliable 3-d small-world network-on-chip. IEEE Transactions on Computer-Aided Design of Integrated Circuits and Systems 36(5), pp. 719–732.

Deb, S., A. Ganguly, P. P. Pande, B. Belzer, and D. Heo (2012). Wireless NoC as interconnection backbone for multicore chips: Promises and challenges. IEEE Journal on Emerging and Selected Topics in Circuits and Systems 2(2), pp. 228–239.

Dong, Y., Y. Wang, Z. Lin, and T. Watanabe (2009). High performance and low latency mapping for neural network into network on chip architecture. In 2009 IEEE 8th International Conference on ASIC, pp. 891–894.

Fettes, Q., M. Clark, R. Bunescu, A. Karanth, and A. Louri (2018). Dynamic voltage and frequency scaling in NoCs with supervised and reinforcement learning techniques. IEEE Transactions on Computers 68(3), pp. 375–389.

Firuzan, A., M. Modarressi, and M. Daneshtalab (2015). Reconfigurable communication fabric for efficient implementation of neural networks. In 2015 10th International Symposium on Reconfigurable Communication-centric Systems-on-Chip (ReCoSoC), pp. 1–8.

Foroutan, S., Y. Thonnart, and F. Petrot (2013). An iterative computational technique for performance evaluation of networks-on-chip. IEEE Transactions on Computers 62(8), pp. 1641–1655.

Gao, W. and P. Zhou (2019). Customized high performance and energy efficient communication networks for AI chips. IEEE Access 7, pp. 69434–69446.

Hazarika, P., P. Kale, S. Jain, and B. Bhowmik (2021). AI technology for NoC performance evaluation. In 2021 IEEE 17th Asia Pacific Conference on Circuits and Systems (APCCAS), pp. 1–4.

Hestness, J., S. W. Keckler, and D. A. Wood (2015). GPU computing pipeline inefficiencies and optimization opportunities in heterogeneous CPU-GPU processors. In 2015 IEEE International Symposium on Workload Characterization, pp. 87–97.

Hou, J., Q. Han, and M. Radetzki (2019). A machine learning enabled long-term performance evaluation framework for NoCs. In 2019 IEEE 13th International Symposium on Embedded Multicore/Many-core Systems-on-Chip (MCSoC), pp. 164–171.

Hu, Y., M. Mettler, D. Mueller-Gritschneder, T. Wild, A. Herkersdorf, and U. Schlichtmann (2020). Machine learning approaches for efficient design space exploration of application-specific NoCs. ACM Transactions on Design Automation of Electronic Systems (TODAES) 25(5), pp. 1–27.

Jain, S., P. Kale, P. Hazarika, and B. Bhowmik (2021). A machine learning enabled NoC performance evaluation. In 2021 IEEE 25th High Performance Extreme Computing (HPEC), pp. 1–6.

Jiang, N., D. U. Becker, G. Michelogiannakis, J. Balfour, B. Towles, D. E. Shaw, J. Kim, and W. J. Dally (2013). A detailed and flexible cycle-accurate network-on-chip simulator. In 2013 IEEE International Symposium on Performance Analysis of Systems and Software (ISPASS), pp. 86–96.

Joseph, J. M. and T. Pionteck (2014). A cycle-accurate network-on-chip simulator with support for abstract task graph modeling. In 2014 International Symposium on System-on-Chip (SoC), pp. 1–6.

Kahng, A. B., B. Li, L.-S. Peh, and K. Samadi (2012). Orion 2.0: A power-area simulator for interconnection networks. IEEE Transactions on Very Large Scale Integration (VLSI) Systems 20(1), pp. 191–196.

Kempa, K., J. Rybczynski, Z. Huang, K. Gregorczyk, A. Vidan, B. Kimball, J. Carlson, G. Benham, Y. Wang, A. Herczynski, et al. (2007). Carbon nanotubes as optical antennae. Advanced Materials 19(3), pp. 421–426.

Kim, K., B. Floyd, J. Mehta, H. Yoon, C.-m. Hung, D. Bravo, T. Dickson, X. Guo, R. Li, and N. Trichy, et al. (2006). Silicon integrated circuits incorporating antennas. In IEEE Custom Integrated Circuits Conference, pp. 473–480.

Kim, R. G., W. Choi, Z. Chen, P. P. Pande, D. Marculescu, and R. Marculescu (2016, July). Wireless NoC and dynamic VFI codesign: Energy efficiency without performance penalty. IEEE Transactions on Very Large Scale Integration (VLSI) Systems 24(7), pp. 2488–2501.

Kumar, A. and B. Talawar (2019). Floorplan based performance estimation of network-on-chips using regression techniques. In 2019 IEEE 5th International Conference for Convergence in Technology (I2CT), pp. 1–6.

Kundu, S. and S. Chattopadhyay (2014). Network-on-Chip: The Next Generation of System-on-Chip Integration. CRC Press, Taylor & Francis Group.

Lee, K., S.-J. Lee, and H.-J. Yoo (2006). Low-power network-on-chip for high-performance SoC design. IEEE Transactions on Very Large Scale Integration (VLSI) Systems 14(2), pp. 148–160.

Lee, J., S. Li, H. Kim, and S. Yalamanchili (2013, December). Design space exploration of on-chip ring interconnection for a CPU-GPU heterogeneous architecture. Journal of Parallel and Distributed Computing 73(12), pp. 1525–1538.

Lin, J.-J., H.-T. Wu, Y. Su, L. Gao, A. Sugavanam, J. E. Brewer, et al. (2007). Communication using antennas fabricated in silicon integrated circuits. IEEE Journal of Solid-State Circuits 42(8), pp. 1678–1687.

Liu, X., W. Wen, X. Qian, H. Li, and Y. Chen (2018). Neu-NoC: A high-efficient interconnection network for accelerated neuromorphic systems. In 2018 23rd Asia and South Pacific Design Automation Conference (ASP-DAC), pp. 141–146.

Lowe-Power, J., A. M. Ahmad, A. Akram, M. Alian, R. Amslinger, M. Andreozzi, A. Armejach, and Others (2020). The gem5 simulator: Version 20.0+, CoRR.

Mello, A., N. Calazans, and F. Moraes (2011). Vieira-de-Mello, Aline. "ATLAS-An Environment for NoC Generation and Evaluation."

Millberg, M., E. Nilsson, R. Thid, S. Kumar, and A. Jantsch (2004). The nostrum backbone-A communication protocol stack for networks on chip. In IEEE 17th International Conference on VLSI Design. Proceedings, pp. 693–696.

Mishra, M. and M. Srivastava (2014). A view of artificial neural network. In 2014 International Conference on Advances in Engineering Technology Research (ICAETR - 2014), pp. 1–3.

Morgan, A. A., A. S. Hassan, M. W. El-Kharashi, and A. Tawfik (2020). NoC2: An efficient interfacing approach for heavily-communicating NoC-based systems. IEEE Access 8, pp. 185992–186011.

Narayana, S. A. (2016). An artificial neural networks based temperature prediction framework for network-on-chip based multicore platform. CoRR abs/1612.04197.

Ogras, U. and R. Marculescu (2006). "It's a small world after all": NoC performance optimization via long-range link insertion. IEEE Transactions on Very Large Scale Integration (VLSI) Systems 14(7), pp. 693–706.

Ogras, U. Y., P. Bogdan, and R. Marculescu (2010). An analytical approach for network-on-chip performance analysis. IEEE Transactions on Computer-Aided Design of Integrated Circuits and Systems 29(12), pp. 2001–2013.

Pagallo, W. B. U. (2020). Advanced Introduction to Law and Artificial Intelligence. Cheltenham, UK: Edward Elgar Publishing.

Painkras, E., L. A. Plana, J. Garside, S. Temple, S. Davidson, J. Pepper, D. Clark, C. Patterson, and S. Furber (2012). Spinnaker: A multi-core system-on-chip for massively-parallel neural net simulation. In Proceedings of the IEEE 2012 Custom Integrated Circuits Conference, pp. 1–4.

Pérez, I., E. Vallejo, M. Moretó, and R. Beivide (2020). BST: A BookSim-based toolset to simulate NoCs with single- and multi-hop bypass. In 2020 IEEE International Symposium on Performance Analysis of Systems and Software (ISPASS), pp. 47–57.

Phing, N. Y., M. M. Warip, P. Ehkan, F. Zulkefli, and R. B. Ahmad (2017). Topology design of extended torus and ring for low latency network-on-chip architecture. Telecommunication Computing Electronics and Control 15(2), pp. 869–876.

Ping, L. B., C. P. Kit, and E. K. Karuppiah (2013). Network latency prediction using high accuracy prediction tree. In ACM Proceedings of the 7th International Conference on Ubiquitous Information Management and Communication, ICUIMC '13, pp. 1–8.

Qian, Z.-L., D.-C. Juan, P. Bogdan, C.-Y. Tsui, D. Marculescu, and R. Marculescu (2015). A support vector regression (SVR)-based latency model for network-on-chip (NoC) architectures. IEEE Transactions on Computer-Aided Design of Integrated Circuits and Systems 35(3), pp. 471–484.

Rafie, M., A. Khadem-Zadeh, and M. Reshadi (2014). Performance improvement of application-specific network on chip using machine learning algorithms. International Journal of High Performance Systems Architecture 5, pp. 71–83.

Ramanujam, R. S., V. Soteriou, B. Lin, and L.-S. Peh (2010). Design of a high-throughput distributed shared-buffer NoC router. In 2010 Fourth ACM/IEEE International Symposium on Networks-on-Chip, pp. 69–78.

Rao, N., A. Ramachandran, and A. Shah (2018). MLNoC: A machine learning based approach to NoC design. In 2018 IEEE 30th International Symposium on Computer Architecture and High Performance Computing (SBAC-PAD), pp. 1–8.

Sant'Ana, A. C., H. M. Medina, K. B. Fiorentin, and F. G. Moraes (2019). Lightweight security mechanisms for MPSoCs. In 2019 32nd Symposium on Integrated Circuits and Systems Design (SBCCI), pp. 1–6.

Savva, A. G., T. Theocharides, and V. Soteriou (2012, January). Intelligent on/off dynamic link management for on-chip networks. Journal of Electrical and Computer Engineering 6(1), p. 1.

Shacham, A., B. G. Lee, A. Biberman, K. Bergman, and L. P. Carloni (2007, August). Photonic NoC for DMA communications in chip multiprocessors. In Symposium on High-Performance Interconnects (HOTI), pp. 29–38.

Silva, J., M. Kreutz, M. Pereira, and M. Da Costa-Abreu (2019). An investigation of latency prediction for NoC-based communication architectures using machine learning techniques. Journal of Supercomputing 75(11), pp. 7573–7591.

Stevens, J. R., A. Ranjan, and A. Raghunathan (2018). AxBA: An approximate bus architecture framework. In 2018 IEEE/ACM International Conference on Computer-Aided Design (ICCAD), pp. 1–8.

Thiem, C. V., and S. Oyanagi (2011). An input buffer architecture for on-chip routers. In 2011 Second International Conference on Networking and Computing, pp. 280–283.

Venkatagiri, R., A. Mahmoud, S. K. S. Hari, and S. V. Adve (2016). Approxilyzer: Towards a systematic framework for instruction-level approximate computing and its application to hardware resiliency. In 2016 49th Annual IEEE/ACM International Symposium on Microarchitecture (MICRO), pp. 1–14.

Wächter, E. W., C. De Bellefroid, K. R. Basireddy, A. K. Singh, B. M. Al-Hashimi, and G. Merrett (2019). Predictive thermal management for energy-efficient execution of concurrent applications on heterogeneous multicores. IEEE Transactions on Very Large Scale Integration (VLSI) Systems 27(6), pp. 1404–1415.

Wang, K., A. Louri, A. Karanth, and R. Bunescu (2019a). High-performance, energy-efficient, fault-tolerant network-on-chip design using reinforcement learning. In 2019 IEEE Design, Automation & Test in Europe Conference & Exhibition (DATE), pp. 1166–1171.

Wang, K., A. Louri, A. Karanth, and R. Bunescu (2019b). IntelliNoC: A holistic design framework for energy-efficient and reliable on-chip communication for manycores. In 2019 ACM/IEEE 46th Annual International Symposium on Computer Architecture (ISCA), pp. 1–12.

Wang, L., X. Wang, and Y. Wang (2017). ABDTR: Approximation-based dynamic traffic regulation for networks-on-chip systems. In 2017 IEEE International Conference on Computer Design (ICCD), pp. 153–160.

Wettin, P., R. Kim, J. Murray, X. Yu, P. P. Pande, A. Ganguly, and D. Heoamlan (2014). Design space exploration for wireless NoCs incorporating irregular network routing. IEEE Transactions on Computer-Aided Design of Integrated Circuits and Systems 33(11), pp. 1732–1745.

Wolf, W., A. A. Jerraya, and G. Martin (2008). Multiprocessor system-on-chip (MPSoC) technology. IEEE Transactions on Computer-Aided Design of Integrated Circuits and Systems 27(10), pp. 1701–1713.

Wu, X., J. Xu, Y. Ye, X. Wang, M. Nikdast, Z. Wang, and Z. Wang (2015). An inter/intra-chip optical network for manycore processors. IEEE Transactions on Very Large Scale Integration (VLSI) Systems 23(4), pp. 678–691.

Xiao, Y., S. Nazarian, and P. Bogdan (2019a). Self-optimizing and self-programming computing systems: A combined compiler, complex networks, and machine learning approach. IEEE Transactions on Very Large Scale Integration (VLSI) Systems 27(6), pp. 1416–1427.

Xiao, S., X. Wang, M. Palesi, A. K. Singh, and T. Mak (2019b). ACDC: An accuracy- and congestion-aware dynamic traffic control method for networks-on-chip. In 2019 Design, Automation Test in Europe Conference Exhibition (DATE), pp. 630–633.

Yin, J., S. Sethumurugan, Y. Eckert, C. Patel, A. Smith, E. Morton, M. Oskin, N. Enright Jerger, and G. H. Loh (2020). Experiences with ML-driven design: A NoC case study. In 2020 IEEE International Symposium on High Performance Computer Architecture (HPCA), pp. 637–648.

Zhang, Q. and Q. Xu (2020). Approxit: A quality management framework of approximate computing for iterative methods. IEEE Transactions on Computer-Aided Design of Integrated Circuits and Systems 39(5), pp. 991–1002.

Zhao, Y., F. Ge, C. Cui, F. Zhou, and N. Wu (2020). A mapping method for convolutional neural networks on network-on-chip. In 2020 IEEE 20th International Conference on Communication Technology (ICCT), pp. 916–920.

Zheng, H. and A. Louri (2019). An energy-efficient network-on-chip design using reinforcement learning. In Proceedings of the 56th Annual Design Automation Conference 2019, pp. 1–6.

7 Modeling, Analysis, and Simulation of Hydrogen Leakage Jet in the Air

Mohamed F. El-Amin

CONTENTS

7.1 INTRODUCTION

Although restricted amounts of fossil fuels are available on Earth, the world's effort to develop alternative energy sources to meet our demands has been slow. Fuels derived from fossil sources often have high energy content and density, which encourages their use in various applications in our daily lives. However, the hunt for cleaner fuels that meet our demands with little impact on the environment has never concluded. Hydrogen has a high energy content and occupies a large volume under normal conditions. As a result, hydrogen must be stored in a compressed form to maximize its energy density. Furthermore, storing hydrogen is far more difficult than storing other fuel gases as hydrogen gas molecules are the tiniest of all other gases and may flow through any material. Hydrogen energy evokes hope due to its high energy potential and clean combustion and fears as it is dangerous. Since May 6, 1937, the Hindenburg tragedy created hydrogen anxiety known as the "Hindenburg syndrome," and hydrogen has been considered excessively dangerous. The best way to assuage fears is to conduct an objective examination of hydrogen dangers to identify and implement preventive and protective measures to prevent similar catastrophes from occurring in the future or minimize the consequences to a bare minimum (see Figure 7.1).

Hydrogen is more difficult to store than other gases due to its characteristics. Hydrogen leaks are hazardous because they can cause a fire if they contact air (see Figure 7.2). Hydrogen is combustible and explosive in various concentrations, 4–75%, and 15–59%, respectively, under typical conditions. Furthermore, hydrogen is easy to ignite and fire could be started by an unnoticed spark or static electrical discharge. As a result, ensuring

DOI: 10.1201/9781003229018-7

FIGURE 7.1 Hindenburg hydrogen balloon disaster (1937).

high-quality leak-proofing is critical for hydrogen plants. Long-term exposure of the container results in hydrogen embrittlement, leading to fracturing and, therefore, hydrogen leakage. Hydrogen leakage may occur for instance in main infrastructures, usage and distribution. Hydrogen leaks can be caused by faulty components, vents, etc. The shape of the leakage source is an important aspect that has a major effect on how leakage flow and mixtures are characterized in the air. There are two different types of hydrogen leakage: flammable quick leaks and flammable slow-leaks. The first type of hydrogen leaking can be described using traditional turbulent jet flame models. This chapter concentrates on the second category concerned with slow hydrogen leaks.

This chapter discusses a variety of hydrogen leakage scenarios. The chapter begins with an overview of the available literature on hydrogen leaks in the atmosphere. The vertical and horizontal hydrogen-free turbulent plumes are discussed.

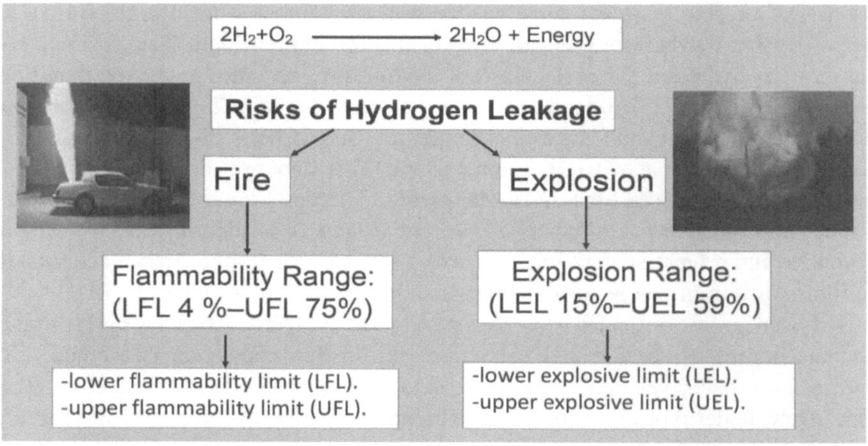

FIGURE 7.2 Hydrogen leakage results classification.

Momentum-dominated and buoyancy-dominated regimes of the hydrogen plume are highlighted. The methodology that has been used in this chapter is the analytical, semi-analytical and numerical methods. Selecting the suitable method depends on the complexity of the governing equations, as shown throughout the chapter. A set of expressions such as velocity-concentration correlation, turbulent-Reynolds-stress, etc., are calculated analytically. Also, we created and analyzed similarity solutions for a buoyant hydrogen jet caused by slow hydrogen leakage in the air. Third, a horizontal hydrogen buoyant jet has been studied with similar findings to the above two categories.

7.2 HYDROGEN LEAKAGES IN OPEN SPACE

There are two types of hydrogen leaking, the momentum-dominated-turbulent-jet (high Froude number) and buoyancy-dominated-turbulent-jet (low Froude number).

7.2.1 Jet and Plume Theory

When fluid leaves a nozzle with initial momentum, it forms a jet or plume. A non-buoyant flow created by a large initial momentum flux is known as a pure jet. The buoyancy flux is the sole cause of the pure plume; however, the initial momentum flux and the buoyancy flux together can form a buoyant jet. Rajaratnam [1], List [2], and Fischer et al. [3] gave comprehensive reviews of jets. Turner [4] and Shabbir and Georg [5] studied a turbulent vertical circular jet and plume behavior. The flow gets turbulent far away from the inlet, a distance approximately equals 10d, such that d is the inlet diameter. The jet source spreads conically far away from the source (Figure 7.3), and

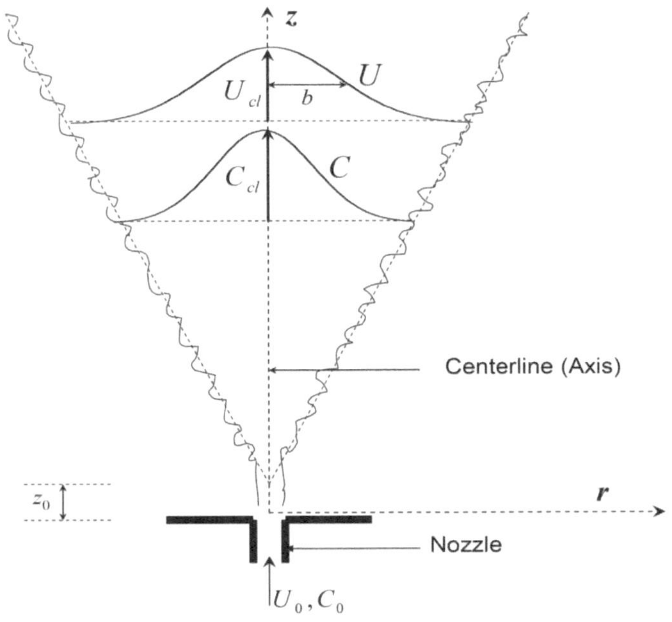

FIGURE 7.3 Turbulent jet.

TABLE 7.1
Mean Variables and Coefficients of Jets and Plumes

	Jet	Plume
Width	$b(z) = c_m(z - z_0)$	$b(z) = c_m(z - z_0)$
Centerline velocity	$U_{cl}(z) = \dfrac{AdU_0}{z - z_0}$	$U_{cl}(z) = \dfrac{AF_0^{1/3}}{(z - z_0)^{1/3}}$
Centerline concentration	$C_{cl}(z) = \dfrac{5.6\sqrt{\pi}\, dC_0}{2(z - z_0)}$	$C_{cl}(z) = \dfrac{9.4Y}{F_0^{1/3}(z - z_0)^{5/3}}$
c_m	0.1–0.13	0.1–0.13
A	5–7	3.9–4.7
Entrainment coefficient	$\alpha = c_m/2$	$\alpha = 5c_m/6$

the flow becomes self-similar after a 10d distance. Although the actual flow is turbulent and three-dimensional, the average mean flow seems steady and two-dimensional.

In Figure 7.3, r is the horizontal distance and z the distance along the centerline. The velocity scale is represented by the mean centerline velocity U_{cl}, while the length scale is the jet width b. The mean flow (similarity) variables are shown in Table 7.1 [3, 5].

where

$$F_0 = Kd^2U_0g\Delta\rho_0 / \rho_\infty, K = \pi / (1+\lambda^2), \lambda = c_m / c_c \tag{7.1}$$

$$Y = QC_0 = \frac{\pi}{4}d^2U_0C_0, Q = \frac{\pi}{4}d^2U_0 \tag{7.2}$$

In the above equations:

c_m momentum spread-rate,
C_0 concentration at the nozzle,
F_0 buonancy flux,
Q volume flux,
U_0 velocity at the nozzle
Y mass flux,
z_0 virtual origin.

In this chapter, we assume $z_0 = 0$, consequently, the mean variables are reduced to the following forms (Table 7.2):

7.2.2 Momentum-Dominated Hydrogen Jet

The circular hydrogen jet is mainly created from circular source, such as pinholes. Hydrogen jet is similar to helium jet at the momentum-dominated region [6]. It has been found that U_{cl} and C_{cl} vary inversely with z, while the jet width behaves as a linear function of z [1–6], thus,

$$U_{cl}(z) = \frac{AdU_0}{(z - z_0)} \tag{7.3}$$

TABLE 7.2
Reduced Mean Variables and Coefficients of Jets and Plumes

	Jet	Plume
Width	$b = 2\alpha z$	$b = (6\alpha/5)z$
Centerline velocity	$U_{cl}(z) = \dfrac{AdU_0}{z}$	$U_{cl}(z) = \dfrac{AF_0^{1/3}}{z^{1/3}}$
Centerline concentration	$C_{cl}(z) = \dfrac{5.6\sqrt{\pi}\, dC_0}{2z}$	$C_{cl}(z) = \dfrac{9.4Y}{F_0^{1/3} z^{5/3}}$

$$C_{cl}(z) = \frac{d^*}{K_C(z - z_0)} = \frac{d(\rho_{H_2} / \rho_{air})^{1/2}}{K_C(z - z_0)} \tag{7.4}$$

$$b(z) = c_m(z - z_0) \tag{7.5}$$

In general, $c_m = 0.1 \sim 0.13$, while $A = 5 \sim 7$ [3, 4]. For the hydrogen jet, $c_m = 0.103$ and $c_c = 0.13$ such that $c_c \neq c_m$. Therefore, the concentration profile can be rewritten in this form:

$$\begin{aligned} C(r,z) &= C_{cl}(z)\exp(-\lambda^2 r^2 / b^2) \\ &= C_{cl}(z)\exp(-\lambda^2 \eta^2) \\ &= C_{cl}(z)\exp(-0.63\eta^2) \end{aligned} \tag{7.6}$$

where:

$$\lambda = c_m / c_c,$$

$$\eta = r / b(z) = r / c_m(z - z_0).$$

The streamwise velocity and concentration behave as a Gaussian distribution [2, 4, 6–8], namely:

$$U(r,z) = U_{cl}(z)\exp\left(-\frac{r^2}{b^2(z)}\right) = U_{cl}(z)\exp(-\eta^2) \tag{7.7}$$

$$C(r,z) = C_{cl}(z)\exp\left(-\lambda^2 \frac{r^2}{b^2(z)}\right) = C_{cl}(z)\exp(-\lambda^2\eta^2) \tag{7.8}$$

The self-similar axisymmetric governing equations are given as [9–12]:

$$V\frac{\partial U}{\partial r} + U\frac{\partial U}{\partial z} + \frac{1}{r}\frac{\partial(r\overline{uv})}{\partial r} = 0 \tag{7.9}$$

$$V\frac{\partial C}{\partial r} + U\frac{\partial C}{\partial z} + \frac{1}{r}\frac{\partial(r\overline{vc})}{\partial r} = 0 \tag{7.10}$$

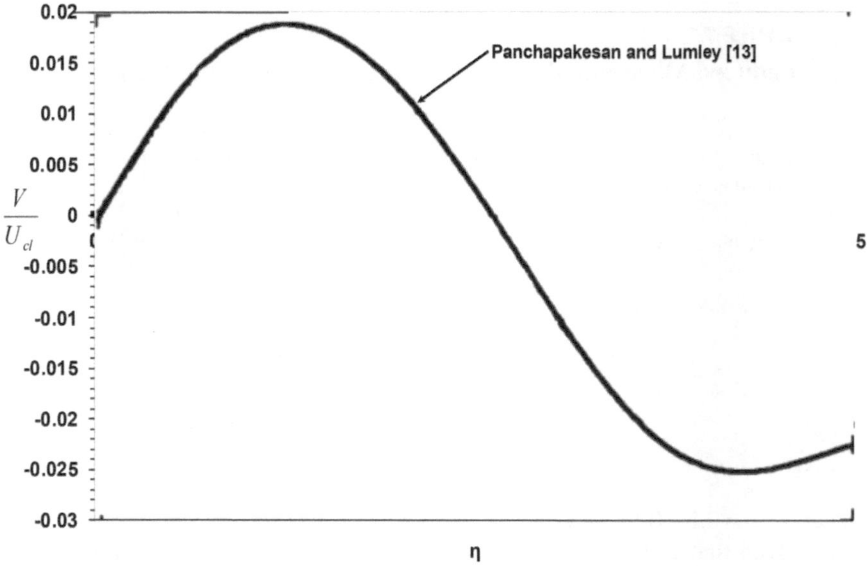

FIGURE 7.4 Validation of the velocity V/U_{cl} [13].

By integration Equation (7.9), one gets the following expression (Figure 7.4):

$$\frac{V}{U_{cl}} = \frac{c_m}{2\eta}\left(-1+\exp(-\eta^2)+2\eta^2\exp(-\eta^2)\right) \tag{7.11}$$

which is a normalized form.

The cross-stream velocity becomes outward close to the centerline and vanishes at $0 \le \eta \le 1.12$ with a maximum value of 0.0173 and approaches zero at $\eta = 1.12$. Also, it becomes negative (inward, $V(\eta)/U_{cl} < 0$) with a minimum of -0.0216 with an asymptote of 0 as $\eta \to \infty$.

From Equations (7.7), (7.9) and (7.11) with integration, the normalized Reynolds stress can be obtained as (Figure 7.5):

$$\frac{\overline{uv}}{U_{cl}^2} = \frac{c_m}{2\eta}\exp(-\eta^2)\left(1-\exp(-\eta^2)\right) \tag{7.12}$$

It is noteworthy that $max(\overline{uv}/U_{cl}^2) = 0.0181$ at $\eta = 0.6$.

Similarly, inserting U, V and C into Equation (7.10), to find the normalized time-averaged velocity-concentration correlation as (Figure 7.6):

$$\frac{\overline{vc}}{U_{cl}^2} = \frac{c_m}{2\eta}\exp(-\lambda^2\eta^2)\left(1-\exp(-\eta^2)\right) \tag{7.13}$$

Note that $\max(\overline{vc}/U_{cl}^2) = 0.021$ at $\eta = 0.7$. More details can be found in [13].

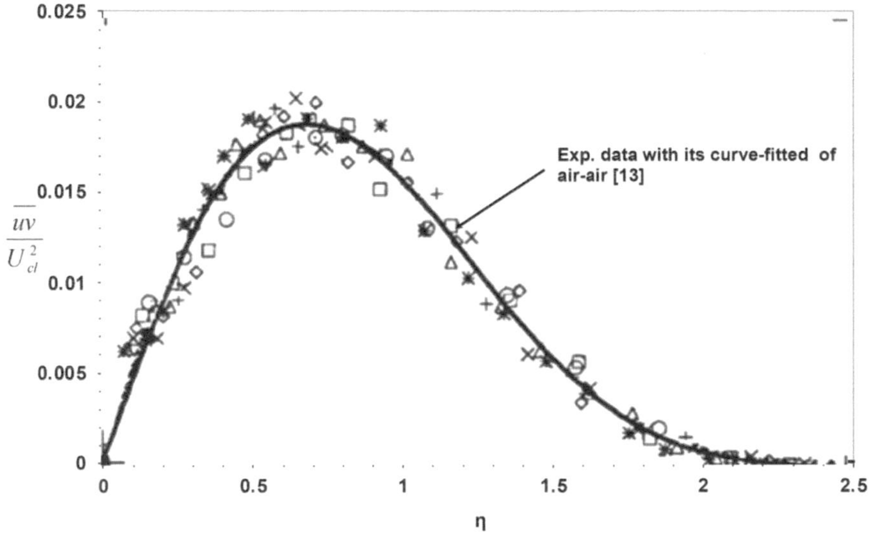

FIGURE 7.5 Comparison of the Reynolds stress for the momentum-dominated case [13].

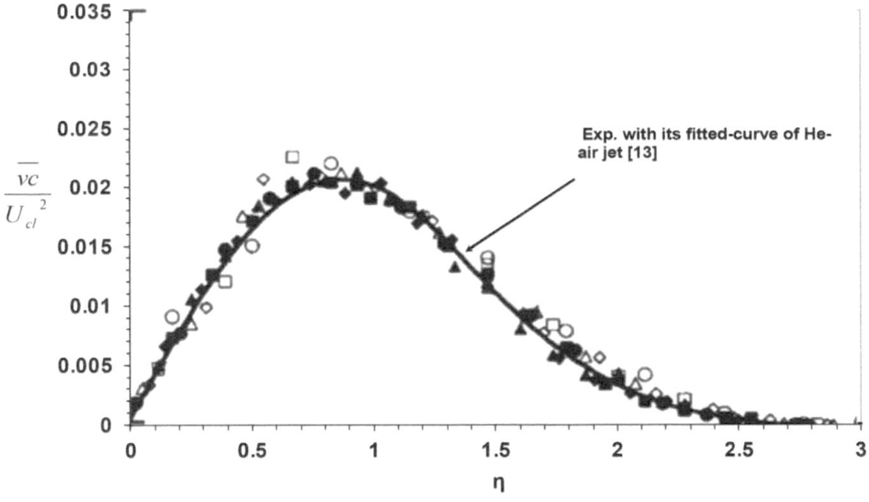

FIGURE 7.6 Validation of the time-averaged velocity-concentration [13].

7.2.3 Buoyant-Dominated Hydrogen Jet

This section obtains and discusses similarity solutions of the hydrogen plume. The density of the H_2-air mixture is a function of concentration. The buoyant jet's polar continuity, momentum, and concentration equations are as follows:

$$\frac{\partial(r\rho V)}{\partial r}+\frac{\partial(r\rho U)}{\partial z}=0 \tag{7.14}$$

$$\frac{\partial(r\rho UV)}{\partial r}+\frac{\partial(r\rho U^2)}{\partial z}+\frac{\partial(r\rho\overline{uv})}{\partial r}=gr(\rho-\rho_\infty) \quad (7.15)$$

$$\frac{\partial(r\rho VC)}{\partial r}+\frac{\partial(r\rho UC)}{\partial z}+\frac{\partial(r\rho\overline{vc})}{\partial r}=0 \quad (7.16)$$

When determining the entrainment rate, the ratio of the ambient densities is also considered. The local entrainment rate consists of entrainment owing to momentum, and the entrainment related to the buoyancy. Holding the ideal gas assumption, one may write:

$$U(r,z)=U_{cl}(z)\exp\left(-\frac{r^2}{b^2(z)}\right) \quad (7.17)$$

$$\rho_\infty-\rho(r,z)=\left(\rho_\infty-\rho_{cl}(z)\right)\exp\left(-\lambda^2\frac{r^2}{b^2(z)}\right) \quad (7.18)$$

$$\rho(r,z)C(r,z)=\rho_{cl}(z)C_{cl}(z)\exp\left(-\lambda^2\frac{r^2}{b^2(z)}\right) \quad (7.19)$$

$$\rho=1/\left([(1/\rho_0)-(1/\rho_\infty)]C+(1/\rho_\infty)\right) \quad (7.20)$$

Following the derivation as in [9, 13], one reaches,

$$\frac{d}{dz}\left(b^2(z)U_{cl}(z)\right)=2\alpha bU_{cl}\left(1-\frac{F_0}{Kgb^2(z)U_{cl}(z)}\right)^{1/2} \quad (7.21)$$

$$\frac{d}{dz}\left(b^2(z)U_{cl}^2(z)\right)=\frac{2F_0}{\lambda^2KU_{cl}(z)}+\frac{2F_0}{(2+\lambda^2)gK}\frac{dU_{cl}}{dz} \quad (7.22)$$

To eliminate F_0 from the preceding equations, we utilize the following non-dimensional transformations [14]:

$$\hat{b}=b\left(\frac{g^3}{F_0^2}\right)^{1/5},\quad \hat{u}=\frac{U_{cl}}{(gF_0)^{1/5}},\hat{z}=z\left(\frac{g^3}{F_0^2}\right)^{1/5},\quad N=\hat{b}^2\hat{u},M=\hat{b}^2\hat{u}^2 \quad (7.23)$$

where the top-hat dimensionless radius, centerline velocity and vertical coordinate are, respectively, $\hat{b}$, $\hat{u}$, $\hat{z}$ [14, 15], and N is the dimensionless mass and M is the dimensionless momentum. Therefore,

$$\frac{dN}{d\hat{z}}=2\alpha M^{1/2}\left(1-\frac{1}{KN}\right)^{1/2} \quad (7.24)$$

$$\frac{dM}{d\hat{z}}=\left(\frac{1}{KN-A}\right)\left(\frac{2N}{\lambda^2M}-2\alpha A\frac{M^{3/2}}{N}\left(1-\frac{1}{KN}\right)^{1/2}\right) \quad (7.25)$$

such that $A=2/(2+\lambda^2)$

Along with the initial conditions,

$$N_0 = \frac{1}{\left(K\left(\rho_\infty - \rho_0\right)/\rho_\infty\right)}, \quad M_0 = \frac{\left(U_0^2/gb_0\right)^{2/5}}{\left(K\left(\rho_\infty - \rho_0\right)/\rho_\infty\right)^{6/5}} \tag{7.26}$$

The above set of Equations (2.24)–(2.26) are treated numerically using the Runge-Kutta method. Figure 7.7 presents validation of theoretical [16] against experimental [17] results. It has been found that the momentum-dominated case has the asymptotic as ($1/C_{cl} \propto z$) while the buoyancy-dominated case has an asymptotic as ($1/C_{cl} \propto z^{5/3}$). More details can be found in El-Amin and Kanayama (2009) [16].

Here, we introduce alternative formula of the concentration equation. The density is defined as,

$$\rho = \frac{1}{[(1/\rho_0)-(1/\rho_\infty)]C+(1/\rho_\infty)} \tag{7.27}$$

and as a function of the mole fraction X becomes,

$$\rho = \rho_\infty(1-X)+\rho_0 X \tag{7.28}$$

Therefore,

$$C = \frac{\rho_0 X}{\rho_\infty(1-X)+\rho_0 X} = \frac{R_\rho X}{1+(R_\rho-1)X} \tag{7.29}$$

such that $R_\rho = \rho_0/\rho_\infty$. Note that $\rho C = \rho_0 X$.

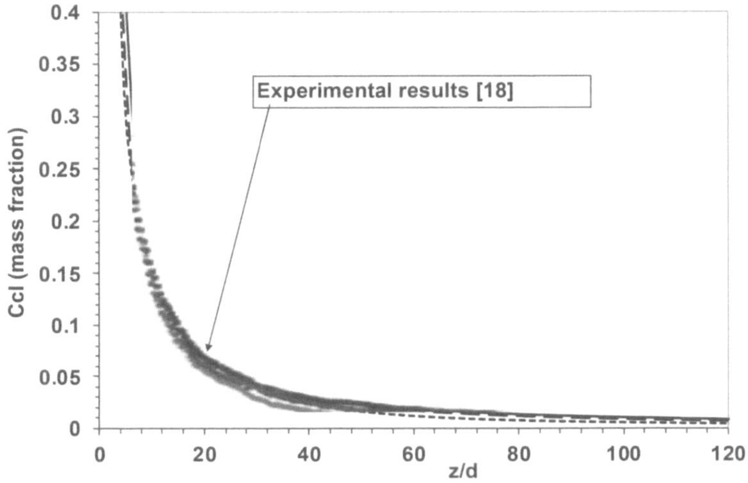

FIGURE 7.7 Centerline mass fraction [16].

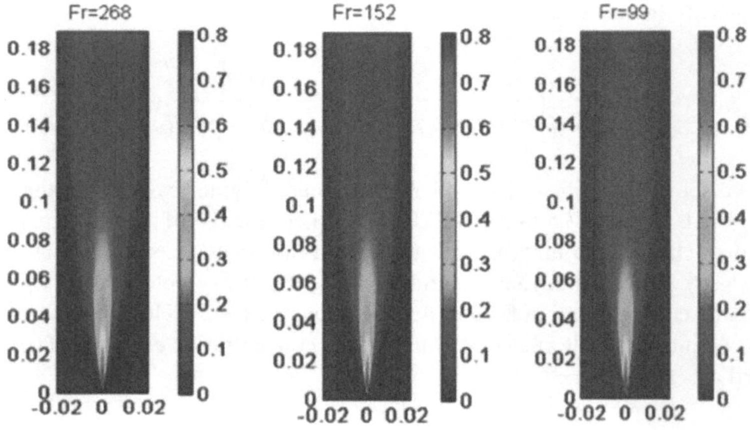

FIGURE 7.8 Mean mole fraction contours for different values of Fr [18].

Therefore, Equation (7.16) may take the form,

$$\frac{\partial(rVX)}{\partial r}+\frac{\partial(rUX)}{\partial z}+\frac{\partial(r\overline{vx})}{\partial r}=0 \tag{7.30}$$

where x is the mole fraction fluctuation. Using the relationships of density and mass/mole fractions, the mole-fraction contours for various values of Fr (Figure 7.8). Figure 7.8 indicates that as Fr increases, the spreading rate increases [18].

7.3 HORIZONTAL HYDROGEN BUOYANT JET

The current section introduces physical investigation using numerical simulation of the hydrogen buoyant jet from a horizontal circular source (Figure 7.9). The jet exits the nozzle of an angle θ_0 with the x-axis, to form the s-axis which is a

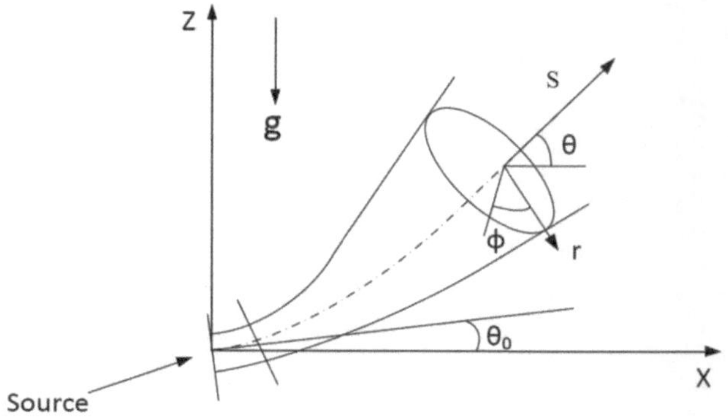

FIGURE 7.9 Schematic diagram of horizontal hydrogen jet [19].

parametric coordinate. The angle θ is located between the s-axis and the horizontal axis, while φ is the azimuthal angle. Therefore, the mean equations can be written as follows:

$$U(s,r) = u_{cl}(s)\exp\left(-\frac{r^2(s)}{b^2(s)}\right) \tag{7.31}$$

$$\rho_\infty - \rho(s,r) = \left(\rho_\infty - \rho_{cl}(s)\right)\exp\left(-\lambda^2\frac{r^2(s)}{b^2(s)}\right) \tag{7.32}$$

$$\rho(s,r)C(s,r) = \rho_{cl}(s)C_{cl}(s)\exp\left(-\lambda^2\frac{r^2(s)}{b^2(s)}\right) \tag{7.33}$$

where:

$U(s,r)$ mean velocity,
$\rho(s,r)$ mean density,
$u_{cl}(s)$ centerline velocity,
$\rho_{cl}(s)$ centerline density.

The mole-fraction profiles, for various Fr, are shown in Figures 7.10 and 7.11. For Fr = 62, the curve is almost linear representing the momentum-dominated case, while for Fr = 31, it is still approximately linear. Thus, the Froude number has a slight effect on the buoyancy.

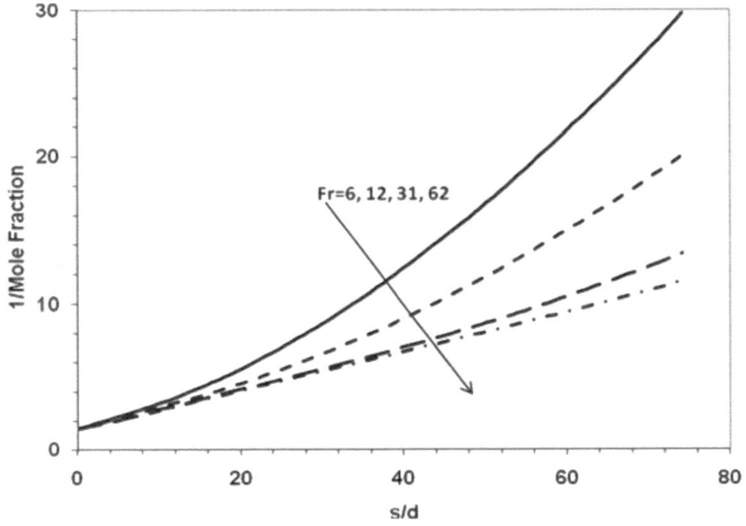

FIGURE 7.10 Hydrogen mole fraction with various Fr [19].

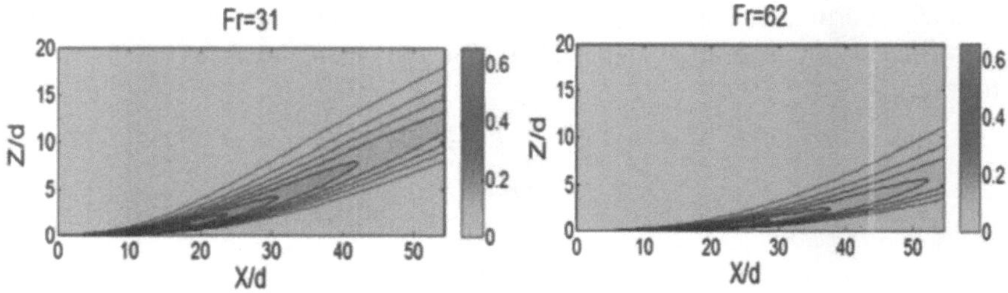

FIGURE 7.11 Mean mole Contours fraction at Fr = 31, 62 [19].

7.4 CONCLUSION

This chapter concerns studying the behavior of hydrogen leakage in the open air. Several scenarios have been considered, with developing analytical models that are solved and verified against experimental data. The research also includes theoretical modeling for hydrogen-air buoyant jets caused by a slow hydrogen leak. For the altered model, numerical solutions have been obtained (ODEs set). The buoyant hydrogen jet model was created, and numerical solutions were achieved. The mean centerline quantities are used to construct similarity solutions. The chapter contains many interesting results; we will highlight the most important ones. For example, standard Gaussian expressions for the mean streamwise velocity/concentration are inserted into the conservative equations to obtain essential quantities.

REFERENCES

[1] N. Rajaratnam. Turbulent jets. Amsterdam, The Netherlands: Elsevier, 1976.
[2] E.J. List, Turbulent jets and plumes, Ann. Rev. Fluid Mech., 14 (1982), pp. 189–212.
[3] H.B. Fischer, E.J. List, R.C.Y. Koh, J. Imberger, N.H. Brooks. Mixing in inland and coastal waters. Cambridge, MA: Academic Press, 1979.
[4] J.S. Turner, Turbulent entrainment: The development of the entrainment assumption, and its application to geophysical flows, J. Fluid Mech., 173 (1986), pp. 431–471.
[5] A. Shabbir, W.K. George, Experiments on a round turbulent plume, J. Fluid Mech., 275 (1994), pp. 1–32.
[6] R.W. Schefer, W.G. Houf, T.C. Williams, Investigation of small-scale unintended releases of hydrogen: Momentum-dominated regime, Int. J. Hydrogen Energy, 33 (21) (2008a), pp. 6373–6384.
[7] A. Agrawal, A.K. Prasad, Integral solution for the mean flow profiles of turbulent jets, plumes, and wakes, ASME J. Fluids Eng., 125 (2003), pp. 813–822.
[8] G.S. Bhat, R. Narasimha, A volumetrically heated jet: Large eddy structure and entrainment characteristics, J. Fluid Mech., 325 (1996), pp. 303–330.
[9] C.J. Chen, W. Rodi. Vertical turbulent buoyant jets: A review of experimental data. Oxford, UK: Pergamon Press; 1980; 11–12.
[10] B. Gebhart, Y. Jaluria, R.L. Mahajan, B. Sammakia. Buoyancy-induced flows and transport. Washington, DC: Hemisphere; 1988; 661–663.
[11] H.A. Becker, H.C. Hottel, G.C. Williams, The nozzle-fluid concentration of the round turbulent free jet, J. Fluid Mech., 30 (1967), pp. 285–303.
[12] N.R. Panchapakesan, J.L. Lumley, Turbulence measurements in axisymmetric jets of air and helium. Part 1. Air jet, J. Fluid Mech., 246 (1993), pp. 197–223.

[13] M.F. El-Amin, H. Kanayama, Integral solutions for selected turbulent quantities of small-scale hydrogen leakage: A non-buoyant jet or momentum-dominated buoyant jet regime, Int. J. Hydrogen Energy, 34 (3) (2009), pp. 1607–1612.
[14] F.R. Steward, Prediction of the height of turbulent diffusion buoyant flames, Combust. Sci. Technol., 2 (1970), pp. 203–212.
[15] A.A. Townsend, The mechanism of entrainment in free turbulent flows, J. Fluid Mech., 26 (1966), pp. 689–715.
[16] M.F. El-Amin, H. Kanayama, Similarity consideration of the buoyant jet resulting from hydrogen leakage, Int. J. Hydrogen Energy, 34 (14) (2009), pp. 5803–5809.
[17] R.W. Schefer, W.G. Houf, T.C. Williams, Investigation of small-scale unintended releases of hydrogen: Buoyancy effects, Int. J. Hydrogen Energy, 33 (17) (2008), pp. 4702–4712.
[18] M.F. El-Amin, Non-Boussinesq turbulent buoyant jet resulting from hydrogen leakage in air, Int. J. Hydrogen Energy, 34 (18) (2009), pp. 7873–7882.
[19] M.F. El-Amin, S. Sun, Horizontal H_2-air turbulent buoyant jet resulting from hydrogen leakage, Int. J. Hydrogen Energy, 37 (4) (2012), pp. 3949–3957.

8 Multi-Objective Interval Assignment Problems and their Solutions Using Genetic Algorithms

Anita Ravi Tailor, Dhiren Pandit, and Jayesh M. Dhodiya

CONTENTS

DOI: 10.1201/9781003229018-8

8.1 INTRODUCTION

Assignment problem (AP) is a distinguished optimization problem wherein the objective is to assign a number of jobs to an equal number of machines/people/facilities with optimum decision parameters. The mathematical formulation of the problem recommends that this is a 0-1 linear programming problem and is vastly degenerate [1, 2]. Nevertheless, in real-life situations, the elements of the effective matrices of the AP should be an imprecise number in place of a deterministic number as decision parameters for one job assigned to one person might vary due to people's limited knowledge associated with the problematic areas, the lack of data, incorrect performance ratings, etc. The inaccurate information of decision parameters may be in the type of interval numbers [3, 4]. Thus, these axes of assessment are usually equipped by objective functions to be optimized in the formation of the Interval Assignment Problem (IAP) model [5–9]. When AP with interval numbers involves multiple decision variables, in this situation, IAP forms a new problem as a multi-objective interval assignment problem (MOIAP). Thus, new methods to solve the MOIAP are needed.

To find the solution of MOIAP, several methods which are available for multi-objective interval linear programming problems can be utilized. In particular, there are hardly any studies on MOIAP existing in the literature. As far as we know, Kagade K. L. et al. [10] have obtained the solution of multi-objective assignment problem (MOAP) with interval cost using the fuzzy method. Shen Kai et al. [11] proposed a solution for solving the MOIAP using the risk attitude parameter (RAP). Kayvan Salehi [12] has developed an approach for solving MOAP with interval parameters using the substitution variables approach. In addition, to transform the multi-objective optimization problem (MOOP) into a single objective optimization problem (SOOP), a weighted min-max method is applied. Tailor and Dhodiya [13] have developed a genetic algorithm(GA)-based hybrid approach for finding solution of MOIAP using estimation theory.

To find the solution of MOIAP, the MOIAP is converted into MOAP by using the RAP, and then MOAP is transformed into a single objective non-linear optimization problem (SONOP) with various realistic constraints, and as a result, this exacting problem becomes an NP-hard" problem. For dealing with such problems, GA is an appropriate approach. Therefore, this chapter provides a GA-based approach for determining MOIAP solutions. This chapter also deliberates the outcomes of the convergence to the efficient solutions (ESs) of MOIAP when decision maker (DM) gives a distinctive aspiration level (AL) and sensitivity analysis in the sense of changing the RAP value. Finally, this chapter compares the developed approach to other approaches.

In this chapter, we study a MOIAP subject to many realistic constraints and propose a GA-based approach to solve it using the RAP. The chapter is organized as follows. Section 8.2 describes and formulates the MOIAP. Section 8.3 presents the details of some preliminaries which are useful to solve MOIAP. Section 8.4 proposes the solution approach for MOIAP using the RAP and also presents the convergence criteria of GA. Numerical illustrations of MOIAP are provided in Section 8.5 to demonstrate the proposed GA-based algorithm and their sensitivity analysis, and the discussion and comparison are discussed in Section 8.6. Finally, conclusion is provided in Section 8.7.

8.2 MATHEMATICAL FORMULATION OF MULTI-OBJECTIVE INTERVAL ASSIGNMENT PROBLEM

The general mathematical formulation of MOIAP is as follows [10, 11, 12, 13]:

Model-1.1:

$$\min Z_k = \sum_{i=1}^{n}\sum_{j=1}^{n}\left[Z_{ijk}^{L}, Z_{ijk}^{U}\right]x_{ij}; k = 1, 2, \ldots m$$

Subject to the constraints:

$$\sum_{j=1}^{n} x_{ij} = 1;\ i = 1, 2, \ldots, n \tag{8.1}$$

$$\sum_{i=1}^{n} x_{ij} = 1;\ j = 1, 2, \ldots, n \tag{8.2}$$

$$x_{ij} = \begin{cases} 1; \text{ if i}^{\text{th}} \text{ job is assigned to j}^{\text{th}} \text{ machine} \\ 0; \text{ otherwise} \end{cases} \tag{8.3}$$

Where $\left[Z_{ijk}^{L}, Z_{ijk}^{U}\right]$ is an interval representing the uncertain parameter for the AP. Here $\left(Z_{ij1}^{L}, Z_{ij2}^{L}, \ldots, Z_{ijk}^{L}\right)$ and $\left(Z_{ij1}^{R}, Z_{ij2}^{R}, \ldots, Z_{ijk}^{R}\right)$ stand for left bound and right bound for Z_k respectively.

8.3 SOME PRELIMINARIES

To convert the MOIAP into MOAP, some preliminaries are useful.

Interval Arithmetic:
For basic concepts and operation of interval $\tilde{x} = \left[x^L, x^U\right] = \left\{x/x^L \le x \le x^U, x \in R\right\}$, where x^L and x^U are denoted as lower limit and upper limit respectively, researchers refer to [11], which plays a significant role in the analysis of MOIAP.

Risk attitude parameter:
For the assignment analysis process in MOIAP to manage the uncertainty, DM is introduced the different parameters/approaches like RAP, order relations between interval, fuzzy technique, substitution variable approach, interval estimation, etc., to convert the interval number into a crisp number [3, 11, 12, 13]. In this section, out of them, RAP is used to convert the interval numbers into crisp numbers and the sensitivity analysis is also discussed by changing the values of RAP.

For positive interval number $\tilde{x} = \left[x^L, x^U\right]$, the interval mapping function is defined as follows [11]:

$$\phi_\varepsilon(x) = m(\tilde{x}) + (\varepsilon)d(\tilde{x}); \tag{8.4}$$

where ε is DM's RAP for uncertain data of real world problem and $|\varepsilon| \leq 0.5$ $m(\tilde{x}) = \dfrac{x^U + x^L}{2}$ and $d(\tilde{x}) = x^U - x^L$ are the center and the difference of an interval $\tilde{x}$ respectively.

For the assignment analysis process, the DM gives RAP. The RAP can be divided into optimistic, most likely, and pessimistic, and the corresponding values for ε are $-0.5 \leq \varepsilon < 0, \varepsilon = 0, 0 < \varepsilon \leq 0.5$ respectively. Once RAP is determined, the interval values are converted into crisp values by equation (8.4).

Aspiration level:

In many MOOPs, it is essential to attain a solution that closely reflects the DM's opinion. To make decisions based on the assortment of value judgments and intricate alterations in the decision making's nature, the AL presents the adequate trading method which cannot work through the significant DM opinion but moreover work with the opinions on important dynamics. AL does not impose any consistency of judgment on DM as DM frequently changes his/her mindset even during the entire decision-making process [14, 15, 16].

Positive Ideal Solution (PIS) and Negative Ideal Solution (NIS):

The PIS and NIS for each objective function are definite as follows. They need to calculate the membership value of all objective functions [6, 13, 16–19].

The PIS and NIS for objective Z_k are:

PIS for objective function	**NIS for objective function**
$Z_k^{\text{PIS}} = \min Z_k$	$Z_k^{\text{NIS}} = \max Z_k$
Subject to constraints (8.1) - (8.3)	Subject to constraints (8.1) - (8.3)

Exponential Membership Function:

In MOOPs, the various ALs of DM are renowned by fuzzy membership functions for the objective functions. Furthermore, the membership function (MF) is listed for portraying the presentation of the ambiguous data, the use of the fuzzy numbers of DM, preferences in the direction of uncertainty, etc. The exponential membership function (EMF) gives a better exhibit than others and gives the litheness to convey the degree of accuracy in parameter values. It additionally reflects reality enhanced than the linear MF [13, 16–17, 19–21].

If Z_k^{PIS} and Z_k^{NIS} are PIS and NIS of objective Z_k, the EMF $\mu_{Z_k}^E$ is defined by

$$\mu_{Z_k}^E(x) = \begin{cases} 1; & \text{if } Z_k \leq Z_k^{\text{PIS}} \\ \dfrac{e^{-S\psi_k(x)} - e^{-S}}{1 - e^{-S}}, & \text{if } Z_k^{\text{PIS}} < Z_k < Z_k^{\text{NIS}} \\ 0; & \text{if } Z_k \geq Z_k^{\text{NIS}} \end{cases} \tag{8.5}$$

where $\psi_k(x) = \frac{z_k - z_k^{\text{PIS}}}{z_k^{\text{NIS}} - z_k^{\text{PIS}}}$ and $S \neq 0$ is SP which summarized by DM so that $\mu_{Z_k}(x) \in [0,1]$. The MF is sternly convex for $S < 0$ and concave for $S > 0$ in $[Z_k^{\text{PIS}}, Z_k^{\text{NIS}}]$.

Genetic algorithm:
Taking into account adaptive evolution and the usual selection of biological systems, the GA is an excellent method of randomly search and large-scale optimization. It is also the most proper method for solving discrete, non-linear, and non-convex global optimization problems in place of any conventional method, as it looks for the optimal solution by simulating the natural evolution process and mimicking the usual chromosome processing work and evaluation principles on genetics [–17, 20–27]. It has demonstrated a number of key benefits, such as high robustness, the most advantageous large-scale convergence and analogous research capacity. In GA, the chromosomes are coded together with the question and the fitness function is used to quantify the chromosome. Gene operators, for instance selection, crossover and mutation, are used to generate new populations [13, 16, 18–19]. It has demonstrated a number of key benefits, such as high robustness, the most advantageous large-scale convergence and analogous research capacity.

Convergence criteria:
If after getting some certain value, i.e. an optimum value, we say that GA is converging. For NP-hard problems, it is not practical for GA to converge to an optimal value on a large scale except you earlier contain a test data set with the best-known solution. Moreover, the dimension of the problem also influences the convergence of the GA. The chromosome size (the number of solutions) can be defined according to the parameters of the problem. Furthermore, it should be noted that increasing chromosome size will affect the speed of GA convergence [16–17, 20].

8.3.1 Formulation of Crisp Multi-Objective Assignment Model using Risk Attitude Parameter

To convert the model-1 into crisp multi-objective assignment model, the RAP is utilized as follows:

Let $Z_k = [z_{ijk}]_{n \times n}$ be the cost (profit and loss) matrix with respect to the objective k, where $z_{ijk} = [z_{ijk}^L, z_{ijk}^U]$ is representing the cost if the person A_i assigned jobs B_j related to the objective k; $i, j = 1,2,...,n$, $k = 1,2,...,m$. This $Z_k = [z_{ijk}]_{n \times n}$ is converted into crisp values $Z_k^{\varepsilon} = [z_{ijk}^{\varepsilon}]_{n \times n}$ by using the equation

$$Z_k^{\varepsilon} = [z_{ijk}^{\varepsilon}] = \frac{(z_{ijk}^U + z_{ijk}^L)}{2} + \varepsilon(z_{ijk}^U - z_{ijk}^L) \tag{8.6}$$

where ε is DM's RAP for uncertain data of real-world problem. For assignment analysis process, the DM gives RAP.

To replicate the disparate scenario using RAP, the MOIAP is changed into crisp MOAP as follows:

Model-2:

$$\min Z_k^{\varepsilon} = \sum_{i=1}^{n}\sum_{j=1}^{n}\left[z_{ijk}^{\varepsilon}\right]$$

Subject to constraints (8.1)–(8.3)

8.4 SOLUTION APPROACH FOR MOIAP WITH RISK ATTITUDE PARAMETER

This section outlines a GA-based approach to find the best ESs of MOIAP. This approach also offers more flexibility in solving MOIAP in terms of a disparate assortment of ALs for all objective functions.

8.4.1 Steps to Find the Efficient Solutions of MOIAP

The subsequent is a gradual depiction of a GA-based approach to find ESs for the MOIAP.

Step-1: According to RAP, define the crisp multi-objective assignment model-2.
Step-2: Convert maximization problem into a minimization form.
Step-3: Find out PIS and NIS for each objective function.
Step-4: Find fuzzy EMF value for Z_K.
Step-5: In this step, MOIAP can be written in the SOOPs, which is as follows.

Model-3:

$$\max W = \prod_{k=1}^{m}\mu_{Z_k^{\varepsilon}} \tag{8.6}$$

Subject to the constraints: (8.1)-(8.3)

$$\mu_{Z_k^{\varepsilon}}(x) - \overline{\mu_{Z_k^{\varepsilon}}}(x) \geq 0;\ k = 1, 2, \ldots m \tag{8.7}$$

Step-6: To deal with the SOOP of model-3, GA is used with various choices of the SP.

8.4.1.1 Encoding of Chromosomes

Chromosome coding is the main task of solving any problem by GA. To create a SOOP solution via GA, the chromosome data formation in the coding space must be well-thought-out, which indicates a solution to the problem. In the chromosome data formation, put all 0's to $n \times n$ genes on a chromosome, and then capriciously select the genes of the chromosome, and put 1 exactly once in each row and column of the chromosome precisely one time to satisfy the constraints (8.1) to (8.3). In chromosomal formation, each string can be alone expressed as in $2^\wedge r;\ r = 0 : N - 1$ way.

8.4.1.2 Evolution of Fitness Function

The fitness function is a significant factor in solving any optimization problem in GA. With the fulfillment of constraints (8.1) to (8.3) and (8.7), the objective function of P-2 is calculated to test the fitness of chromosomes.

8.4.1.3 Selection

The selection operator is used to decide how to preferentially select the chromosomes of the current population and treat them as parents to support the next genetic operation, that is, the crossover/mutation/next population with the maximum fitness. To find MOIAP's solutions, tournament selection is used because of their capabilities and straightforward execution. In this selection, chromosomes are subjectively preferred as of the population and judged by each other. The most suitable (winner) chromosome is the first choice for the next generation, while others are not eligible. This process continues in anticipation of the population size and the number of winners is equal [13, 16–20].

8.4.1.4 Crossover

Crossover is the process of finding solutions to both parents and turning them into a child. For MOIAP, partially matched crossover (PMX) is used to breed new offspring. In the PMX, two chromosomes are linked and two crossover points are selected at random. These two crossover points provide a toning alternative and are used to transform a cross by swapping positions one by one [16, 18–19].

8.4.1.5 Threshold Construction

To keep the population diverse after the crossing, a threshold is established for generating a MOIAP solution. During this stage, certain chromosomes are selected from the parent-child set to sustain the new iteration. To formulate a threshold, it is first necessary to sort the total population in ascending/descending order proportionate to its objective function values and then the encoded entity string is selected from all groups. According to these objective functional values, the population is divided into four categories: above $\mu + 3 * \sigma$, among $\mu + 3 * \sigma$ and μ, among μ and $\mu - 3 * \sigma$, and below $\mu - 3 * \sigma$. So, the proficient string cannot be lost, where σ and μ denote the standard deviation and expected value of the objective function values of parent-childhood population respectively [13, 16–20].

8.4.1.6 Mutation

Here, swap mutation operator is used among the many available mutation operators to enhance the misplaced genetic materials and also used to arbitrarily distribute genetic information. In this mutation, any two points in a string are selected and the subsequent values are exchangedbetweenthe points.

8.4.1.7 Termination Criteria

A GA is executed on a specified number of iterations until the end state is reached, i.e., the best fitness condition of the population has remained unchanged for a number of generations.

As a result, two cases are carried out to find the SOOP's solution using GA:

i. Without mutation
ii. With mutation

In the above cases, GA converged towards SOOP's optimal solution, and lastly ES of MOIAP at different RAP with disparate SPs and ALs.

If the DM agreed with the achievable solution, consider it to be an ideal solution and stop the solution process, otherwise modify the SP, AL and restate the steps 1–6 until it reaches an acceptable ES.

8.5 NUMERICAL EXAMPLES AND RESULT ANALYSIS

In this section, we present the MOIAPs and their solutions by using GA-based approach.

8.5.1 MOIAP-1

To justify the proposed methodology, MOIAP has been referred to article [11, 13], which is given in Table 8.1.

Consider the AP of the matches between development teams and the new product tasks. In this problem, there are four development teams (M_1, M_2, M_3, M_4) and four new products (N_1, N_2, N_3, N_4) to be developed. In the assignment analysis procedure, three objectives should be considered as: (1) Development period (Z_1): The greater the development period is, the more resources of company are occupied, (2) Fund investment (Z_2): The development of a new product will consume huge fund and

TABLE 8.1
Profit and Loss Matrices under Each Objective (Data from Kai et al (2011) [11] and Tailor and Dhodiya (2016) [13])

		M_1	M_2	M_3	M_4
Z_1	N_1	[120,140]	[110,150]	[100,160]	[105,160]
	N_2	[60,100]	[65,95]	[70,90]	[65,90]
	N_3	[180,220]	[190,210]	[160,240]	[170,230]
	N_4	[60,80]	[60,80]	[50,90]	[65,75]
Z_2	N_1	[20,25]	[19,27]	[18,27]	[21,24]
	N_2	[11,13]	[9,12]	[10,12]	[9,11]
	N_3	[25,23]	[26,30]	[27,31]	[27,30]
	N_4	[5,9]	[6,7]	[4,8]	[4,9]
Z_3	N_1	[0.7,0.9]	[0.8,0.9]	[0.7,1]	[0.6,1]
	N_2	[0.8,0.9]	[0.7,0.9]	[0.8,1]	[0.6,1]
	N_3	[0.6,0.8]	[0.6,1]	[0.8,0.9]	[0.6,0.9]
	N_4	[0.7,0.9]	[0.7,0.8]	[0.7,1]	[0.7,1]

TABLE 8.2
PIS and NIS for Z_1, Z_2 and Z_3 at $\varepsilon = -0.5$, $\varepsilon = 0$ and $\varepsilon = 0.5$

		Objectives		
α – Level	Solutions	Z_1	Z_2	Z_3
ε = −0.5	PIS	385	56	0.1
	NIS	445	65	0.6
ε = 0	PIS	477.5	66.5	0.2
	NIS	482.5	70.5	0.5
ε = 0.5	PIS	515	74	0.1
	NIS	580	80	0.4

(3) Expected effect (Z_3): The question of whether the new product can meet customer demand is significant in the development of the product. Except the development effect is profit-type objectives with a value in the form of an interval between 0 and 1.

Table 8.2 gives the PIS and NIS for each objective function of model-2 at different RAPs $\varepsilon = -0.5$, $\varepsilon = 0$ and $\varepsilon = 0.5$. For the uniformity, we convert maximize quality function into minimizing form and then find the solution.

Using PIS and NIS value for Z_1, Z_2 and Z_3 in model-3, an equivalent crisp model for $\varepsilon = -0.5, 0$ and 0.5 can be formulated respectively for MOIAP-1 as follows:

$$\max w = \mu_{Z_1}.\mu_{Z_2}.\mu_{Z_3}$$

Subject to the constraints: (8.1)–(8.3) and (8.7); $i, j = 1,2,3,4;\ k = 1,2,3$
where,
For $\varepsilon = -0.5$

$$\mu_{Z_1} = \frac{\exp\left(-S\begin{pmatrix}120x_{11}+110x_{12}+100x_{13}+105x_{14}+60x_{21}+65x_{22}\\ +70x_{23}+65x_{24}+180x_{31}+190x_{32}+160x_{33}+170x_{34}\\ +60x_{41}+60x_{42}+50x_{43}+65x_{44}-384\end{pmatrix}/61\right)-\exp(-S)}{(1-\exp(-S))}$$

$$\mu_{Z_2} = \frac{\exp\left(-S\begin{pmatrix}20x_{11}+19x_{12}+18x_{13}+21x_{14}+11x_{21}+9x_{22}\\ +10x_{23}+9x_{24}+25x_{31}+26x_{32}+27x_{33}+27x_{34}\\ +5x_{41}+6x_{42}+4x_{43}+4x_{44}-56\end{pmatrix}/9\right)-\exp(-S)}{(1-\exp(-S))}$$

$$\mu_{Z_3} = \frac{\exp\left(-S\begin{pmatrix}0.1x_{11}+0x_{12}+0.1x_{13}+0.2x_{14}+0x_{21}+0.1x_{22}\\ +0x_{23}+0.2x_{24}+0.2x_{31}+0.2x_{32}+0x_{33}+0.2x_{34}\\ +0.1x_{41}+0.1x_{42}+0.1x_{43}+0.1x_{44}-0.1\end{pmatrix}/0.5\right)-\exp(-S)}{(1-\exp(-S))}$$

For $\varepsilon = 0$

$$\mu_{Z_1} = \frac{\exp\left(-S\left(\begin{pmatrix}130x_{11}+130x_{12}+130x_{13}+132.5x_{14}+80x_{21}+80x_{22}\\ +80x_{23}+77.5x_{24}+200x_{31}+200x_{32}+200x_{33}\\ +200x_{34}+70x_{41}+70x_{42}+70x_{43}+70x_{44}-477.5\end{pmatrix}/5\right)\right)-\exp(-S)}{(1-\exp(-S))}$$

$$\mu_{Z_2} = \frac{\exp\left(-S\left(\begin{pmatrix}22.5x_{11}+23x_{12}+22.5x_{13}+22.5x_{14}+12x_{21}+10.5x_{22}\\ +11x_{23}+10x_{24}+28.5x_{31}+28x_{32}+29x_{33}+28.5x_{34}\\ +7x_{41}+6.5x_{42}+6x_{43}+6.5x_{44}-66.5\end{pmatrix}/4\right)\right)-\exp(-S)}{(1-\exp(-S))}$$

$$\mu_{Z_3} = \frac{\exp\left(-S\left(\begin{pmatrix}0.1x_{11}+0.05x_{12}+0.05x_{13}+0.1x_{14}+0.05x_{21}+0.1x_{22}\\ +0x_{23}+0.1x_{24}+0.2x_{31}+0.1x_{32}+0.05x_{33}+0.15x_{34}\\ +0.1x_{41}+0.15x_{42}+0.05x_{43}+0.05x_{44}-0.2\end{pmatrix}/0.3\right)\right)-\exp(-S)}{(1-\exp(-S))}$$

For $\varepsilon = 0.5$

$$\mu_{Z_1} = \frac{\exp\left(-S\left(\begin{pmatrix}140x_{11}+150x_{12}+160x_{13}+160x_{14}+100x_{21}+95x_{22}\\ +90x_{23}+90x_{24}+220x_{31}+210x_{32}+240x_{33}\\ +230x_{34}+80x_{41}+80x_{42}+90x_{43}+75x_{44}-515\end{pmatrix}/65\right)\right)-\exp(-S)}{(1-\exp(-S))}$$

$$\mu_{Z_2} = \frac{\exp\left(-S\left(\begin{pmatrix}25x_{11}+27x_{12}+27x_{13}+24x_{14}+13x_{21}+13x_{22}\\ +13x_{23}+11x_{24}+32x_{31}+30x_{32}+31x_{33}+30x_{34}\\ +9x_{41}+7x_{42}+8x_{43}+9x_{44}-74\end{pmatrix}/6\right)\right)-\exp(-S)}{(1-\exp(-S))}$$

$$\mu_{Z_3} = \frac{\exp\left(-S\left(\begin{pmatrix}0.1x_{11}+0.1x_{12}+0x_{13}+0x_{14}+0.1x_{21}+0.1x_{22}\\ +0x_{23}+0x_{24}+0.2x_{31}+0x_{32}+0.1x_{33}+0.1x_{34}\\ +0.1x_{41}+0.2x_{42}+0x_{43}+0x_{44}-0.1\end{pmatrix}/0.3\right)\right)-\exp(-S)}{(1-\exp(-S))}$$

Here, we find the ESs for model-3 using the various SPs and ALs that are given by the DM for $\varepsilon = -0.5, 0$ and 0.5.

Some of combinations of the SPs and its corresponding ALs are shown in Table 8.3.

According to different values of RAP, the assignment plans for MOIAP are noted in Table 8.4 with distinct values of the SPs and ALs which are stated by the DM.

TABLE 8.3
Different Values of SPs and ALs

Case	Shape Parameter	Aspiration Level
1	(-5, -5, -5)	(0.9, 0.85, 0.7)
2	(-3, -3, -3)	(0.8, 0.7, 0.85)
3	(-2, -2, -2)	(0.7, 0.65, 0.75)
4	(-1, -1, -1)	(0.6, 0.5, 0.7)

TABLE 8.4
Summary Results at ε = –0.5, ε = 0 and ε = 0.5

α	Case	Degree of Satisfaction	Membership Values $(\mu_{Z_{1j}}, \mu_{Z_{2j}}, \mu_{Z_{3j}})$	Objective Values Z_1, Z_2, Z_3	Optimum Allocations
$\varepsilon = -0.5$	1	0.9442	(0.9912, 0.9442, 0.9567)	(395, 60, 2.9)	$M_1 \to N_2, M_2 \to N_4, M_3 \to N_3, M_4 \to N_1$
	2	0.7750	(0.9660, 0.7750, 1.0000)	(395, 61, 3.1)	$M_1 \to N_2, M_2 \to N_1, M_3 \to N_3, M_4 \to N_4$
	3	0.6811	(0.9381, 0.6811, 1.0000)	(395, 61, 3.1)	$M_1 \to N_2, M_2 \to N_1, M_3 \to N_3, M_4 \to N_4$
	4	0.5676	(0.8945, 0.5676, 1.0000)	(395, 61, 3.1)	$M_1 \to N_2, M_2 \to N_1, M_3 \to N_3, M_4 \to N_4$
$\varepsilon = 0$	1	0.9241	(1.0000, 1.0000, 0.9241)	(477.5, 66.5, 3.25)	$M_1 \to N_1, M_2 \to N_4, M_3 \to N_2, M_4 \to N_3$
	2	0.8176	(0.8176, 0.8910, 0.9660)	(480, 68, 3.35)	$M_1 \to N_1, M_2 \to N_3, M_3 \to N_2, M_4 \to N_4$
	3	0.7311	(0.7311, 0.8252, 0.9381)	(480, 68, 3.35)	$M_1 \to N_1, M_2 \to N_3, M_3 \to N_2, M_4 \to N_4$
	4	0.6225	(0.6225, 0.7325, 0.8945)	(480, 68, 3.35)	$M_1 \to N_1, M_2 \to N_3, M_3 \to N_2, M_4 \to N_4$
$\varepsilon = 0.5$	1	0.9853	(0.9853, 1.0000, 1.0000)	(530, 74, 3.9)	$M_1 \to N_1, M_2 \to N_4, M_3 \to N_2, M_4 \to N_3$
	2	0.9477	(0.9477, 1.0000, 1.0000)	(530, 74, 3.9)	$M_1 \to N_1, M_2 \to N_4, M_3 \to N_2, M_4 \to N_3$
	3	0.9082	(0.9082, 1.0000, 1.0000)	(530, 74, 3.9)	$M_1 \to N_1, M_2 \to N_4, M_3 \to N_2, M_4 \to N_3$
	4	0.9853	(0.9853, 1.0000, 1.0000)	(530, 74, 3.9)	$M_1 \to N_1, M_2 \to N_4, M_3 \to N_2, M_4 \to N_3$

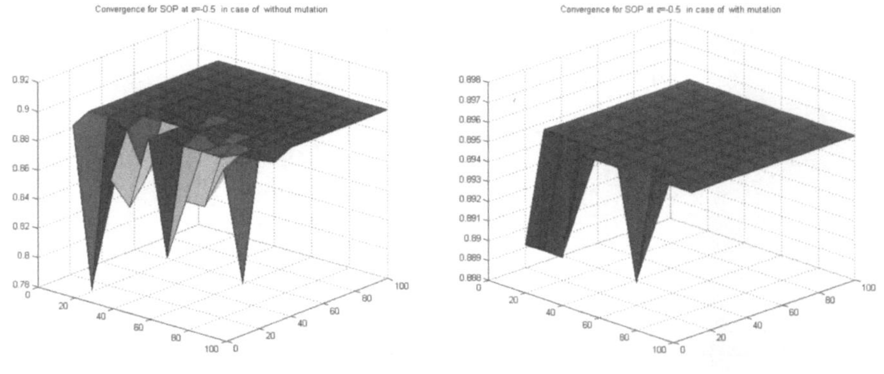

FIGURE 8.1 Convergence rate of GA at ε = –0.5 for case-1

8.5.1.1 Convergence Rate of GA for MOIAP-1

The convergence rate of GA for MOIAP-1 is accessed for case-1 of Table 8.3 at different RAPs $\varepsilon = -0.5$, $\varepsilon = 0$ and $\varepsilon = 0.5$ as shown in Figures 8.1, 8.2 and 8.3. To find the convergence rate, we have used different number of populations and iterations with the values of max $W = \Pi_{i=1}^{3} Z_i$.

In both the cases as in cases of without and with mutation operator, GA converges after 40 populations and 60 iterations as well as 20 populations and 20 iterations at $\varepsilon = -0.5$ respectively for case-1 of Table 8.3, which is shown in Figure 8.1. Similarly, Figures 8.2 and 8.3 show the GA converges after 30 populations, 20 iterations and 70 populations, 100 iterations in case of without mutation operator as well as 10 populations, 10 iteration and 40 populations, 60 iterations in case of with the mutation operator at $\varepsilon = 0$ and $\varepsilon = 0.5$ respectively. For other cases of Table 8.3, the convergence rate of GA almost remains the unchanged. Figures 8.1, 8.2 and 8.3 also provide alternate options for the DM as required.

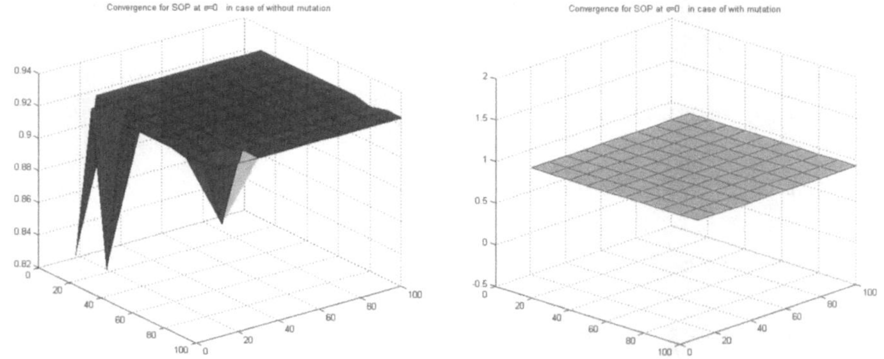

FIGURE 8.2 Convergence rate of GA at ε = 0 for case-1

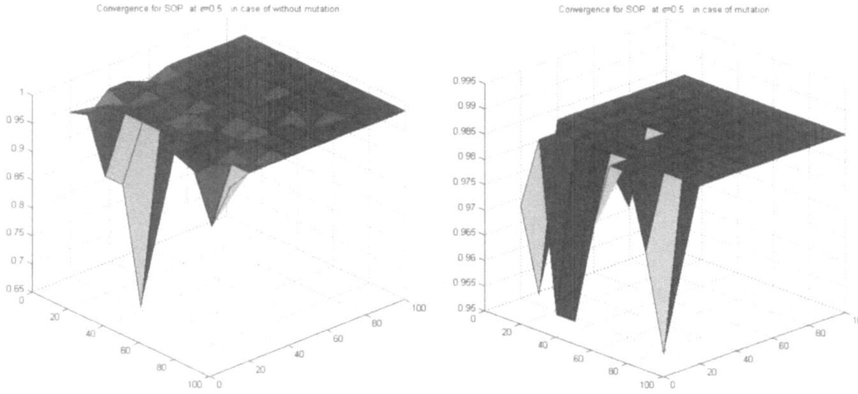

FIGURE 8.3 Convergence rate of GA at ε = 0.5 for case-1.

Figures 8.4, 8.5 and 8.6 indicate the ESs of each objectives as (395, 60, 2.9), (395, 61, 3.1), (395, 61, 3.1) and (395, 61, 3.1) at ε = –0.5, (477.5, 66.5, 3.25), (480, 68, 3.35), (480, 68, 3.35) and (480, 68, 3.35) at ε = 0 and (530, 74, 3.9), (530, 74, 3.9), (530, 74, 3.9) and (530, 74, 3.9) at ε = 0.5 for (-5, -5, -5), (-3, -3, -3), (-2, -2, -2) and (-1, -1, -1), SPs and its corresponding (0.9, 0.85, 0.7), (0.8, 0.7, 0.85), (0.7, 0.65, 0.75) and (0.6, 0.5, 0.7) ALs respectively.

Figures 8.7, 8.8 and 8.9 show the degree of satisfaction levels of each objective of MOIAP-1 at ε = –0.5, ε = 0 and ε = 0.5 for (-5, -5, -5), (-3, -3, -3), (-2, -2, -2) and (-1, -1, -1), SPs and its corresponding (0.9, 0.85, 0.7), (0.8, 0.7, 0.85), (0.7, 0.65, 0.75) and (0.6, 0.5, 0.7) ALs respectively.

8.5.1.2 Sensitivity Analysis

Sensitivity analysis is used here to inform the impact of uncertainty in the DM's judgment.

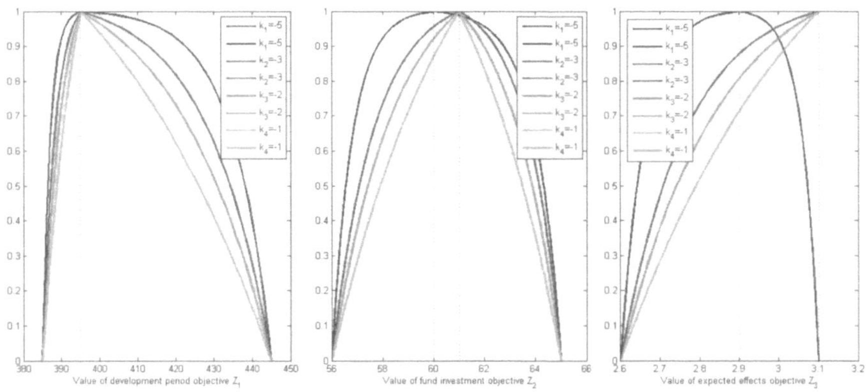

FIGURE 8.4 ESs of MOIAP-1 at ε = –0.5 for different SPs and ALs.

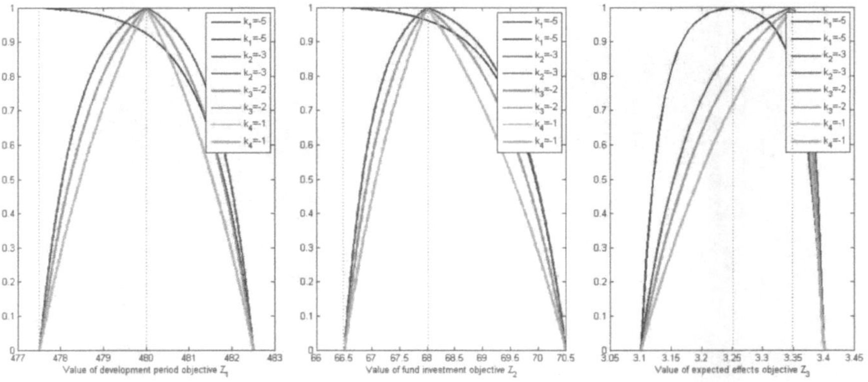

FIGURE 8.5 ESs of MOIAP-1 at ε = 0 for different SPs and ALs.

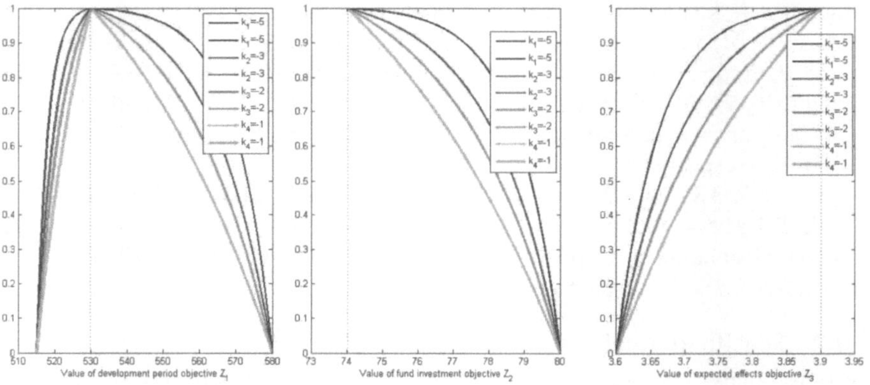

FIGURE 8.6 ESs of MOIAP-1 at ε = 0.5 for SPs and ALs.

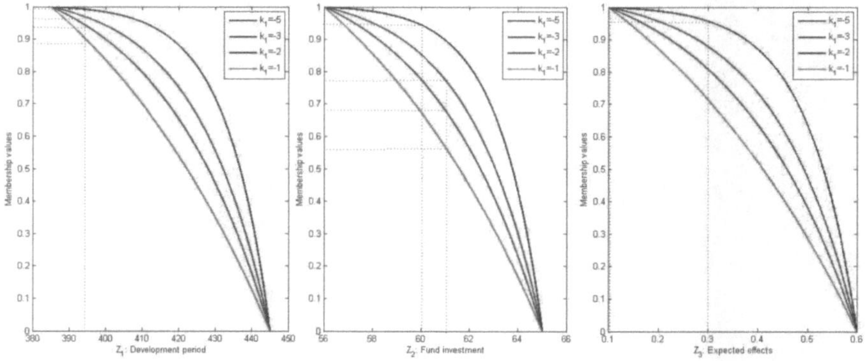

FIGURE 8.7 The degree of satisfaction of the MOIAP-1 at ε = –0.5.

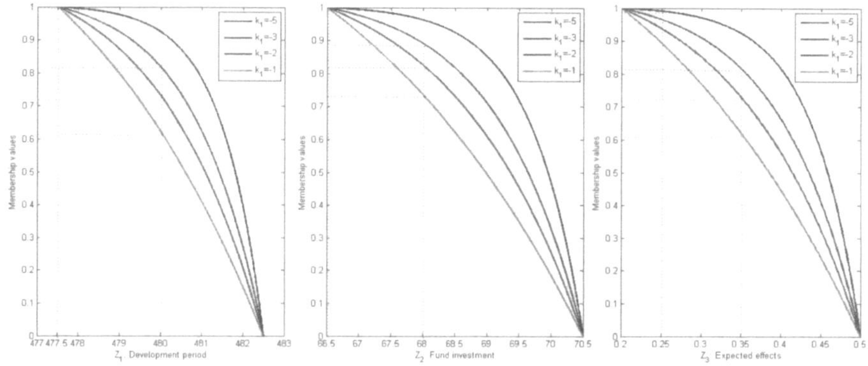

FIGURE 8.8 The degree of satisfaction of the MOIAP-1 at $\varepsilon = 0$.

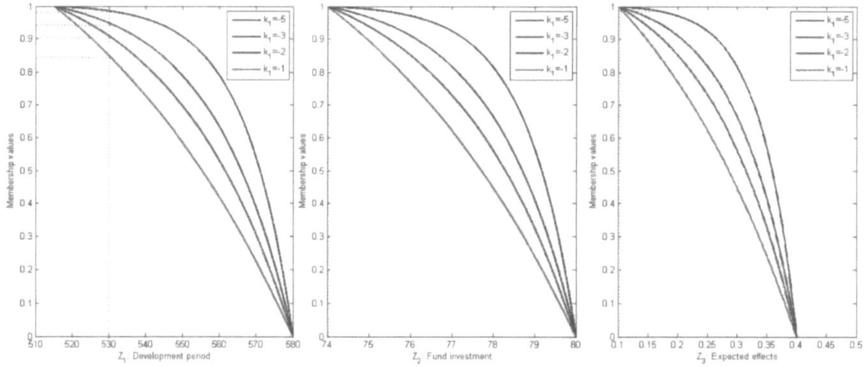

FIGURE 8.9 The degree of satisfaction of the MOIAP-1 at $\varepsilon = 0.5$.

8.5.1.2.1 Sensitivity Analysis with Respect to ε for Different SPs and its Corresponding ALs

For MOIAP, the sensitivity analysis relating to ε for different SPs and its corresponding ALs are discussed in Tables 8.5 and 8.6.

Table 8.5 shows the sensitivity analysis of assignment plans at the different values of RAPs for the (−5, −5, −5) SP and (0.9, 0.85, 0.7) AL.

For the purpose of observing sensitivity analysis of RAPs, normalized values of $z_1^{\varepsilon}, z_2^{\varepsilon}, \& z_3^{\varepsilon}$ are required. Here, normalized values are obtained by dividing all the objective function by its maximum number and sensitivity of the solution with respect to ε is described in Figure 8.10.

From the Figure 8.10, when 'ε' moves from left side to the average value of the interval, values of Z_1, Z_2 and Z_3 increase. When 'ε' moves from 0 to 0.1, value of Z_2 decreases and Z_1 and Z_3 increases, while 'ε' moves from 0.1 to 0.5, values of Z_1, Z_2 and Z_3 increase. Table 8.5 and Figure 8.10 show the nature of the objective functions of MOIAP under different values of RAPs. Figure 8.10 also indicates that, under the

TABLE 8.5
ε and its Corresponding Solutions with (-5, -5, -5) SP and (0.9, 0.85, 0.7) AL

ε	Degree of Satisfaction	Membership Values	$\left(Z_1^{\varepsilon}, Z_2^{\varepsilon}, Z_3^{\varepsilon}\right)$	The Assignment Solutions
-0.5	0.9442	(0.9912, 0.9442, 0.9567)	(395, 60, 2.9)	$M_1 \to N_2, M_2 \to N_4, M_3 \to N_3, M_4 \to N_1$
-0.4	0.9433	(0.9926, 0.9433, 0.9588)	(411.5, 61.8, 2.98)	$M_1 \to N_2, M_2 \to N_4, M_3 \to N_3, M_4 \to N_1$
-0.3	0.9419	(0.9948, 0.9419, 0.9612)	(428, 63.6, 3.06)	$M_1 \to N_2, M_2 \to N_4, M_3 \to N_3, M_4 \to N_1$
-0.2	0.9397	(0.9983, 0.9397, 0.9640)	(444.5, 65.4, 3.14)	$M_1 \to N_2, M_2 \to N_4, M_3 \to N_3, M_4 \to N_1$
-0.1	0.9007	(1.0000, 0.9007, 0.9672)	(461, 67.2, 3.22)	$M_1 \to N_2, M_2 \to N_4, M_3 \to N_3, M_4 \to N_1$
0	0.9241	(1.0000, 1.0000, 0.9241)	(477.5, 66.5, 3.25)	$M_1 \to N_1, M_2 \to N_4, M_3 \to N_2, M_4 \to N_3$
0.1	0.9752	(0.9973, 1.0000, 0.9752)	(488, 68, 3.38)	$M_1 \to N_1, M_2 \to N_4, M_3 \to N_2, M_4 \to N_3$
0.2	0.9862	(0.9914, 1.0000, 0.9862)	(498.5, 69.5, 3.51)	$M_1 \to N_1, M_2 \to N_4, M_3 \to N_2, M_4 \to N_3$
0.3	0.9883	(0.9883, 1.0000, 0.9929)	(509, 71, 3.64)	$M_1 \to N_1, M_2 \to N_4, M_3 \to N_2, M_4 \to N_3$
0.4	0.9865	(0.9865, 1.0000, 0.9972)	(519.5, 72.5, 3.77)	$M_1 \to N_1, M_2 \to N_4, M_3 \to N_2, M_4 \to N_3$
0.5	0.9853	(0.9853, 1.0000, 1.0000)	(530, 74, 3.9)	$M_1 \to N_1, M_2 \to N_4, M_3 \to N_2, M_4 \to N_3$

values of ε between "-0.1 to 0", the value of Z_2 decreases, which provided an opportunity to take decision for finding the better alternatives to DM.

Table 8.6 shows the sensitivity analysis of assignment plans at the different values of RAPs for the (-3, -3, -3) SP and (0.8, 0.7, 0.85) AL.

From Figure 8.11, when 'ε' moves from -0.5 to 0.4, the values of Z_1 and Z_2 increase and value of Z_3 decreases. When 'ε' moves from -0.4 to the average value of the interval, values of Z_1, Z_2 and Z_3 increase. When 'ε' moves from 0 to 0.1, value of Z_2 remains same

TABLE 8.6
ε and its Corresponding Solutions at (-3, -3, -3) SP and (0.8, 0.7, 0.85) AL

ε	Degree of Satisfaction	Membership Values	$\left(Z_1^{\varepsilon}, Z_2^{\varepsilon}, Z_3^{\varepsilon}\right)$	The Assignment Solutions
-0.5	0.7750	(0.9660, 0.7750, 1.0000)	(395, 61, 3.1)	$M_1 \to N_2, M_2 \to N_1, M_3 \to N_3, M_4 \to N_4$
-0.4	0.8519	(0.9709, 0.8519, 0.8829)	(411.5, 61.8, 2.98)	$M_1 \to N_2, M_2 \to N_4, M_3 \to N_3, M_4 \to N_1$
-0.3	0.8881	(0.9786, 0.8943, 0.8881)	(428, 63.6, 3.06)	$M_1 \to N_2, M_2 \to N_4, M_3 \to N_3, M_4 \to N_1$
-0.2	0.8452	(0.9925, 0.8452, 0.8942)	(444.5, 65.4, 3.14)	$M_1 \to N_2, M_2 \to N_4, M_3 \to N_3, M_4 \to N_1$
-0.1	0.7796	(1.0000, 0.7796, 0.9013)	(461, 67.2, 3.22)	$M_1 \to N_2, M_2 \to N_4, M_3 \to N_3, M_4 \to N_1$
0	0.8176	(0.8176, 0.8910, 0.9660)	(480, 68, 3.35)	$M_1 \to N_1, M_2 \to N_3, M_3 \to N_2, M_4 \to N_4$
0.1	0.9205	(0.9884, 1.0000, 0.9205)	(488, 68, 3.38)	$M_1 \to N_1, M_2 \to N_4, M_3 \to N_2, M_4 \to N_3$
0.2	0.9503	(0.9668, 1.0000, 0.9503)	(498.5, 69.5, 3.51)	$M_1 \to N_1, M_2 \to N_4, M_3 \to N_2, M_4 \to N_3$
0.3	0.9569	(0.9569, 1.0000, 0.9702)	(509, 71, 3.64)	$M_1 \to N_1, M_2 \to N_4, M_3 \to N_2, M_4 \to N_3$
0.4	0.9513	(0.9513, 1.0000, 0.9880)	(519.5, 72.5, 3.77)	$M_1 \to N_1, M_2 \to N_4, M_3 \to N_2, M_4 \to N_3$
0.5	0.9477	(0.9477, 1.0000, 1.0000)	(530, 74, 3.9)	$M_1 \to N_1, M_2 \to N_4, M_3 \to N_2, M_4 \to N_3$

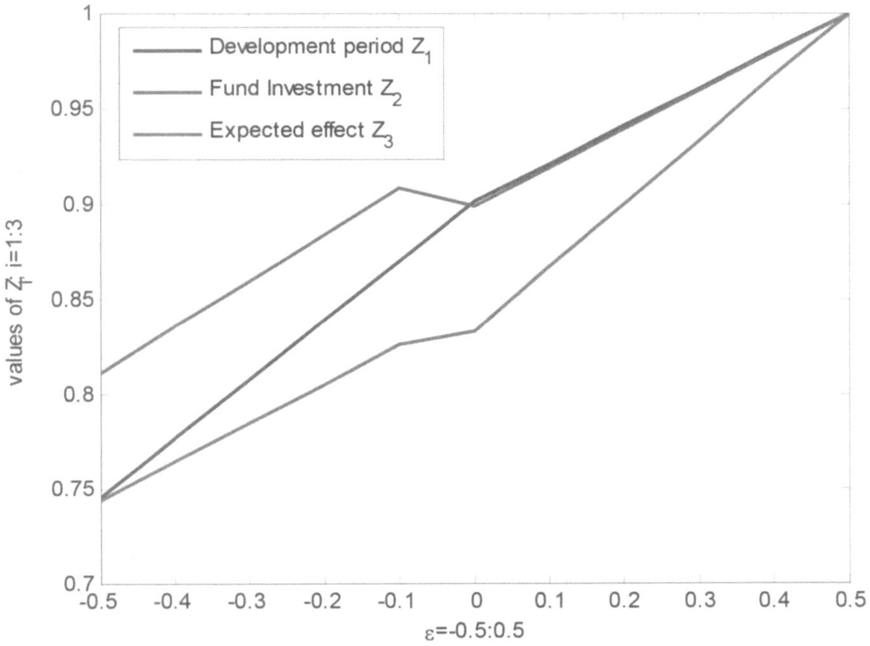

FIGURE 8.10 Sensitivity of different values of ε and corresponding solutions at (-5, -5, -5) SP and (0.9, 0.85, 0.7) AL.

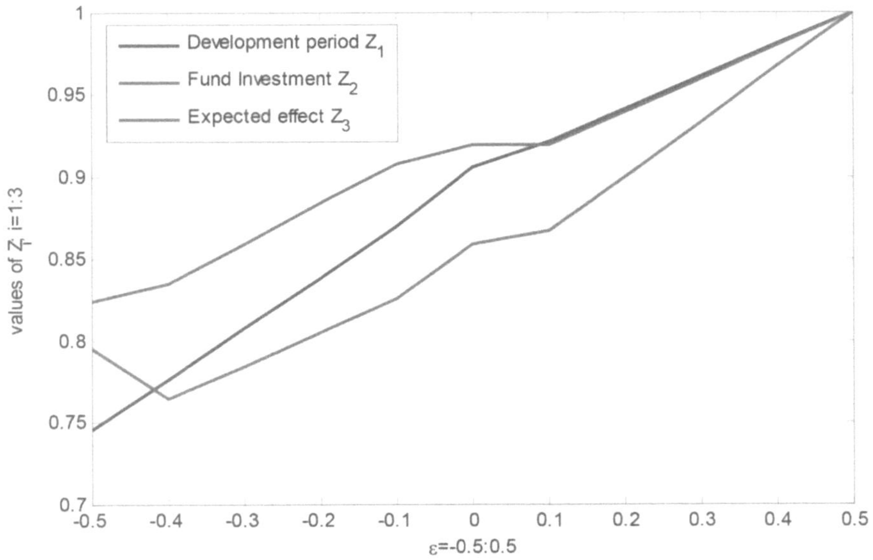

FIGURE 8.11 Sensitivity of different values of ε and corresponding solutions at (-3, -3, -3) SP and (0.8, 0.7, 0.85) AL.

TABLE 8.7
ε and its Corresponding Solutions at (-5, -5, -5) SP and (0.9, 0.85, 0.7) AL

ε	Degree of Satisfaction	Membership Values	$(Z_1^\varepsilon, Z_2^\varepsilon, Z_3^\varepsilon)$	The Assignment Solutions
-0.5	0.9442	(1.0000, 0.9442, 0.9567)	(477.5, 60, 2.9)	$M_1 \to N_2, M_2 \to N_4, M_3 \to N_3, M_4 \to N_1$
-0.4	0.9442	(1.0000, 0.9442, 0.9588)	(477.5, 60, 2.98)	$M_1 \to N_2, M_2 \to N_4, M_3 \to N_3, M_4 \to N_1$
-0.3	0.9442	(1.0000, 0.9442, 0.9612)	(477.5, 60, 3.06)	$M_1 \to N_2, M_2 \to N_4, M_3 \to N_3, M_4 \to N_1$
-0.2	0.9442	(1.0000, 0.9442, 0.9640)	(477.5, 60, 3.14)	$M_1 \to N_2, M_2 \to N_4, M_3 \to N_3, M_4 \to N_1$
-0.1	0.9442	(1.0000, 0.9442, 0.9672)	(477.5, 60, 3.22)	$M_1 \to N_2, M_2 \to N_4, M_3 \to N_3, M_4 \to N_1$
0	0.9442	(1.0000, 0.9442, 0.9709)	(477.5, 60, 3.30)	$M_1 \to N_2, M_2 \to N_4, M_3 \to N_3, M_4 \to N_1$
0.1	0.9752	(1.0000, 0.9862, 0.9752)	(477.5, 58, 3.38)	$M_1 \to N_3, M_2 \to N_4, M_3 \to N_2, M_4 \to N_1$
0.2	0.9862	(1.0000, 0.9862, 0.9862)	(477.5, 58, 3.51)	$M_1 \to N_3, M_2 \to N_4, M_3 \to N_2, M_4 \to N_1$
0.3	0.9862	(1.0000, 0.9862, 0.9929)	(477.5, 58, 3.64)	$M_1 \to N_3, M_2 \to N_4, M_3 \to N_2, M_4 \to N_1$
0.4	0.9862	(1.0000, 0.9862, 0.9972)	(477.5, 58, 3.77)	$M_1 \to N_3, M_2 \to N_4, M_3 \to N_2, M_4 \to N_1$
0.5	0.9862	(1.0000, 0.9862, 1.0000)	(477.5, 58, 3.9)	$M_1 \to N_3, M_2 \to N_4, M_3 \to N_2, M_4 \to N_1$

and at the same time Z_1 and Z_3 increase while 'ε' moves from 0.1 to 0.5, values of Z_1, Z_2 and Z_3 increase. Figure 8.11 also provides an opportunity to take decision for finding the better alternatives to DM.

8.5.1.2.2 Sensitivity Analysis with Respect to the Different Values of ε for each Objective at (-5, -5, -5) SP and (0.9, 0.85, 0.7) AL

The sensitivity analysis relating to the different ε values for each objective at (-5, -5, -5) SP and (0.9, 0.85, 0.7) AL are shown in the Tables 8.7, 8.8 and 8.9.

TABLE 8.8
ε and its Corresponding Solutions at (-5, -5, -5) SP and (0.9, 0.85, 0.7) AL

ε	Degree of Satisfaction	Membership Values	$(Z_1^\varepsilon, Z_2^\varepsilon, Z_3^\varepsilon)$	The Assignment Solutions
-0.5	0.9180	(0.9180, 0.9862, 0.9567)	(535, 58, 2.9)	$M_1 \to N_2, M_2 \to N_3, M_3 \to N_1, M_4 \to N_4$
-0.4	0.9180	(0.9180, 0.9846, 0.9567)	(535, 60.2, 2.9)	$M_1 \to N_2, M_2 \to N_3, M_3 \to N_1, M_4 \to N_4$
-0.3	0.9180	(0.9180, 0.9821, 0.9567)	(535, 62.4, 2.9)	$M_1 \to N_2, M_2 \to N_3, M_3 \to N_1, M_4 \to N_4$
-0.2	0.9180	(0.9180, 0.9776, 0.9567)	(535, 64.6, 2.9)	$M_1 \to N_2, M_2 \to N_3, M_3 \to N_1, M_4 \to N_4$
-0.1	0.8705	(0.9936, 0.9677, 0.8705)	(515, 66.4, 2.8)	$M_1 \to N_1, M_2 \to N_3, M_3 \to N_2, M_4 \to N_4$
0	0.8705	(0.9936, 0.9625, 0.8705)	(515, 68, 2.8)	$M_1 \to N_2, M_2 \to N_3, M_3 \to N_1, M_4 \to N_4$
0.1	0.8705	(0.9936, 0.9650, 0.8705)	(515, 69.6, 2.8)	$M_1 \to N_2, M_2 \to N_3, M_3 \to N_1, M_4 \to N_4$
0.2	0.8705	(0.9936, 0.9669, 0.8705)	(515, 71.2, 2.8)	$M_1 \to N_2, M_2 \to N_3, M_3 \to N_1, M_4 \to N_4$
0.3	0.8705	(0.9936, 0.9685, 0.8705)	(515, 72.8, 2.8)	$M_1 \to N_2, M_2 \to N_3, M_3 \to N_1, M_4 \to N_4$
0.4	0.8705	(0.9936, 0.9698, 0.8705)	(515, 74.4, 2.8)	$M_1 \to N_2, M_2 \to N_3, M_3 \to N_1, M_4 \to N_4$
0.5	0.8705	(0.9936, 0.9709, 0.8705)	(515, 76, 2.8)	$M_1 \to N_2, M_2 \to N_3, M_3 \to N_1, M_4 \to N_4$

TABLE 8.9
ε and its Corresponding Solutions at (-5, -5, -5) SP and (0.9, 0.85, 0.7) AL

ε	Degree of Satisfaction	Membership Values	$\left(Z_1^{\varepsilon}, Z_2^{\varepsilon}, Z_3^{\varepsilon}\right)$	The Assignment Solutions
-0.5	0.9241	(0.9241, 0.9831, 1.0000)	(415, 67.5, 3.9)	$M_1 \to N_3, M_2 \to N_4, M_3 \to N_2, M_4 \to N_1$
-0.4	0.9304	(0.9304, 0.9831, 1.0000)	(427.5, 67.5, 3.9)	$M_1 \to N_3, M_2 \to N_4, M_3 \to N_2, M_4 \to N_1$
-0.3	0.9401	(0.9401, 0.9831, 1.0000)	(440, 67.5, 3.9)	$M_1 \to N_3, M_2 \to N_4, M_3 \to N_2, M_4 \to N_1$
-0.2	0.9567	(0.9567, 0.9831, 1.0000)	(452.5, 67.5, 3.9)	$M_1 \to N_3, M_2 \to N_4, M_3 \to N_2, M_4 \to N_1$
-0.1	0.9709	(0.9709, 0.9831, 1.0000)	(465, 67.5, 3.9)	$M_1 \to N_3, M_2 \to N_4, M_3 \to N_2, M_4 \to N_1$
0	1	(1.0000, 1.0000, 1.0000)	(477.5, 66.5, 3.9)	$M_1 \to N_1, M_2 \to N_4, M_3 \to N_2, M_4 \to N_3$
0.1	0.9973	(0.9973, 1.0000, 1.0000)	(488, 66.5, 3.9)	$M_1 \to N_1, M_2 \to N_4, M_3 \to N_2, M_4 \to N_3$
0.2	0.9914	(0.9914, 1.0000, 1.0000)	(498.5, 66.5, 3.9)	$M_1 \to N_1, M_2 \to N_4, M_3 \to N_2, M_4 \to N_3$
0.3	0.9883	(0.9883, 1.0000, 1.0000)	(509, 66.5, 3.9)	$M_1 \to N_1, M_2 \to N_4, M_3 \to N_2, M_4 \to N_3$
0.4	0.9865	(0.9865, 1.0000, 1.0000)	(519.5, 66.5, 3.9)	$M_1 \to N_1, M_2 \to N_4, M_3 \to N_2, M_4 \to N_3$
0.5	0.9853	(0.9853, 1.0000, 1.0000)	(530, 66.5, 3.9)	$M_1 \to N_1, M_2 \to N_4, M_3 \to N_2, M_4 \to N_3$

Table 8.7 shows the sensitivity analysis for $\varepsilon = 0$, $\varepsilon = -0.5$, and $\varepsilon = -0.5:0.5$ of Z_1, Z_2 and Z_3 objectives at (-5, -5, -5) SP and (0.9, 0.85, 0.7) AL respectively.

Figure 8.12 indicates that when 'ε' moves from left side to the average value of the interval, values of Z_1 and Z_2 remain same and at the same time Z_3 increases. When 'ε' moves from 0 to 0.1, value of Z_1 remains same; at the same time Z_2 and Z_3 decreases and increase respectively. When 'ε' moves from 0.1 to 0.5, values of Z_1 and Z_2 remain same and z_3 increases. Figure 8.12 also indicates that under the values of ε between "0 to 0.1", value of Z_2 decreases, which provides an opportunity to take decision for finding the best alternatives to DM as per their requirement.

Table 8.8 shows the sensitivity analysis for $\varepsilon = 0.5$, $\varepsilon = -0.5:0.5$ and $\varepsilon = -0.5$ of Z_1, Z_2 and Z_3 at (-5, -5, -5) SP and (0.9, 0.85, 0.7) AL respectively.

Figure 8.13 indicates that when 'ε' moves from left side of the interval to -0.2, the values of Z_1 and Z_3 remain same and at the same time Z_2 increases. When 'ε' moves from -0.2 to -0.1, values of Z_1 and Z_3 decrease and Z_2 increases, while 'ε' moves from the average value to the right side of the interval, values of Z_1 and Z_3 remain same and Z_2 increases. Figure 8.13 also indicates that under the values of ε between "-0.2 to -0.1", value of Z_1 and Z_3 decreases, which provides an opportunity to take decision for finding the better alternatives to DM.

Table 8.9 discussed the sensitivity analysis for $\varepsilon = -0.5:0.5$, $\varepsilon = 0$ and $\varepsilon = 0.5$ of Z_1, Z_2 and Z_3 at (-5, -5, -5) $(-5, -5, -5)$ SP and (0.9, 0.85, 0.7) AL respectively.

Figure 8.14 indicates that when 'ε' moves from left side of the interval to -0.1, the values of Z_3 remain same, and at the same time Z_1 and Z_2 increases and decreases respectively. When 'ε' moves from right side of the average value of the interval, values of Z_2 and Z_3 remain same and Z_1 increases.

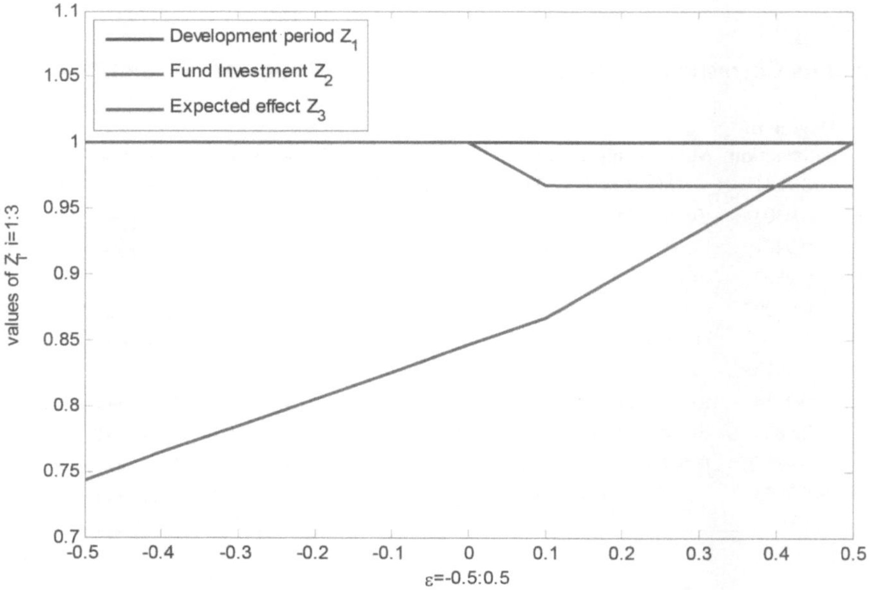

FIGURE 8.12 Sensitivity analysis of different values of ε (ε = 0, ε = –0.5, and ε = –0.5 : 0.5) for each objective and its corresponding solutions.

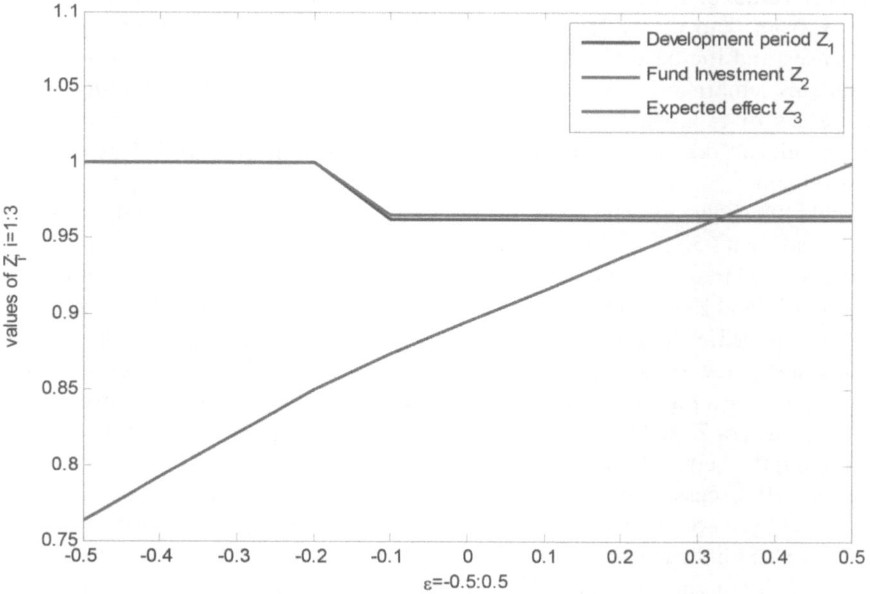

FIGURE 8.13 Sensitivity analysis of different values of ε (ε = 0.5, ε = –0.5 : 0.5, ε = –0.5) for each objective and corresponding solutions.

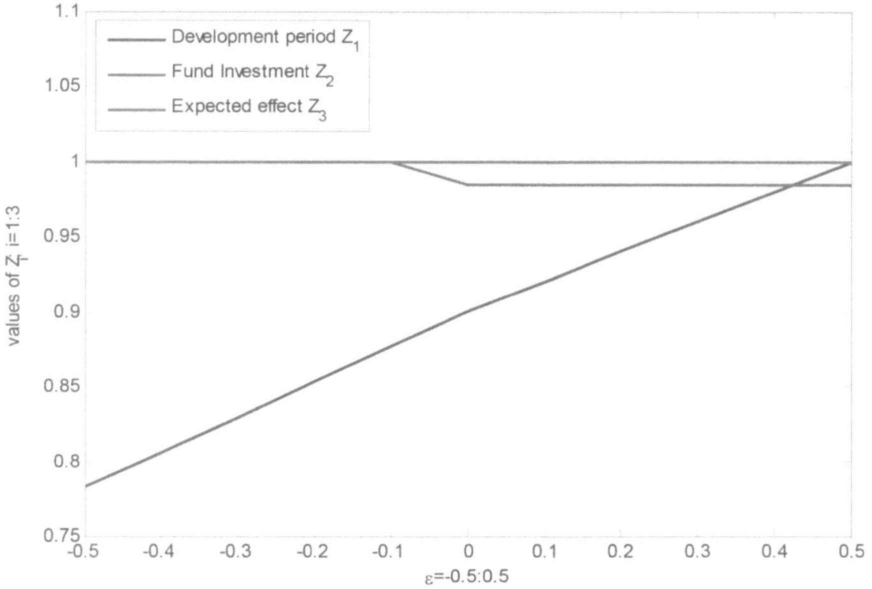

FIGURE 8.14 Sensitivity analysis of different values of ε (ε = –0.5 : 0.5, ε = 0 and ε = 0.5) for each objective and corresponding solutions.

8.5.2 MOIAP-2

To justify the GA-based approach, the cost and time interval objective AP has been referred from the article [10, 13], which is defined as follows:

$$Min\, z^1 = [1,3]x_{11} + [5,9]x_{12} + [4,8]x_{13} + [7,10]x_{21} + [2,6]x_{22} + [3,5]x_{23} + [7,11]x_{31} + [3,5]x_{32} + [5,7]x_{33}$$

$$Min\, z^2 = [3,5]x_{11} + [2,4]x_{12} + [1,5]x_{13} + [4,6]x_{21} + [7,10]x_{22} + [9,11]x_{23} + [4,8]x_{31} + [3,6]x_{32} + [1,2]x_{33}$$

Subject to:

$$\sum_{j=1}^{3} x_{ij} = 1; \; i = 1,2,3$$

$$\sum_{i=1}^{3} x_{ij} = 1; \; j = 1,2,3$$

Table 8.10 gives the PIS and NIS for Z_1 and Z_2 of model-2 at different RAPs ε = –0.5, ε = 0 and ε = 0.5.

Using PIS and NIS value for Z_1 and Z_2 in model-3, an equivalent crisp model for ε = –0.5, 0 and 0.5 can be formulated respectively for MOIAP-2 as follows:

$$\max w = \mu_{Z_1} . \mu_{Z_2}$$

TABLE 8.10
PIS and NIS of Z_1 and Z_2 of MOIAP-2 at $\varepsilon = -0.5$, $\varepsilon = 0$ and $\varepsilon = 0.5$

		Objectives	
α – Level	Solutions	Z_1	Z_2
$\varepsilon = -0.5$	PIS	7	7
	NIS	17	17.5
$\varepsilon = 0$	PIS	10	9.5
	NIS	21.5	19
$\varepsilon = 0.5$	PIS	13	12
	NIS	26	23

Subject to the constraints: (8.1)–(8.3) and (8.7); $i, j = 1,2,3; k = 1,2$
where,
For $\varepsilon = -0.5$

$$\mu_{Z_1} = \frac{\exp\left(-S\left(\begin{array}{l}1x_{11}+5x_{12}+4x_{13}+7x_{21}+2x_{22}+3x_{23}\\+7x_{31}+3x_{32}+5x_{33}-7\end{array}\right)/10\right)-\exp(-S)}{\left(1-\exp(-S)\right)}$$

$$\mu_{Z_2} = \frac{\exp\left(-S\left(\begin{array}{l}3x_{11}+2x_{12}+1x_{13}+4x_{21}+7x_{22}+9x_{23}\\+4x_{31}+3x_{32}+1x_{33}-7\end{array}\right)/10.5\right)-\exp(-S)}{\left(1-\exp(-S)\right)}$$

For $\varepsilon = 0$

$$\mu_{Z_1} = \frac{\exp\left(-S\left(\begin{array}{l}2x_{11}+7x_{12}+6x_{13}+8.5x_{21}+4x_{22}+4x_{23}\\+9x_{31}+4x_{32}+6x_{33}-10\end{array}\right)/11.5\right)-\exp(-S)}{\left(1-\exp(-S)\right)}$$

$$\mu_{Z_2} = \frac{\exp\left(-S\left(\begin{array}{l}4x_{11}+3x_{12}+3x_{13}+5x_{21}+8.5x_{22}+10x_{23}\\+6x_{31}+4.5x_{32}+1.5x_{33}-9.5\end{array}\right)/9.5\right)-\exp(-S)}{\left(1-\exp(-S)\right)}$$

For $\varepsilon = 0.5$

$$\mu_{Z_1} = \frac{\exp\left(-S\left(\begin{array}{l}3x_{11}+9x_{12}+8x_{13}+10x_{21}+6x_{22}+5x_{23}\\+11x_{31}+5x_{32}+7x_{33}-13\end{array}\right)/7\right)-\exp(-S)}{\left(1-\exp(-S)\right)}$$

$$\mu_{Z_2} = \frac{\exp\left(-S\left(\begin{array}{l}5x_{11}+4x_{12}+5x_{13}+6x_{21}+10x_{22}+11x_{23}\\+8x_{31}+6x_{32}+2x_{33}-12\end{array}\right)/11\right)-\exp(-S)}{\left(1-\exp(-S)\right)}$$

TABLE 8.11
Different values of SPs and ALs

Case	Shape parameter	Aspiration level
Case-1:	(-5, -5)	(0.95, 0.9)
Case-2:	(-3, -3)	(0.85, 0.75)
Case-3:	(-2, -2)	(0.75, 0.7)
Case-4:	(-1, -1)	(0.65, 0.6)

Here, we find the ESs for model-3 using the various SPs and ALs that are given by the DM for $\varepsilon = -0.5, 0$ and 0.5.

Each combination of the SPs and its corresponding distinct ALs are shown in Table 8.11.

According to RAPs, the assignment plans for MOIAP-2 are noted in Table 8.12 with disparate values of the SPs and ALs which are stated by the DM.

8.5.2.1 Convergence Rate of GA for MOIAP-2

The convergence rate of GA for MOIAP-2 is obtained for case-1 of Table 8.11 at disparate RAPs $\varepsilon = -0.5$, $\varepsilon = 0$ and $\varepsilon = 0.5$ as shown in Figures 8.15, 8.16 and 8.17 respectively. To find the convergence rate, we have used disparate number of iterations and populations with the values of max $W = \Pi_{i=1}^{2} Z_i$.

In both the cases as in cases of without and with mutation operator, the GA converges after 20 populations and 10 iterations at $\varepsilon = -0.5$ for case-1 of Table 8.11,

TABLE 8.12
Summary Results at $\varepsilon = -0.5$, $\varepsilon = 0$ and $\varepsilon = 0.5$

α	Case	Degree of Satisfaction	Membership Values $(\mu_{Z_{1j}}, \mu_{Z_{2j}})$	Objective Values Z_1, Z_2	Optimum Allocations
$\varepsilon = -0.5$	1	0.9241	(0.9956, 0.9241)	(8, 11)	$x_{11} = x_{22} = x_{33} = 1$
	2	0.8176	(0.9817, 0.8176)	(8, 11)	$x_{11} = x_{22} = x_{33} = 1$
	3	0.7311	(0.9653, 0.7311)	(8, 11)	$x_{11} = x_{22} = x_{33} = 1$
	4	0.6225	(0.9388, 0.6225)	(8, 11)	$x_{11} = x_{22} = x_{33} = 1$
$\varepsilon = 0$	1	0.9343	(0.9906, 0.9343)	(12, 14)	$x_{11} = x_{22} = x_{33} = 1$
	2	0.8354	(0.9641, 0.8354)	(12, 14)	$x_{11} = x_{22} = x_{33} = 1$
	3	0.7529	(0.9349, 0.7529)	(12, 14)	$x_{11} = x_{22} = x_{33} = 1$
	4	0.6474	(0.8895, 0.6474)	(12, 14)	$x_{11} = x_{22} = x_{33} = 1$
$\varepsilon = 0.5$	1	0.9409	(0.9853, 0.9409)	(16, 17)	$x_{11} = x_{22} = x_{33} = 1$
	2	0.8457	(0.9477, 0.8457)	(16, 17)	$x_{11} = x_{22} = x_{33} = 1$
	3	0.7680	(0.9082, 0.7680)	(16, 17)	$x_{11} = x_{22} = x_{33} = 1$
	4	0.6651	(0.8489, 0.6651)	(16, 17)	$x_{11} = x_{22} = x_{33} = 1$

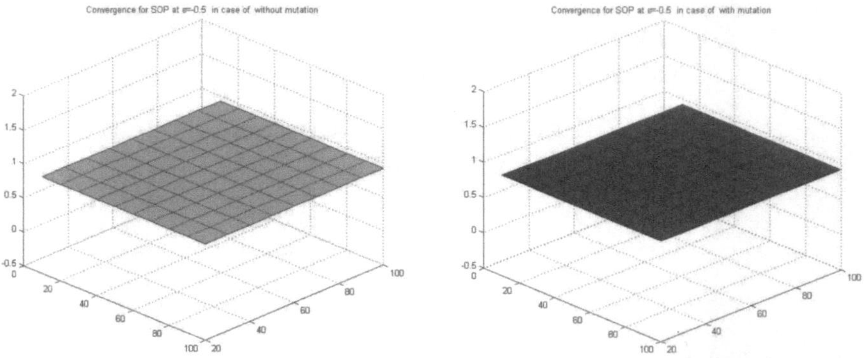

FIGURE 8.15 Convergence rate of GA at ε = –0.5 for case-1.

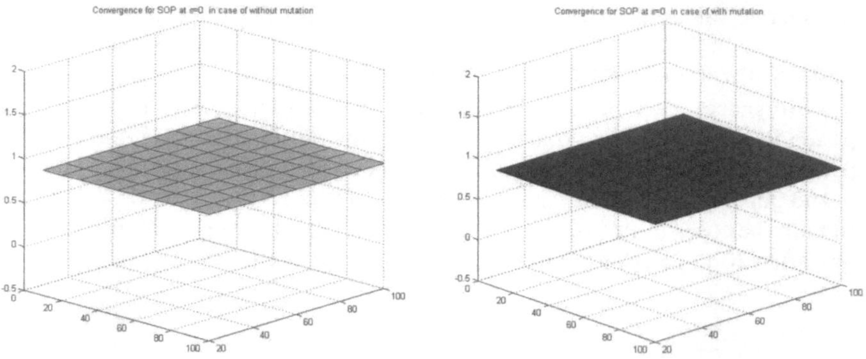

FIGURE 8.16 Convergence rate of GA at ε = 0 for case-1.

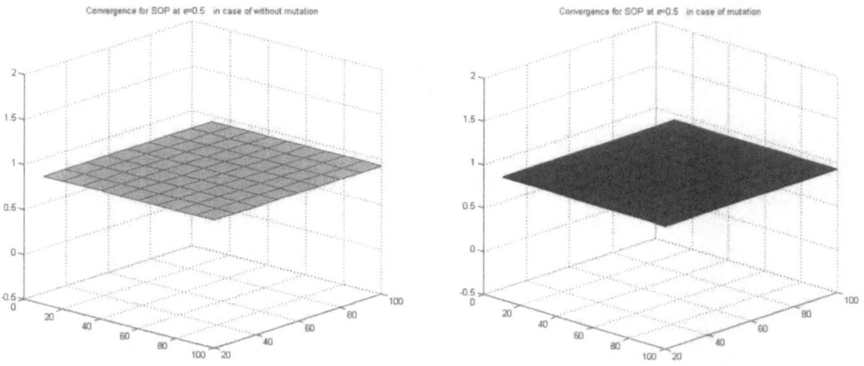

FIGURE 8.17 Convergence rate of GA at ε = 0.5 for case-1.

TABLE 8.13
ε and its Corresponding Solutions at (-5, -5) Shape Parameter and (0.95, 0.9) Aspiration Level

ε	Degree of Satisfaction	$\mu_{Z_1^{\varepsilon}}$	$\mu_{Z_2^{\varepsilon}}$	Z_1^{ε}	Z_2^{ε}	The Assignment Solutions
-0.5	0.9241	0.9956	0.9241	8	11	$x_{11} = x_{22} = x_{33} = 1$
-0.4	0.9266	0.9946	0.9266	8.8	11.6	$x_{11} = x_{22} = x_{33} = 1$
-0.3	0.9288	0.9876	0.9288	9.6	12.2	$x_{11} = x_{22} = x_{33} = 1$
-0.2	0.9308	0.9927	0.9308	10.4	12.8	$x_{11} = x_{22} = x_{33} = 1$
-0.1	0.9327	0.9916	0.9327	11.2	13.4	$x_{11} = x_{22} = x_{33} = 1$
0	0.9343	0.9906	0.9343	12	14	$x_{11} = x_{22} = x_{33} = 1$
0.1	0.9359	0.9896	0.9359	12.8	14.6	$x_{11} = x_{22} = x_{33} = 1$
0.2	0.9373	0.9885	0.9373	13.6	15.2	$x_{11} = x_{22} = x_{33} = 1$
0.3	0.9386	0.9874	0.9386	14.4	15.8	$x_{11} = x_{22} = x_{33} = 1$
0.4	0.9398	0.9864	0.9398	15.2	16.4	$x_{11} = x_{22} = x_{33} = 1$
0.5	0.9409	0.9853	0.9409	16	17	$x_{11} = x_{22} = x_{33} = 1$

which is shown in Figure 8.15. Similarly, Figures 8.16 and 8.17 indicate that the GA converges after 20 populations and 10 iterations in both cases at $\varepsilon = 0$ and $\varepsilon = 0.5$ respectively. For other cases of Table 8.14, the convergence rate of GA almost remains unchanged.

8.5.2.2 Sensitivity Analysis

8.5.2.2.1 Sensitivity Analysis with Respect to ε for Different SPs and its Corresponding ALs

For MOIAP-2, the sensitivity analysis relating to ε for (-5, -5) SP and (0.95, 0.9) AL is discussed in Table 8.13.

From Figure 8.18, when 'ε' moves from -0.5 to 0.5, the values of Z_1 and Z_2 increase. Similarly, for the (-3, -3), (-2, -2) and (-1, -1) SPs and their corresponding ALs (0.85, 0.75), (075, 0.7) and (0.65, 0.6), when 'ε' moves from -0.5 to 0.5, the values of Z_1 and Z_2 increase.

8.5.2.2.2 Sensitivity Analysis of Different Values of ε for each Objective at (-5, -5) SP and (0.95, 0.9) AL

The sensitivity analysis of the different ε values for each objective at = (-5, -5) SP and (0.95, 0.9) AL are shown in Table 8.14.

Table 8.14 shows the sensitivity analysis for $\varepsilon = -0.5$ and $\varepsilon = -0.5 : 0.5$ of Z_1 and Z_2 at (-5, -5) SP and (0.95, 0.9) AL respectively.

From Figure 8.19, it is clear that when 'ε' moves from left side of the interval to 0.5, the values of Z_1 remain same and at the same time Z_2 increases.

From Tables 8.4 and 8.12, it is finalized that as the values of ε increase, the degree of satisfaction of objectives increases. The resulting solutions and assignment plans (Tables 8.4 and 8.15) with disparate SPs in EMF for MOIAP provide the distinct

TABLE 8.14
ε and its Corresponding Solutions at (-5, -5) SP and (0.95 0.9) AL

ε	Degree of Satisfaction	$\mu_{Z_1^\varepsilon}$	$\mu_{Z_2^\varepsilon}$	Z_1^ε	Z_2^ε	The Assignment Solutions
-0.5	0.9241	0.9956	0.9241	8	11	$x_{11} = x_{22} = x_{33} = 1$
-0.4	0.9266	0.9956	0.9266	8	11.6	$x_{11} = x_{22} = x_{33} = 1$
-0.3	0.9288	0.9956	0.9288	8	12.2	$x_{11} = x_{22} = x_{33} = 1$
-0.2	0.9308	0.9956	0.9308	8	12.8	$x_{11} = x_{22} = x_{33} = 1$
-0.1	0.9327	0.9956	0.9327	8	13.4	$x_{11} = x_{22} = x_{33} = 1$
0	0.9343	0.9956	0.9343	8	14	$x_{11} = x_{22} = x_{33} = 1$
0.1	0.9359	0.9956	0.9359	8	14.6	$x_{11} = x_{22} = x_{33} = 1$
0.2	0.9373	0.9956	0.9373	8	15.2	$x_{11} = x_{22} = x_{33} = 1$
0.3	0.9386	0.9956	0.9386	8	15.8	$x_{11} = x_{22} = x_{33} = 1$
0.4	0.9398	0.9956	0.9398	8	16.4	$x_{11} = x_{22} = x_{33} = 1$
0.5	0.9409	0.9956	0.9409	8	17	$x_{11} = x_{22} = x_{33} = 1$

scenarios to DM for allotment strategy in an uncertain context. Tables 8.4, 8.12 and Figures 8.4–8.9 show that the proposed GA-based approach gives litheness and large amount of details in the sensation of altering the SPs. If the DM is dissatisfied with the resulting allocation plans, it is possible to develop more plans by varying the values ε as well as values of SPs in EMFs.

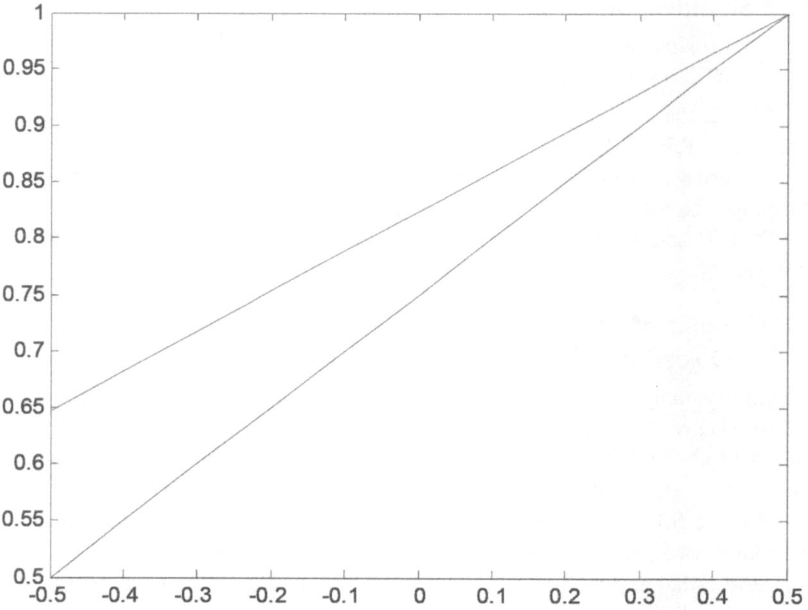

FIGURE 8.18 Sensitivity of different values of ε and corresponding solutions at (-5, -5) SP and (0.95, 0.9) AL.

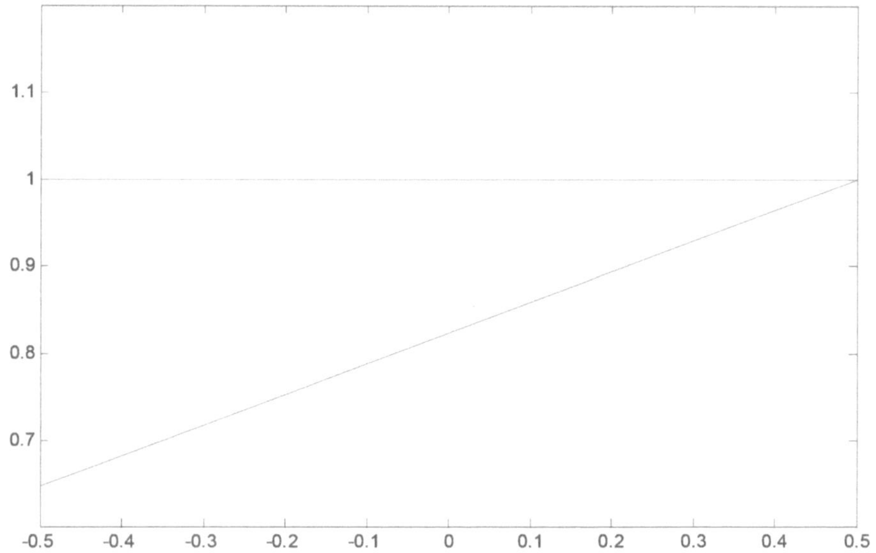

FIGURE 8.19 Sensitivity analysis of different values of ε (ε = 0.5, ε = −0.5 : 0.5) for each objective and corresponding solutions.

8.6 DISCUSSION AND COMPARISON WITH SHEN KAI'S APPROACH

In order to continue looking at the developed approach, we compare the results obtained from the GA-based approach with an existing approach of Shen Kai et al. [11] presented in Tables 8.15 and 8.16. Tables 8.15 and 8.16 show the comparison

TABLE 8.15
Comparison of Results of Problem-1 Obtained by Different Approach

ε	Shen Kai et al. Approach [11]	Shape Parameter	Aspiration Level	GA-Based Approach with ε (RAP)
-0.5	$Z_1 = 395$	(-5, -5, -5)	0.9, 0.85, 0.7	(395, 60, 2.9)
	$Z_2 = 61$	(-3, -3, -3)	0.8, 0.7, 0.85	(395, 61, 3.1)
	$Z_3 = 3.1$	(-2, -2, -2)	0.7, 0.65, 0.75	(395, 61, 3.1)
		(-1, -1, -1)	0.6, 0.5, 0.7	(395, 61, 3.1)
0	$Z_1 = 480$	(-5, -5, -5)	0.9, 0.85, 0.7	(477.5, 66.5, 3.25)
	$Z_2 = 70.5$	(-3, -3, -3)	0.8, 0.7, 0.85	(480, 68, 3.35)
	$Z_3 = 3.4$	(-2, -2, -2)	0.7, 0.65, 0.75	(480, 68, 3.35)
		(-1, -1, -1)	0.6, 0.5, 0.7	(480, 68, 3.35)
0.5	$Z_1 = 515$	(-5, -5, -5)	0.9, 0.85, 0.7	(530, 74, 3.9)
	$Z_2 = 76$	(-3, -3, -3)	0.8, 0.7, 0.85	(530, 74, 3.9)
	$Z_3 = 3.9$	(-2, -2, -2)	0.7, 0.65, 0.75	(530, 74, 3.9)
		(-1, -1, -1)	0.6, 0.5, 0.7	(530, 74, 3.9)

TABLE 8.16
Comparison of Results of Problem-2 Obtained by Different Approach

ε	Shen Kai et al. Approach [11]	Shape Parameter	Aspiration Level	GA-Based Approach with ε
-0.5	$Z_1 = 8$	(-5, -5)	0.95, 0.9	$Z_1 = 8$, $Z_2 = 11$
	$Z_2 = 11$	(-3, -3)	0.85, 0.75	$Z_1 = 8$, $Z_2 = 11$
		(-2, -2)	0.75, 0.7	$Z_1 = 8$, $Z_2 = 11$
		(-1, -1)	0.65, 0.6	$Z_1 = 8$, $Z_2 = 11$
0	$Z_1 = 12$	(-5, -5)	0.95, 0.9	$Z_1 = 12$, $Z_2 = 14$
	$Z_2 = 14$	(-3, -3)	0.85, 0.75	$Z_1 = 12$, $Z_2 = 14$
		(-2, -2)	0.75, 0.7	$Z_1 = 12$, $Z_2 = 14$
		(-1, -1)	0.65, 0.6	$Z_1 = 12$, $Z_2 = 14$
0.5	$Z_1 = 16$	(-5, -5)	0.95, 0.9	$Z_1 = 16$, $Z_2 = 17$
	$Z_2 = 17$	(-3, -3)	0.85, 0.75	$Z_1 = 16$, $Z_2 = 17$
		(-2, -2)	0.75, 0.7	$Z_1 = 16$, $Z_2 = 17$
		(-1, -1)	0.65, 0.6	$Z_1 = 16$, $Z_2 = 17$

of the solutions of problem-1 and problem-2 which are obtained by GA-based approach using RAP as well as estimation theory and Shen Kai et al. [11] approach respectively.

From Tables 8.15 and 8.16, GA-based approach with RAP can effectively handle uncertainty (interval number) in the objective function of MOIAP. The GA-based approach with the RAP provides an effective and situational solution, with sensitivity analysis to DM and multiple alternatives for the DM to make a decision as part of the decision-making process.

8.7 CONCLUSION

The GA-based approach provided the solutions of MOIAP with RAPs using an EMF with sensitivity analysis. In this chapter, RAP is used to convert the interval number into crisp number and AL is used to closely reflect the DM's opinion. It also discusses how GA's leading operators are affected on the MOIAPs. The GA-based approach also provides all possible solutions in each situation, which is very helpful for DMs to make decisions based on the situation.

REFERENCES

1. Bufardi, Ahmed. "On the efficiency of feasible solutions of a multicriteria assignment problem." *The Open Operational Research Journal* 2, no. 1 (2008): 25–28.
2. Pentico, David W. "Assignment problems: A golden anniversary survey." *European Journal of Operational Research* 176, no. 2 (2007): 774–793.
3. Majumdar, Sarangam. "Interval linear assignment problems." *Universal Journal of Applied Mathematics*1, no. 1 (2013): 14–16.

4. Oliveira, C., and C. H. Antunes. "An interactive method of tackling uncertainty in interval multiple objective linear programming." *Journal of Mathematical Sciences* 161, no. 6 (2009): 854–866.
5. Chanas, Stefan, and Dorota Kuchta. "Multiobjective programming in optimization of interval objective functions—a generalized approach." *European Journal of Operational Research* 94, no. 3 (1996): 594–598.
6. Bitran, Gabriel R. "Linear multiple objective problems with interval coefficients." *Management Science* 26, no. 7 (1980): 694–706.
7. Ida, M. "Necessary efficient test in interval multiobjective linear programming." In *Proceedings of the Eighth International Fuzzy Systems Association World Congress*, pp. 500–504. 1999.
8. Ida, M. "Efficient solution generation for multiple objective linear programming and uncertain coefficients." *Proceedings of the 8th Bellman Continuum*, pp. 132–136. 2000.
9. Ida, Masaaki. "Efficient solution generation for multiple objective linear programming based on extreme ray generation method." *European Journal of Operational Research* 160, no. 1 (2005): 242–251.
10. Kagade, K. L., and V. H. Bajaj. "Fuzzy method for solving multi-objective assignment problem with interval cost." *Journal of Statistics and Mathematics* 1, no. 1 (2010): 1.
11. Kai, Shen, Liu Yang, Jia Lu, and Fan Zhi-ping. "An interval multiobject assignment method based on decision-maker's risk attitude." In *2011 International Conference on E-Business and E-Government (ICEE)*, pp. 1–4. IEEE, 2011.
12. Salehi, Kayvan. "An approach for solving multi-objective assignment problem with interval parameters." *Management Science Letters* 4, no. 9 (2014): 2155–2160.
13. Tailor, Anita Ravi, and Jayesh M. Dhodiya. "A genetic algorithm based hybrid approach to solve multi-objective interval assignment problem by estimation theory." *Indian Journal of Science and Technology* 9, no. 35 (2016): 0974–5645.
14. Nakayama, Hirotaka. "Aspiration level approach to interactive multi-objective programming and its applications." In P. M. Pardalos, Y. Siskos, and C. Zopounidis (Eds.), *Advances in multicriteria analysis*, pp. 147–174. Springer, Boston, MA, 1995.
15. Nakayama, Hirotaka, Yeboon Yun, and Min Yoon. (Eds.) "Interactive programming methods for multi-objective optimization." In *Sequential approximate multiobjective optimization using computational intelligence*, pp. 17–43. Springer, Berlin, Heidelberg, 2009.
16. Tailor, Anita R., and Jayesh M. Dhodiya. "Multi-objective Assignment Problems and Their Solutions by Genetic Algorithm." In K. Tajeddini, S. Patnaik, and V. Jain (Eds.), *Computational Management*, pp. 409–428. Springer, Cham, 2021.
17. Dhodiya, Jayesh M., and Anita Ravi Tailor. "Genetic algorithm based hybrid approach to solve uncertain multi-objective COTS selection problem for modular software system." *Journal of Intelligent & Fuzzy Systems* 34, no. 4 (2018): 2103–2120.
18. Tailor, Anita Ravi, and Jayesh M. Dhodiya. "Genetic algorithm based hybrid approach to solve multi-objective assignment problem." *International Journal of Innovative Research in Science, Engineering and Technology* 5, no. 1 (2016): 524–535.
19. Tailor, Anita Ravi, and Jayesh M. Dhodiya. "Genetic algorithm based hybrid approach to solve optimistic, most-likely and pessimistic scenarios of fuzzy multi-objective assignment problem using exponential membership function." *Journal of Advances in Mathematics and Computer Science* 17, no. 2 (2016): 1–19.
20. Dhodiya, Jayesh M., and Anita Ravi Tailor. "Genetic algorithm based hybrid approach to solve fuzzy multi-objective assignment problem using exponential membership function." *Springer Plus* 5, no. 1 (2016): 1–29.
21. Gupta, Pankaj, Mukesh K. Mehlawat, and Garima Mittal. "A fuzzy approach to multicriteria assignment problem using exponential membership functions." *International Journal of Machine Learning and Cybernetics* 4, no. 6 (2013): 647–657.

22. Rajan, K. "Adaptive techniques in genetic algorithm and its applications." PhD diss, Mahatma Gandhi University (2013).
23. Ratli, Mustapha, Mansour Eddaly, Bassem Jarboui, Sylvain Lecomte, and Saïd Hanafi. "Hybrid genetic algorithm for bi-objective assignment problem." In *Proceedings of 2013 International Conference on Industrial Engineering and Systems Management (IESM)*, pp. 1–6. IEEE, 2013.
24. Sahu, Anshuman, and Rudrajit Tapadar. "Solving the Assignment Problem using Genetic Algorithm and Simulated Annealing." In *IMECS*, pp. 762–765. 2006.
25. Sivanandam, N., and S. N. Deepa. *Introduction to Genetic Algorithms*. Springer-Verlag, Heidelberg, Berlin, 2008.
26. Wu, Bin, Xuyan Tu, and Jian Wu. "Generalized self-adaptive genetic algorithms." *Journal of University of Science and Technology Beijing English Edition* 7, no. 1 (2000): 72–75.
27. Younas, Irfan. *"Using Genetic Algorithms for Large Scale Optimization of Assignment, Planning and Rescheduling Problems."* PhD diss., KTH Royal Institute of Technology, 2014.

9 Evolutionary Approaches in Engineering Applications

S. Tilva and J. Dhodiya

CONTENTS

9.1 INTRODUCTION

The optimization scenario is a fundamental process in various organizations, leaders, and planning devices and applications. An advancement issue by and large figures out how to glance by the best individual from all the achievable individuals called an ideal individual. It is also known as an optimal solution. The investigation and application in the enhancement area are viewed as a singular objective. Simultaneously, feasible errands consider something like at least two or more than two objections. A couple restricting or conflicting objections is customary in many issues, making the headway issues further enthralling to enlighten. In these fields, various and much of the time, conflicting targets ought to be achieved. It is called the

DOI: 10.1201/9781003229018-9

multi-objective optimization (MOO) problem. The solution has been maneuvered into the thought of experts for quite a while.

There are three objectives to pursue after settling MOO problems: (1) Demonstration of convergence, (2) Various assortment in the individual, and (3) Dispersed consistency in individuals. In addition, the obtained non-dominated individuals should be as near to as possible the Pareto ideal of the MOO problems. In single-objective optimization (SOO) problems, this goal resembles concurrent at worldwide ideal. Commonly, uncountable arrangement sets are lying on the Pareto ideal. But finite individuals are made all through the interaction of improvement. In addition, to diminish the computational cost, the amount of made individuals should be limited. Regardless, the greatest possible chance of choice should be set up on a Decision Maker (DM). Consequently, an especially scattered estimate set is mentioned, which is a target that includes itself of two requests: (1) the set that is just about as broad as could be anticipated in light of the current situation and (2) a dispersion that is pretty much as consistently partitioned as sensibly anticipated. Pareto ideal can be discontinuous, so every one of the things considered, a decisively uniform course of individuals is unreasonable. Regardless, the non-dominated individuals should be distributed on the Pareto ideal's total space and remake a major bend of the Pareto ideal as adequately as could be anticipated in light of the current situation. These necessities don't have a partner in SOO problems since simply a solitary ideal individual is delivered in light of everything.

MOO problems are tackled by both techniques, particularly evolutionary (developed) and classical (old-style) techniques. In old-style techniques, we generally substitute all objectives into an identical objective that implies the MOO problem changes over into the SOO problem. The identical objective is to look for the individual who amplifies/limits that single goal along with the system's actual limitations. The optimization individual achieves alone worth that profits a compromise between all objectives. The craftsmanship in this cycle is to produce the capacity to achieve this optimal compromise individual.

Changing the MOO problem into an SOO problem is regularly done by storing each objective in a weighted capacity or changing everything aside from one of the objections into constraints. This approach to managing disentangling MOO problem has numerous limitations: (1) by the previous data, collect the general significance of the objections, and limitations on objections provide the constraints; (2) from an accumulated work, infer simply a solitary individual; (3) efficient arrangements between objections don't obtain easily; (4) mostly an individual isn't feasible.

This essential theory of optimization isn't, now, deserving of systems with various conflicting objections. All the while, experts might need to realize each potential individual of the multitude of objections. For this present reality, it is known as an analysis of trade-offs. In a real situation, there are numerous cases of interest to execute the analysis of trade-offs. For example, they are organizing appropriated controllers while diminishing costs; both are conflicted objectives. By automobiles that venture to every part of the city at a specific time, anticipating the base fuel and comfort is a MOO problem. Restricting the functioning cost of the profession, even though taking consistent other factors, is conflicting. Hence it is a MOO problem [1–3]. Various real-world problems are identified as the MOO problem, like CPM, task issue, transportation issue, COTS determination issue, etc. [4–7].

The rest of the chapter is organized as follows. In Section 9.2, we discuss the concept of MOO, the model formulation of MOO, and some basic definitions used in MOO. Section 9.3 provides a brief description of many trendy evolutionary approaches with the conclusion in Section 9.4. At last, we include some references.

9.2 MULTI-OBJECTIVE OPTIMIZATION (MOO)

The goal of MOO is to find efficient individuals and give them to the DM. Then, the DM will choose an arrangement set as per their requirements. Additional prerequisites or guidelines are recommended beforehand or after the DM's inquiry cycle, which can be very helpful. However, we will look at the traditional circumstance. So, no previous information could be provided to DM.

9.2.1 Model of MOO

Suppose we need to purchase the lounge chair. For that, here we are considering the two factors: (1) Price assigned for a lounge chair, and (2) Comforting behavior of a lounge chair. Here, both of the factors are opposite. Thus, conflicting objections occur. Since our budget is low for a lounge chair, we couldn't feel that much comforted from a lounge chair. On the vice-versa, if we give priority to the comforting behavior of a lounge chair, then the price must be high. We need to search for a solution that provides the price in our range and desires comfort level. This scenario is known as a model of MOO.

9.2.2 Formulation of MOO

The general MOO problem is defined as follows:

$$\text{Min } Z(m) = \left[z_1(m), z_2(m), \ldots, z_p(m)\right]^T$$

Sub. to cons.,

$$g_s(m) \geq 0, \quad s = 1, 2, \ldots, a.$$
$$h_t(m) = 0, \quad t = 1, 2, \ldots, b.$$
$$m_u^L \leq m_u \leq m_u^U, \quad u = 1, 2, \ldots, c.$$

Where m_u^L and m_u^U denote the lower and upper limits of the u-th choice factor (decision variable). a represents the sum of inequality constraints; b represents the sum of equality constraints. The individual (solution) m_u that satisfies the $(a+b)$ constraints is known as feasible individual. The feasible region is the collection of all the feasible individuals and denoted as Ω. The generalized form of the MOO problem is always minimization in nature. Whenever the real-world problem is formulated in a maximization nature, it can be easily converted into minimization by the duality principle.

9.2.3 Basic Definitions

Definition 1:

Consider two individuals $m, n \in \mathbb{R}^k$, then m **dominated** n and denoted as $m \prec n$, when $z_j(m) \le z_j(n) \quad \forall\, j = 1,\ldots,P$ & $z_j(m) < z_j(n)$ for at least one j.

Definition 2:

The individual $m \in M \subset \mathbb{R}^k$ is **non-dominated** with respect to M, when 6 another m' in $M \ni z(m') \prec z(m)$.

Definition 3:

The individual $m \in M \subset \mathbb{R}^k$ is **weakly non-dominated** with respect to M, when 6 another m' in $M \ni m_u^L$.

Definition 4:

The individuals $m^* \in S \subset \mathbb{R}^k$ is **Pareto optimal** when m^* is non-dominated with respect to S, where S is feasible region.

Definition 5:

The **Pareto optimal** set O^* is defined as: $O^* = \{m^* \in S \mid m^* \; is \; Pareto-optimal\}$.

Definition 6:

The **Pareto front** f^* is defined as:$f^* = \{f(m^*) \in \mathbb{R}^k \mid m^* \in O^*\}$.

9.3 BRIEF PRESENTATION OF THE EVOLUTIONARY ALGORITHMS (EAS)

9.3.1 Genetic Algorithm (GA)

The GA [8] is one of the most seasoned and most popular nature-based optimization techniques. For the GA, solution space searching invites the natural procedure that are available in the surrounding. So, Darwinian theory for the evolution of species is taking place. The population has termed a chromosome that addresses a solution space for the issue. The issue being tackled is defined by the goal work. Contingent upon how "great" the given population is fit to a goal work, the worth that addresses caliber is assigned to it. The worth is alluded to as the population's fitness. It is a fundamental assessing component. The chromosomes which have higher fitness values get more chances for selection in the next iteration. There are three operators: selection, crossover, and mutation.

9.3.2 Genetic Programing (GP)

GP [9] is a particular type of GA that works on a particular kind of arrangement, utilizing hybrid operators of genetic. Koza [9] created the GP. It always tries to search a path for making the program codes automatically when criteria of evolution are properly known. As the program is solution searching, the code of potential solutions evolved in the pattern of the tree rather than linear chromosomes widely distributed in GA. The principal loop of GP is the same as that of GA, as GP is apart from GA as in the scheme of coding. The genetic operators represent considerable authority in dealing with trees like, and crossover.

9.3.3 Differential Evolution Algorithm (DEA)

The DEA is a kind of evolutionary approach helpful principally for the capacity enhancement in constant search space. Also, a form of DEA for combinatorial issues has also been discussed [10]; Storn and Price [11] have examined the DEA's essential form. The primary benefits of DEA over a conventional GA are: It provides less computational effort and complexity, has efficient memory usage, and is easy to apply [12]. Introduced there, DEA advances an issue by n choice factors. CR parameter denotes the crossover rate, and F parameter increases the value of specific choice factor [12] ($x_{i,j}$ denotes the worth of j-th choice factor variable put away in i-th individual in the population). The step-wise description of the working scenario of parameters is obtained from the literature [13]. The principal thought of a DEA is associated by figuring out the contrast among both the people picked arbitrarily from the population. (The DEA searches the objective gradient inside a provided region—not at a solitary point.) Thus, the DEA always tries to provide globally optimal solutions, precluding the solution sticking scenario at local extrema [12]. Two decades of DEA improvement brought about numerous modifications.

9.3.4 Evolution Strategies (ESs)

The ESs are distinctive when contrasted with the GAs, primarily in the choice system. In the GA, a parental population makes the succeeding iteration by picking people contingent upon their value of fitness by holding a consistent population size. ESs formulate a transitory population that is different from the parental population (contingent upon the expected parameters μ and λ). In this progression, the worth of fitness isn't significant. People of the transitory population go through mutation and crossover. The expected number of the best people is chosen for the upcoming iteration of the population. The traditional GA works on double vectors, while ESs work on the vectors of the floating point numbers. ES$(1+1)$, ES$(\lambda+\mu)$, and ES(λ,μ) are the essential kinds of ESs.

> **Evolution strategy ES(1 + 1)**
> It is the oldest approach; only one individual x is evolved. The initial individual x is randomly generated. In each iteration, only one new individual y is created. The crossover operator does not exist, and the mutation operator creates the individual y by adding a randomly generated number to each gene of the individual x. The normal distribution N with a mean value equal to zero and a standard deviation equal to one is used. The value of i-th gene in the individual y is computed as follows: $y_i = x_i + r \cdot N_i(0,1)$, r is a parameter determining the mutation range. Based on the fitness value of individuals x and y, the better one is selected for the new generation and becomes a new individual x. Parameter r undergoes adaptation by the so-called rule of 1/5 successes. According to this rule, the best results are obtained when the relation R between successful mutations and all mutations equals 1/5. When during k successive generations, the relation R is higher than 1/5, then the value of the r parameter is increased. When the relation R is lower than 1/5, then the value of the r parameter is decreased. The parameter r does not change when the relation R is equal to 1/5 [14].

Evolution strategy ES(λ + μ)
This is an extension of the ES(1 + 1). The ES(λ + μ) has a self-adaptive mutation range, which replaces the 1/5 success rule implemented in ES(1 + 1). In the ES(λ + μ), each individual in the population contains additional chromosomes r, consisting of standard deviation values for each gene. These values are used during the mutation procedure. The crossover operator operates before the mutation. Both chromosomes (consisting of the value of variables and the value of r parameters) undergo mutation and crossover processes [14].

Evolution strategy ES(λ, μ)
This type of ES is used more often than ES(λ + μ). The operation of both algorithms is almost identical. The only difference is that in the ES(λ,μ), the new population $P(t)$ is created using only the best individuals from the "children" population $M(t)$. In this case, μ has to be greater than λ. Such selection gives the advantage of ES(λ,μ) over the ES(λ + μ); in the latter, the population can be dominated by one much better individual than the others. The values of standard deviations r are not well-tuned. The ES(λ,μ) does not have this disadvantage because the individuals from the parental population $P(t-1)$ are not copied to the new generation $P(t)$.

Today, the covariance matrix adaptation evolution strategy (CMA-ES) is perceived as a state-of-the-art ES [15, 16]. Several variants of CMA-ES were developed [16] to enhance the efficiency or robustness by different techniques. In the CMA-ES algorithm, the adaptation of the population size or other parameters has been presented in earlier papers [17]. The CMA-ES algorithm employs global weighted recombination for both strategy and object variables, adapts the full covariance matrix for mutation, and, in general, is based on the scheme of the ES(λ,μ). The CMA-ES algorithm can handle poorly scaled functions, and its performance remains invariant under rotation of the search space [15].

9.3.5 Evolutionary Programming (EP)

The EP was created as a device for finding punctuation of an obscure language. When EP was developed as the mathematical optimization technique, it turned out to be more famous. The EP is like the ES(λ + μ), however, keeping one fundamental contrast [14]. The updated population of EP is formulated by changing each individual of the parental population. At the same time, in the ES(λ + μ), each individual has a similar likelihood for choosing in the temporary population on which the hereditary tasks are executed. The recently made and the parent population both have similar sizes (λ = μ). At last, the new iteration of the population is made utilizing the positioning choice of the people from both the population.

9.3.6 Multi-Objective Genetic Algorithm (MOGA)

The MOGA is a first Pareto-based transformative methodology developed by Fonseca and Fleming in 1993 [18]. MOGA used the idea of niching mechanism and Pareto-based rank to rouse the inquiry toward the genuine Pareto front and take the variety from a population. Each individual has been given a place addressed as a component of the number of individuals dominating through it. Assume the nd^s is a total number of individuals dominating a particular individual y at a s

generation, then, at that place, the situation at s of y individual is described as below:

$$pos^s(y) = 1 + nd^s \tag{9.1}$$

Through a ranking technique, all the non-dominated individuals are allocated at rank 1. The strategy of MOGA is as per the individual's situation and the normal fitness worth of the population. The methodology for ascertaining the fitness esteem is portrayed underneath. At first, the population is sorted through rank. From that point forward, the fitness esteem is given to each individual by the insertion of best to the most exceedingly awful situation by some predefined work. Finally, the individuals who gave a similar rank got the normal worth of the fitness. That ensures each individual who has a similar rank is inspected with an indistinct recurrence. This information is used to keep up reliable worldwide fitness of the population with a reasonable proportion of particular weight. Also, MOGA utilizes a niching idea and uses a parameter as the radius of niche. It is symbolized as σ_{rd}. For producing the uniform dissemination of the surmised Pareto front, the system of niching is done on the goal space. The individuals involved under the space of niching span are given punishment in their fitness esteem. Although in MOGA, the relegated fitness esteem relies upon the Pareto predominant idea, the arrangement may have the very position that might not have the equivalent fitness esteem. That may make an unfortunate tendency for a particular space of search space. Particularly, MOGA may be powerful by a Pareto front calculation, alongside an individual's thickness on the pursuit space. Moreover, the niching component gives grace toward the individuals with lower ranks over those with higher ranks when these last are more packed.

9.3.7 Niche Pareto Genetic Algorithm (NPGA)

The NPGA developed by Horn et al. [19] is different from prior developed MOEAs in an administrator of determination. The MOGA uses the relative determination strategy; instead, this methodology uses the selection of binary tournaments. In the selection of tournaments, arbitrarily, both individuals z and y are picked from a parent population J. A while later, both individuals are inspected reliant upon Pareto strength, with each individual of a subjectively picked sub-population I whose size is id, where $id \ll |J|$. When anybody from both individuals is non-dominated for every individual of sub-population and another individual is predominant by something like a single individual, then, at that point, the non-dominated individual is prevented. The niching mechanism is utilized to pick an individual from z and y, whichever lies at the least crowded areas in situations where neither one nor both individuals dominate through the individuals of sub-population I.

The cycle of NPGA is identified with a worth of σ_{rd} alongside the id. From the mathematical outcomes revealed by Horn et al. [19], we can presume that the population size is more prominent than id. Though, when id is exorbitantly colossal, the non-dominated individual is underscored all around; still, its intricacy will be high. On the other hand, when id is close to nothing, then, at that point, the verification of non-control could be rambunctious that it can't zero in on the non-dominated individual sufficiently. Moreover, *jd* depends on the number of objectives that are to enhance.

9.3.8 Non-Dominated Sorting Genetic Algorithm (NSGA)

NSGA [20] relies upon the method of non-dominated arrangement. This procedure portrays the arrangement population set into numerous ranks. The strategy of non-dominated figuring out begins looking through the non-dominated individuals from the population. Every one of these individuals is allotted at rank one and offers the biggest dummy fitness value. These individuals are then erased among the population; on the other hand, non-dominated individuals are found from the excess population. Besides, the non-dominated individuals of this time are relegated to the second rank, and the spurious worth of fitness is allocated more modest than the previous. This methodology is rehashed till all the individuals of the population have been ranked. To keep up with the variety of individuals, the niching mechanism is utilized for choice space instead of a target space to decrease the worth of fitness as indicated by the worth of σ_{rd}. The partaking in each rank is refined through tallying a benefit of dividing capacity between both individuals, m and n, in a comparable situation as below:

$$Sh_{e_{mn}} = \begin{cases} 1-\left(\dfrac{e_{mn}}{\sigma_{rd}}\right)^2, & \text{if } e_{mn} < \sigma_{rd}, \\ 0, & \text{otherwise.} \end{cases} \tag{9.2}$$

Where e_{mn} indicates the Euclidean distance for n and m. A while later, the above benefit of sharing capacity for each individual is included in their individual rank for creating the niche count parameter. At last, the worth of shared fitness for each individual is determined by dividing the worth of dummy fitness by niche count. Nice individuals are focused on always over the remaining individuals. In this way, the new individuals closer to the non-dominated individuals are prioritized. The niching system gives a way to deal with spreading non-dominated individuals over a Pareto front. Simultaneously, a greater affectability in regards to a parameter σ_{rd} gives a lesser useful execution of NSGA.

9.3.9 Non-Dominated Sorting Genetic Algorithm-II (NSGA-II)

The NSGA-II is the updated variant of NSGA [21, 22]. The recognizable features of NSGA-II are a diversity technique, elitist technique, less computation complexity and it doesn't require any other parameter. The fundamental idea of NSGA-II is described as follows. In NSGA-II, initially generate O_o by using a GA on a self-assertively detailed parent population J_o. The principal generation of NSGA-II is particular different from the first iteration grant. At first, both the J_s and O_s are merged and form T_s where $|J_s| = |O_s| = M$. From that point forward, the arranging of non-dominated individuals is performed to portray the entire T_s. When arranging the non-rule individual is refined, the T_s turns out to be comparatively apportioned into a couple of classes as

NSGA. Then, at that point, the new parent population J_{s+1} is figured by the individuals of the best non-dominated fronts, one by one. As the size of the absolute population is $2*M$, it probably won't lie in M spaces as it is the size of the new population J_{s+1}. When the final allowed front is taken, it may comprise more individuals than the extra spaces open in J_{s+1}. Maybe then erasing the arbitrary individuals from the last front, NSGA-II uses the niching mechanism to pick individuals from the last front in the least crowded region. Additionally, for each situating stage, the crowding distance value (CDV) is determined through ED from surrounding individuals from one or the other side of the individual on each goal work. For keeping up with the limit individuals, these last give the infinite CD. The idea of CD esteem was examined later on alongside non-dominated sorting and some more.

9.3.10 Teaching-Learning-Based Optimization Algorithm (TLBO)

The TLBO was developed by Rao et al. [23] in 2011, which needn't bother with any specific algorithm parameters for its working. A methodology communicates two fundamental ways for refreshing: (1) Instructor (known as educator phase) and (2) Relationship with various understudies (known as understudy phase). Here, the understudy's bunching is considered as a population, along with a particular sub. proposed to understudies considered as different choice factors of SOO/MOO problems. An understudy's result is related to the fitness assessment of thought about an issue. From the whole population, the best arrangement set is appointed as an instructor. Every choice factor is a boundary related to the given issue's objectives. Moreover, the best arrangement sets are the best assessment of objectives.

The working process of TLBO is divided into two parts, named as Educator stage and Learner stage. The process is described as follows [24].

Educator Phase: The first portion of the TLBO is the Teacher stage, where understudies learn through the instructor. At this stage, a teacher endeavors to extend the class' general result in the subject instructed by that individual is reliant upon their latent capacity. For any generation t, expect that there is n no. of sub. (i.e., choice factors), m no. of understudies (i.e., population size, $l = 1,2,\ldots,m$) and $M_{s,t}$ is normal for understudies in a specific sub. s $(s = 1,2,\ldots,n)$. A best broad result $Y_{total-lbest,t}$ taken for each subject to obtain an entire population of understudies with a result of best understudy $lbest$. The variety among the new mean result of each subject and the educator's equal result for each subject is given as follows:

$$VarM_{s,l,t} = r_t(Y_{s,lbest,t} - T_f M_{s,t}) \tag{9.3}$$

Where r_t stands for arbitrary number that lies between 0 and 1. $Y_{s,lbest,t}$ is the outcome of the best student in sub. s. T_f stands for teaching factor and its value of T_f can be 1/2. The value of T_f is defined as follows,

$$T_f = round[1 + rand(0,1)\{2-1\}] \tag{9.4}$$

The assessment of T_f isn't given as a contribution to a methodology, and the methodology self-assertively picks its value from Equation (9.4). After coordinating a few assessments on various benchmark capacities, it is seen that the methodology accomplishes great results if the assessment T_f is someplace in the scope of 1 and 2. Regardless, the methodology is set up to execute much better if the assessment T_f can be 1 or 2. To make a simpler methodology, T_f is prescribed to consider as 1 or 2, depending upon Equation (9.4) actions.

Because of a $Var\,M_{s,l,t}$ the recent individual is refreshed in the educator phase by using the following equation.

$$Y'_{s,l,t} = Y_{s,l,t} + VarM_{s,l,t} \tag{9.5}$$

Where $Y'_{s,l,t}$ addresses refreshed assessment of $Y_{s,l,t}$. $Y'_{s,l,t}$ is viewed as regardless of whether it gives better regard. Toward the finish of the educator stage, each considered regard is kept up because they contribute to the student phase. The student phase depends on the educator phase.

Student Phase: The next part of the TLBO is the student phase, where understudies further develop their relational abilities. An understudy relates indiscriminately with various understudies to work on their abilities. An understudy learns more when another understudy has more data over them. Accepting m as the populace size, the learning system of this phase is explained beneath.

Arbitrarily select two students A and B with the end goal that $Y'_{total-A,t} \neq Y'_{total-B,t}$

$$\text{If } Y'_{total-B,t} < Y'_{total-A,t} \text{ then } Y''_{s,A,t} = Y'_{s,A,t} + r_t(Y'_{s,A,t} - Y'_{s,B,t}), \tag{9.6}$$

$$\text{If } Y'_{total-A,t} < Y'_{total-B,t} \text{ then } Y''_{s,A,t} = Y'_{s,A,t} + r_t(Y'_{s,B,t} - Y'_{s,A,t}). \tag{9.7}$$

$Y''_{s,A,t}$ is taken if it gives a better value.

Equations (6) and (7) are for maximizing problem. For the minimizing problem, Equations (8) and (9) are used.

$$\text{If } Y'_{total-B,t} < Y'_{total-A,t} \text{ then } Y''_{s,A,t} = Y'_{s,A,t} + r_t(Y'_{s,B,t} - Y'_{s,A,t}). \tag{9.8}$$

$$\text{If } Y'_{total-B,t} < Y'_{total-A,t} \text{ then } Y''_{s,A,t} = Y'_{s,A,t} + r_t(Y'_{s,B,t} - Y'_{s,A,t}). \tag{9.9}$$

9.3.11 Non-dominated Sorting Teaching-Learning-Based Optimization (NSTLBO)

The NSTLBO is created for managing the MOO problem. The augmentation of TLBO is called NSTLBO. The NSTLBO goes under the back way to manage MOO problems and keeps up different arrangement sets. NSTLBO considers the educator stage and student stage, similar to the TLBO. Also, to manage different objectives efficaciously and speedily, the NSTLBO incorporates crowding distance computations and a non-dominate arranging approach.

At first, start with the initial random populations' formulation as J number of individuals. Then, arranging the initial populations and ranking relies upon the idea of non-dominance (ND) and constrained dominant (CD). The commonness from the individuals is chosen as below: The need is provided to CD and then subsequently to ND, and a short time later, a CDV of the individuals. An understudy who has the highest raised rank is picked as an instructor of the class. Suppose if two or more than two understudies have a comparable situation, the understudy with the highest raised CDV is assigned as a class educator. Thus, an instructor is picked from a sparse search space (SS) area. After that, understudies are modified as depicted in the educator phase. After that, modified understudies are combined with the underlying understudies to generate $2J$ arrangement sets (understudies). Once more, all understudies are sorted and ranked according to the CDV, ND, and CD of the understudies. The understudy who has poor rank is seen as worst compared to different understudies. If both the understudies stand firm on a comparative situation, then preference is given to the understudy whose CDV is higher. Then, J number of best understudies are picked from the updated ranking list along with the concept of CDV. These understudies are also modified through the learner phase.

The understudy's rank is concluded by their CDV, ND, and CD. The understudy whose rank is lowest raised rank is seen as worst compared to a next understudy. If both the understudies stand firm on a comparative situation, then the understudy with a bigger CDV is seen as better compared to the next. All new understudies are getting together with an old understudy and overarranged in achieving the student stage. From the new ranking and CDV, J number of nice understudies are picked, and they are directly refreshed subject to an educator phase for the next generation [24, 25].

Non-dominated Sorting of the Population

Here, the populations are arranged at numerous ranks as indicated by an idea of dominance described as below: the y_s individual is known as dominant. y_t individual gave the y_s individual put out concerning each objective. Furthermore the y_s individual is stringently acceptable over y_t individual defined for a basic single objective. If this condition is not fulfilled, then y_t individual can't be dominated by the y_s individual.

To J individuals, individuals who aren't dominated by any of the J individuals are called the non-dominated individuals. Each non-dominated individual perceived in a first arranging cycle is decided as the first rank and deleted from the J individual. The remainder of individuals in J individuals are again arranged, and the procedure is reiterated until each individual of the J individuals is arranged and ranked. For constrained MOO problems, the idea of CD is used.

Crowding Distance Value (CDV)

The fundamental idea of CDV is to compute the density of individuals on every side of an individual m. CDV assigned for each individual of the population. Accordingly, a CD (cd_m) is designed as the normal distance of a couple of individuals on everywhere of the individual m is determined for every N objectives. For an individual m, its CDV can be searched toward a front f, described as follow: Step 1: Find a number of individuals situated on front f as $k = |f|$. To each individual m has given $cd_m = 0$. Stage 2: For each objective $n = 1, 2, \ldots, N$, arrange an individual in a most noticeably awful request of z_n. Stage 3: For $n = 1, 2, \ldots, N$, give greatest CDV to restrict individuals ($cd_1 = cd_k = \infty$), and for the leftover individuals $s = 2$ *to* $(k-1)$, CDV is determined as below:

$$cd_s = cd_s + \frac{z_n^{s+1} - z_n^{s-1}}{z_n^{max} - z_n^{min}} \tag{9.10}$$

Where, s denotes the individual order, z_n gives n-th objective value, z_n^{max} and z_n^{min} are the maximum and minimum population value of n-th objective.

Crowding-Comparison Operator

The crowding comparison operator is used for searching an individual that is superior to succeeding one. It is set up on both valuable variables: (1) Non-dominate rank and (2) CDV of every individual s of the population. It is signified as $\prec_n$ notation and it is depicted as follows: $\prec_n$, if $(Rank_s < Rank_t)$ or $[(Rank_s = Rank_t)$ and $(cd_s > cd_t)]$. It means s and t both individuals that have to differentiate non-dominated ranks, the individual with a superior or inferior rank, is focused on. Though, assuming that the two individuals have a similar position $(Rank_s = Rank_t)$, by then, the individual arranged in the lesser crowded region $(cd_s > cd_t)$ is picked.

Constraint-Dominance (CD)

The idea of CD is described as follows: Take the individuals s and t. An individual s is known as constrained-dominant individual t when the recommended occasions happen. An individual t isn't feasible, and the s individuals are feasible. The individuals s and t are infeasible, yet the individuals have less constrained violations as compared to the individual t. The individuals s and t are feasible, and additionally, the individual t is dominated through the individual s. Subsequently, the CD concept gives guarantee that the feasible individuals achieve good rank than the infeasible individuals. For infeasible individuals, the individuals who have less constrained violation are assigned for the good rank than the remaining infeasible individuals. For feasible individuals, the non-dominated individuals got good rank than the dominant individuals.

9.3.12 Jaya Algorithm (JA)

The JA was developed by Rao [26] in 2016. It is not difficult to apply and needn't bother with a change of specific algorithm parameters. In JA, haphazardly introduced individuals are planned between the choice variable's lower and furthest cut-off points. From that point forward, every individual's choice factors are refreshed using Equation (11). Assume z as an objective which is upgraded. Suppose that v is number of choice factors. The worth of target work identified with the best individual is signified as z_{best}. Correspondingly, the worth of target work identified with the most exceedingly terrible individual is signified as z_{worst}.

$$V_{s+1,t,u} = V_{s,t,u} + R_{s,t,1}\left(V_{s,t,B} - \left|V_{s,t,u}\right|\right) - R_{s,t,2}\left(V_{s,t,W} - \left|V_{s,t,u}\right|\right) \tag{9.11}$$

Where B and W denotes the most noticeably best and terrible individuals from the current population. The file of choice factor, emphasis, and an individual is addressed as t, s, and u separately. $V_{s,t,u}$ defines the t-th choice factor of u-th individual at s-th cycle. The arbitrary numbers are signified as $R_{s,t,1}$ and $R_{s,t,2}$. The range of arbitrary numbers is [0,1], and it acts as a scaling factor, providing the great variety in individuals. In this manner, JA endeavors to move the worth of a target capacity of each individual toward the best individual through inspiring the choice factors. When the choice factors are refreshed, there is an examination among a refreshed one and the relating old one. Whichever offers a benefit is conveyed further, and individuals are performing in this way. In every emphasis, the individuals draw nearer to a superior individual by the JA. Furthermore, the individual moves a long way from the most exceedingly terrible individuals. This way, a fair fortifying and widening of the inquiry space are refined. The JA reliably endeavors to draw closer to advance (i.e., showing up at a nice individual) and endeavors to evade frustration (i.e., moving perpetually from a most noticeably terrible individual). The JA attempts to get effective by showing up as a nice individual, and consequently, it is known as Jaya means win/trump in Sanskrit.

There are additionally a couple of varieties of JA that are received in writing. They are named as SAMPEJA [27], SAMPJA [28], QOJA [29], and SAJA [30].

9.3.13 Multi-Objective Jaya Algorithm (MOJA)

MOJA was developed by Rao et al. [31] in 2017 for handling MOO problems. To tackle MOO problems, the JA is extended to the MOJA. As per Equation (11), the MOJA individuals are elevated moreover as in the JA. However, for managing various objectives quickly and effectively, the MOJA is coordinated with the idea of ND ranking and CD.

In the SOO problem, it is easy to see which one individual is acceptable than another set identified with the goal capacity's comparing esteem. In MOJA, most exceedingly awful and best individuals are searched by contrasting a position furnished with the individuals. The situation of individuals relies upon the idea of CDV, ND, and CD.

At first, the arbitrary starting population is a planned number of individuals. Then, this population is arranged and ranked to every arrangement set according to the ND and CD. This underlying population is then placed and allocated to every individual by the idea of CD and ND. A predominance among individuals is essentially concluded by the idea of CD, from that point forward, CDV and ND of individuals. The individual assigned a more noteworthy rank (rank = 1) is seen as good compared to succeeding individuals. In any case, on the off chance that two or more individuals hold a similar rank, the individual with a more noteworthy CDV precedes the other individual. This ensures the individual is browsed in the inadequate space of the SS. An individual who has the greatest lower rank is picked as the most exceedingly awful individual. Additionally, an individual with the most raised rank (rank = 1) is picked as a nice individual. When the most exceedingly awful and nice individuals are picked, then, at that point, the individuals are refreshed by Equation (11).

When each individual is refreshed, then, at that point, the refreshed population is joined by the underlying population. Thus, the $2J$ individuals are obtained. Once more, the combined population is arranged and given a rank as indicated by the idea of CDV, ND, and CD. With the assistance of CDV and updated ranking, the J quantity of better individuals is picked. The CDV and their rank settle the individual's predominance between the individuals. The individual that consists of a more noteworthy rank is taken before the other individuals. Assuming some individuals have a similar rank, the individual who has more noteworthy CD is liked over the

TABLE 9.1
List of Distinct Evolutionary Approaches

Approach, References	Author	Year	Description
GA [8]	Holland	1992	Based on the natural evolution process, best individuals survive
GP [9]	Koza	1992	With the help of natural selection, the populations are genetically bred
DEA [11]	Storn and Price	1997	Population utilized differential mutation operator for perturb vectors
ES [38]	Rechenberg	1973	Search operators use the primarily mutation and selection
EP [39]	Fogel et al.	1966	Works on species and micro-level evolution process
MOGA [18]	Fonseca and Fleming	1993	Inspired from the Pareto-based ranking and mechanism of niching
NPGA [19]	Horn et al.	1994	Works on the principle of binary tournament selection
NSGA [20]	Srinivas and Deb	1994	Inspired by the technique of non-dominated sorting
NSGA-II [21]	Deb et al.	2000	Extended version of NSGA
TLBO [23]	Rao et al.	2011	Works like the mechanism of learning and teaching process
NSTLBO [25]	Rao et al.	2018	Extension of TLBO with maintaining a diverse set of solutions
JA [26]	Rao	2016	Inspired from the independent algorithm specific parameters scenario
MOJA [31]	Rao et al.	2017	Posterior version of JA for solving MOOP

other individuals. The MOJA searches the goal capacity's worth for every individual simply a solitary time at each emphasis. Subsequently, the absolute no. of capacity assesses required by the MOJA = no. of emphases * Size of population.

There are likewise a few varieties of MOJA that are received in writing. The list of varieties are named as Multi-target Quasi-oppositional JA (MOQOJA) [32], Improved JA (IJA) [33], Binary JA (BJA) [34], and elitist JA (EJA) [35].

In like manner, JA, TLBO, and Rao are three straightforward calculations for tackling constraint and unconstraint MOO problem [36, 37].

At last, Table 9.1 shows the summary result analysis of all these approaches.

9.4 CONCLUSION

This chapter describes the introduction and basics of MOO problems. Nowadays, evolutionary approaches are trendy in research nature. Here, we have discussed the most famous evolutionary approaches over the last 5–7 decades. Every year there is a new version of these approaches developed. From that point forward, the calculations work autonomously with specific boundary which is the forward assessment of MOEA. There has been more worry about hybridization developmental methodologies, which consolidate the first streamlining approach for various plans to improve its quality. The consequence of the procured systems by and large reliably gives favored results over the first ones.

REFERENCES

1. Stadler, Wolfram, ed. *Multicriteria Optimization in Engineering and in the Sciences.* Vol. 37. Springer Science & Business Media, Boston, MA, 1988.
2. Tabucanon, Mario T. *Multiple Criteria Decision Making in Industry.* Vol. 8. Elsevier Science Limited, Amsterdam, 1988.
3. Coello, CA Coello, D. A. Van Veldhuizen, and G. B. Lamont. *Evolutionary Algorithms for Solving Multi-Objective Problems.* Kluwer Academic Publishers, Norwell, MA, 2002.
4. Tilva, S., and J. Dhodiya. "Hybrid JAYA algorithm for solving multi-objective 0-1 integer programming problem." *International Journal of Engineering and Advanced Technology* 9, no. 2 (2019): 4867–4871.
5. Dhodiya, Jayesh M., and Anita Ravi Tailor. "Genetic algorithm based hybrid approach to solve fuzzy multi-objective assignment problem using exponential membership function." *Springer Plus* 5, no. 1 (2016): 1–29.
6. Gen, Mitsuo, Yinzhen Li, and Kenichi Ida. "Solving multi-objective transportation problem by spanning tree-based genetic algorithm." *IEICE Transactions on Fundamentals of Electronics, Communications and Computer Sciences* 82, no. 12 (1999): 2802–2810.
7. Dhodiya, Jayesh M., and Anita Ravi Tailor. "Genetic algorithm based hybrid approach to solve uncertain multi-objective COTS selection problem for modular software system." *Journal of Intelligent & Fuzzy Systems* 34, no. 4 (2018): 2103–2120.
8. Holland, John Henry. *Adaptation in Natural and Artificial Systems: An Introductory Analysis with Applications to Biology, Control, and Artificial Intelligence.* MIT Press, Cambridge, 1992.
9. Koza, John R. *Genetic Programming: On the Programming of Computers by Means of Natural Selection.* Vol. 1. MIT Press, Cambridge, 1992.
10. Li, Hong, Li Zhang, and Yongchang Jiao. "Discrete differential evolution algorithm for integer linear bi-level programming problems." *Journal of Systems Engineering and Electronics* 27, no. 4 (2016): 912–919.
11. Storn, Rainer, and Kenneth Price. "Differential evolution–a simple and efficient heuristic for global optimization over continuous spaces." *Journal of Global Optimization* 11, no. 4 (1997): 341–359.
12. Slowik, Adam. "Application of an adaptive differential evolution algorithm with multiple trial vectors to artificial neural network training." *IEEE Transactions on Industrial Electronics* 58, no. 8 (2010): 3160–3167.
13. Das, Swagatam, Sankha Subhra Mullick, and Ponnuthurai N. Suganthan. "Recent advances in differential evolution–an updated survey." *Swarm and Evolutionary Computation* 27 (2016): 1–30.
14. Rutkowski, Leszek. *Computational Intelligence: Methods and Techniques.* Springer-Verlag, Berlin, Heidelberg, 2008.
15. Hansen, Nikolaus, and Andreas Ostermeier. "Completely derandomized self-adaptation in evolution strategies." *Evolutionary Computation* 9, no. 2 (2001): 159–195.
16. Loshchilov, Ilya. "A computationally efficient limited memory CMA-ES for large scale optimization." In *Proceedings of the 2014 Annual Conference on Genetic and Evolutionary Computation*, pp. 397–404. 2014.
17. Liao, Tianjun, Marco A. Montes de Oca, and Thomas Stützle. "Computational results for an automatically tuned CMA-ES with increasing population size on the CEC'05 benchmark set." *Soft Computing* 17, no. 6 (2013): 1031–1046.
18. Fonseca, Carlos M., and Peter J. Fleming. "Genetic algorithms for multi-objective optimization: Formulation, discussion and generalization." *ICGA* 93, no. July (1993), 416–423.

19. Horn, Jeffrey, Nicholas Nafpliotis, and David E. Goldberg. "A niched Pareto genetic algorithm for multi-objective optimization." In *Proceedings of the first IEEE Conference on Evolutionary Computation. IEEE World Congress on Computational Intelligence*, pp. 82–87. IEEE, 1994.
20. Srinivas, Nidamarthi, and Kalyanmoy Deb. "Multi-objective optimization using non-dominated sorting in genetic algorithms." *Evolutionary Computation* 2, no. 3 (1994): 221–248.
21. Deb, Kalyanmoy, Samir Agrawal, Amrit Pratap, and Tanaka Meyarivan. "A fast elitist non-dominated sorting genetic algorithm for multi-objective optimization: NSGA-II." In *International Conference on Parallel Problem Solving from Nature*, pp. 849–858. Springer, Berlin, Heidelberg, 2000.
22. Deb, Kalyanmoy, Amrit Pratap, Sameer Agarwal, and T. A. M. T. Meyarivan. "A fast and elitist multi-objective genetic algorithm: NSGA-II." *IEEE Transactions on Evolutionary Computation* 6, no. 2 (2002): 182–197.
23. Rao, R. Venkata, Vimal J. Savsani, and D. P. Vakharia. "Teaching–learning-based optimization: A novel method for constrained mechanical design optimization problems." *Computer-Aided Design* 43, no. 3 (2011): 303–315.
24. Rao, R. Venkata. "Teaching-learning-based optimization algorithm." In *Teaching Learning Based Optimization Algorithm*, pp. 9–39. Springer, Cham, 2016.
25. Rao, R. Venkata, Dhiraj P. Rai, and Joze Balic. "Multi-objective optimization of machining and micro-machining processes using non-dominated sorting teaching–learning-based optimization algorithm." *Journal of Intelligent Manufacturing* 29, no. 8 (2018): 1715–1737.
26. Rao, R. "Jaya: A simple and new optimization algorithm for solving constrained and unconstrained optimization problems." *International Journal of Industrial Engineering Computations* 7, no. 1 (2016): 19–34.
27. Rao, R. Venkata, and Ankit Saroj. "An elitism-based self-adaptive multi-population Jaya algorithm and its applications." *Soft Computing* 23, no. 12 (2019): 4383–4406.
28. Rao, R. Venkata, and Ankit Saroj. "A self-adaptive multi-population based Jaya algorithm for engineering optimization." *Swarm and Evolutionary Computation* 37 (2017): 1–26.
29. Rao, R. Venkata, and Dhiraj P. Rai. "Optimisation of welding processes using quasi-oppositional-based Jaya algorithm." *Journal of Experimental & Theoretical Artificial Intelligence* 29, no. 5 (2017): 1099–1117.
30. Rao, R. V., and K. C. More. "Design optimization and analysis of selected thermal devices using self-adaptive Jaya algorithm." *Energy Conversion and Management* 140 (2017): 24–35.
31. Rao, R. Venkata, Dhiraj P. Rai, and Joze Balic. "A multi-objective algorithm for optimization of modern machining processes." *Engineering Applications of Artificial Intelligence* 61 (2017): 103–125.
32. Warid, Warid, Hashim Hizam, Norman Mariun, and Noor Izzri Abdul Wahab. "A novel quasi-oppositional modified Jaya algorithm for multi-objective optimal power flow solution." *Applied Soft Computing* 65 (2018): 360–373.
33. Azizipanah-Abarghooee, Rasoul, Payman Dehghanian, and Vladimir Terzija. "Practical multi-area bi-objective environmental economic dispatch equipped with a hybrid gradient search method and improved Jaya algorithm." *IET Generation, Transmission & Distribution* 10, no. 14 (2016): 3580–3596.
34. Prakash, Tapan, V. P. Singh, S. Singh, and S. Mohanty. "Binary Jaya algorithm based optimal placement of phasor measurement units for power system observability." *Energy Conversion and Management* 140 (2017): 34–35.
35. Rao, R. Venkata, and Ankit Saroj. "Multi-objective design optimization of heat exchangers using elitist-Jaya algorithm." *Energy Systems* 9, no. 2 (2018): 305–341.

36. Rao, R. "Rao algorithms: Three metaphor-less simple algorithms for solving optimization problems." *International Journal of Industrial Engineering Computations* 11, no. 1 (2020): 107–130.
37. Rao, R. Venkata, and Rahul B. Pawar. "Constrained design optimization of selected mechanical system components using Rao algorithms." *Applied Soft Computing* 89 (2020): 106141.
38. Rechenberg, Ingo. "Evolution sstrategie—Optimierung technischer Systeme nach Prinzipien der biologischen Information." (1973).
39. Fogel, Lawrence J. "AJ owens, and MJ Walsh." *Artificial Intelligence through Simulated Evolution*. Wiley Publishing, New York, NY, 1966.

10 Quality Prediction in Vertical Centrifugal Casting Using Criterion Function

Kamar Mazloum and Dr. Amit Sata

CONTENTS

10.1 INTRODUCTION

The centrifugal casting process involves pouring the molten metal at a proper temperature into revolving mould and then allowing it to solidify. Centrifugal casting is less difficult than other gravity poured castings because there are neither unconnected gates nor risers. Centrifugal casting process is most widely employed for production of pipes, cylinders, liner brake drums, flywheel and other parts. Vertical centrifugal casting is commonly used for cylindrical shapes where aspect ratio (ratio of diameter to height) is in the range of 1–3. Typical vertical centrifugal casting experimental setup is shown in Figure 10.1. Industrial castings of ASTM A413 (LM6) and ASTM A356 (LM25) manufactured by vertical centrifugal casting process are widely used in several industrial sectors including chemical, food and marine. These industrial castings often face quality issues related to occurrence of defects such as porosity, blow holes, pin holes, hard spots, etc. Occurrence of porosity is mainly due to variations in either solidification or filling related process parameters. Presence of shrinkage porosity in castings is shown in Figure 10.2.

DOI: 10.1201/9781003229018-10

FIGURE 10.1 Vertical centrifugal casting experimental setup.

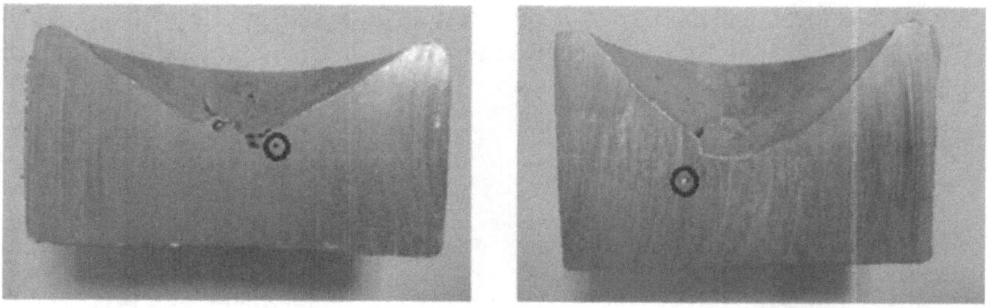

FIGURE 10.2 Shrinkage porosity in vertical centrifugal castings.

The defects, if not prevented, influence the performance of casting and its service life (Plant and Hu 1992). Relatively large amount of effort is required in repairing the defective castings to avoid their expensive rejection and recycling. Repair or recycling of the defective castings also adds to manufacturing cost, and represents wastage of foundry resources. Hence there is a need to predict, analyze and identify the exact causes of the defects to prevent their occurrence. Relevant previous work in the direction of predicting occurrence of defects in metal casting is reviewed here.

The next section provides a detailed review of previous work carried out in the direction of defect prevention using various approaches. Subsequent sections focus on overall steps that have been followed to develop modified criterion function as well as establishment of threshold values to predict shrinkage porosity in vertical centrifugal castings. The last section summarizes the contributions and limitations of the research work, as well as directions for further work.

10.2 PREVIOUS WORK

Given a set of casting design, material composition and process parameters, researchers have attempted to predict the occurrence of casting defects using appropriate mathematical models. Such models also help in analyzing and identifying the parameters responsible for actual occurrence of a defect. There are broadly two approaches for this purpose: data-based methods including Artificial Neural Network (ANN) and statistical analysis (curve fitting, regression, etc.) or simulation of physical phenomena involved in casting process (flow, solidification and cooling of molten metal in mold).

ANN is a data-driven approach, unlike casting simulation, which is a model-driven approach (Partheepan, Sehgal, and Pandey 2011). It is considered as one of the most widely used data-driven approaches in the domain of manufacturing (Sahoo, Zuo, and Tiwari 2012). In the last 20 years, several researchers have explored the application of ANN for prediction of defects in casting. The ANN can learn from examples, and has powerful capabilities to classify and recognize (Zhang, Patuwo, and Hu 1998). It can establish functional relationships from experimental data even when the correlations are difficult to find or describe scientifically (Sata and Ravi 2017). The modelling of ANN for a given application essentially involves selection of the number of layers, number of neurons in hidden layer, activation function, training algorithm, stopping criterion, and performance criterion. ANN offers a promising approach to predict defects in metal casting. However, one of main limitations of ANN is that the relationship between inputs and outputs cannot be readily explained in scientific terms.

Statistical techniques have been widely employed to develop empirical models in engineering domain. These techniques mainly include simple regression, multiple regression, and multivariate regression, which can be used to predict a single dependent variable from a single or multiple independent variables by determining best-fit model (linear or non-linear) on scatter plot of the experimented data (Rao 2005). An important aspect of statistical techniques used for prediction of defects is the extreme care needed in collecting the data as it affects the accuracy of prediction. The collected data should be free of redundant information. This is difficult when the volume of data is huge as it is in case of metal casting (Sata 2016).

Computer simulation of metal casting has accelerated in the last few years with the development of numerical methods and availability of increasingly powerful computers. Casting simulation essentially involves modelling of the physical phenomena such as flow, heat transfer, solidification, phase transformation, and stress/strain behaviour of castings (Jolly 2002). These physical phenomena are expressed in the form of governing equations based on the conservation of mass (continuity equation), momentum (Navier-Stoke) and energy (Eulerian multi-model) for flow, heat transfer and solidification (Lewis and Ravindran 2000), along with metallurgical models such as Johnson-Mehl-Avrami-Kolmorgorov (JMAK) equation for iso-thermal phase transformation

(Lusk, Krauss, and Jou 1995); mixture law, heat conduction and diffusion equation, and constitutive equation for stress-strain behaviour at different temperature (Totten, Howes, and Inoue 2002).

The governing equations are generally differential in nature, and can be solved using appropriate analytical methods (in the case of simple shapes) or numerical methods (in the case of complex shaped industrial castings). Various numerical solution techniques include Finite Element Method (FEM), Finite Difference Method, Finite Volume Method, Gradient Vector Method, and Boundary Element Method. The computations provide the temporal values of metal velocity and temperature, which are used along with various metallurgical models to predict the defects, microstructure and mechanical properties of the casting.

The numerical methods FEM, FDM and FVM presented above are referred to as *physics-based approaches*, since they are based on mathematical modelling (with appropriate assumptions) of the physical phenomena related to filling, solidification and phase change. The accuracy of prediction using casting simulation mainly depends on appropriate inputs, related to the values of thermo-physical properties of the cast metal and mold material (at different temperatures) and boundary conditions at various interfaces (metal-mold, mold-air, etc.).

Casting simulation mostly employed criterion function (simple rule that correlates the local conditions to the tendency to form pores) method to predict formation of porosity. The criterion function usually uses various solidification parameters (thermal gradient, cooling rate, etc.) or filling related parameter (velocity of fluid) resulting from solving suitable governing equations by correct numerical method. Casting simulation representing solidification related phenomena usually results in information related to temporal temperature for domain (casting) considered for simulation. Other information such as thermal gradient, and cooling rate can also be derived using information related to temporal temperature and discretization (type of elements, size of elements, number of elements, etc.). These inputs are further used in computing values of criterion function across discretized domain considered for simulation. These values of criterion function are mapped with experimental results that have already occurred with defects. This will provide information related to values of criterion function below which defects are observed in castings, and this is referred as threshold value.

Application of criterion function for prediction of shrinkage got considerable attention due to its simplicity in predicting formation of defect in castings. Various criterion function developed and their threshold value (value of criterion function below which defect is likely to occur) used for prediction of defects are summarized in Table 10.1.

However, a particular criterion function is driven by specific process and alloy that further drives its threshold value. Also, validity of criterion function for predicting occurrence of shrinkage porosity in LM6 and LM25 alloys manufactured by vertical centrifugal castings is yet to be explored especially considering effect of variation in geometry as well as mold rotation on formation of porosity.

Systematic steps followed for developing threshold value of criterion function to predict formation of shrinkage porosity in LM6 as well as LM25 alloys are discussed next followed by experimentations and their results.

TABLE 10.1
Summary of Criterion Function and Their Threshold Value (Sutaria 2015)

Criterion Function	Threshold Value	Developer
G	1.3–2.6°C/cm 0.2–0.4°C/cm	Bishop et al.
$G/(dT/dt)^{1/2}$	1	Niyama et al.
$Gts^{2/3}/Vs$	1–3	Lee et al. (LCC)
$FRN = \frac{n\mu\Delta T}{\rho_L G V_s \beta D^2}$	% porosity = $23.9 + 9.6 \times 10^{-6}(FRN) + 9.1 \times 10^{-13}(FRN)^2$	Suri et al. (FRN)
$1/ts^m\ Vs^n$	1.52	F. Chisea (FCC)
$N_y^* = \frac{G\lambda_2\sqrt{\Delta P_{cr}}}{\sqrt{\mu_l\beta\Delta T_f\left(\frac{dT}{dt}\right)}}$	610 (0.01%) 137 (0.1%) 211 (0.01%) 23 (0.1%) 776 (0.01%) 99 (0.1%)	Carlson et al. (Dimensionless Niyama, N_y^*)

Note: G: Temperature gradient Vs: Solidification velocity
ts: Local solidification time; C_λ: Material constant
dT/dt: Cooling rate; ΔP_{cr}: Critical pressure drop
β: Total solidification shrinkage; ΔT_f: freezing range
μ_l: Liquid dynamic viscosity; r: Cooling rate

10.3 SYSTEMATIC STEPS

As discussed in the previous section, it is essential to establish and modify existing criterion function to predict formation of shrinkage porosity using casting simulation as existing criterion functions are limited to particular metal-process combination. In turn, prediction of shrinkage porosity in LM6 as well as LM25 alloys also requires establishing the threshold values as these values are not available. The overall steps followed to develop modified criterion function as well as establishment of threshold values to predict shrinkage porosity are as follows.

Step 1: Selecting appropriate benchmark shape of casting that has tendency of defect
Step 2: Planning and performing the experiments
Step 3: Simulating benchmark casting by imposing appropriate boundary conditions
Step 4: Computing values of criterion function
Step 4: Superimposing experimental results on simulated results to establish threshold value
Step 5: Modifying existing criterion function considering effect of geometry and rotation of mold
Above steps are elaborately explained here.

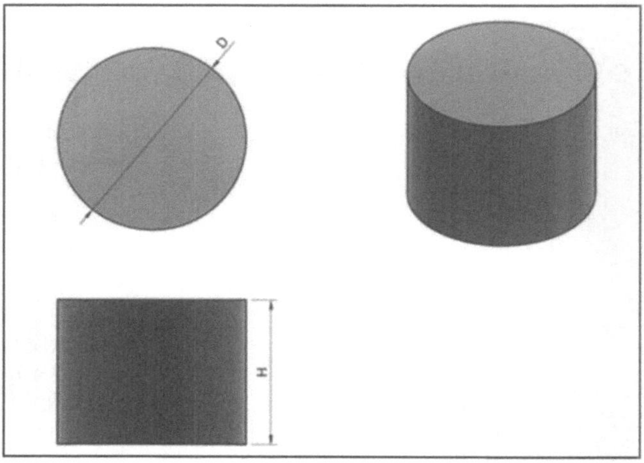

FIGURE 10.3 Benchmark shape vertical centrifugal casting.

10.3.1 Selecting Appropriate Benchmark Shape of Casting

It is known that vertical centrifugal castings are usually manufactured with aspect ratio in the range of 1–3; however, castings with aspect ratio in the range of 1–2 are widely used in various applications (Divya et al. 2018). The benchmark shape with aspect ratio of 1–2 in the steps of 0.25 with cylindrical shape is considered for further proceedings (Figure 10.3).

The vertical centrifugal casting experimental setup was developed using the fundamentals of smart manufacturing. Smart melting furnace with bottom pouring facility was also developed to melt and pour LM6 and LM25 alloys. This experimental setup also helped in Internet of Things (IoT)-enabled process monitoring as well as controlling of casting setup. The detailed methodology adopted for development of IoT-enabled vertical centrifugal casting setup was discussed elsewhere (Divya et al. 2018). This experimental setup is shown in Figure 10.4.

Various metallic molds have been prepared by varying inner diameter and outer diameter. The dimension of outer diameter of various molds was fixed from the values of inner diameter in order to maintain same thickness (difference in dimension of outer and inner diameter of mold) of mold. Capacity of support to hold molds during experimentation was also considered while selecting dimensions of mold. The height of mold was selected based on maximum possible distance between bottom of melting furnace and opening of mold for pouring, and was kept constant. The detailed dimensions of these molds are shown in Table 10.2.

10.3.2 Planning and Performing Experiments

Process parameters and geometric parameters related to vertical centrifugal castings are varied in different ranges (Table 10.3). These parameters were set for two different alloys including LM6 and LM25.

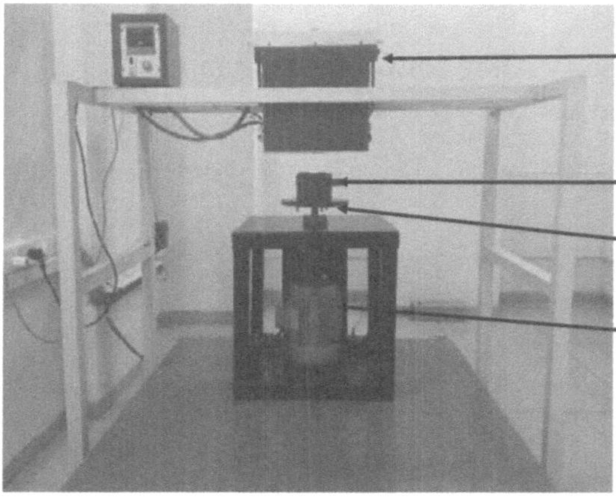

FIGURE 10.4 SMART vertical centrifugal casting setup (Divya et al. 2018).

TABLE 10.2
Dimensions of Metallic Molds

No	Inner Diameter (D_i), mm	Outer Diameter (D_o), mm	Height (H), mm	Aspect Ratio
1	70	120	70	1
2	87.5	137.5	70	1.25
3	105	155	70	1.5
4	122.5	172.5	70	1.75
5	140	190	70	2

TABLE 10.3
Variations in Parameters for Experiments

Alloys	Aspect Ratio, *AR* (D/H)	Rotation of Mold, N_{mold} (rpm)	Pouring Temperature, T_{pour} (°C)
LM6	1	50	700
and	1.25	75	725
LM25	1.5	100	750
	1.75	125	775
	2	150	800

TABLE 10.4
L25 Orthogonal Array of Experimentations for LM6 and LM25

Experiment No	AR	N_{mold} (rpm)	T_{pour} (°C)
1	1	50	700
2	1	75	725
3	1	100	750
4	1	125	775
5	1	150	800
6	1.25	50	800
7	1.25	75	700
8	1.25	100	725
9	1.25	125	750
10	1.25	150	775
11	1.5	50	750
12	1.5	75	775
13	1.5	100	800
14	1.5	125	700
15	1.5	150	725
16	1.75	50	775
17	1.75	75	800
18	1.75	100	700
19	1.75	125	725
20	1.75	150	750
21	2	50	800
22	2	75	700
23	2	100	725
24	2	125	750
25	2	150	775

Design of Experiment (DoE) technique adopted for planning of experiments, and *L25* orthogonal array (OA) has been adopted in order to plan experiments for each alloy (Table 10.4). Experiments were conducted using experimental setup shown in Figure 10.4. In total, 25 experiments were conducted for each alloy to get results of possible occurrence of shrinkage porosity. The representative experimental castings, presence of shrinkage porosity, and sectional view of experimental castings are shown in Figure 10.5(a), (b) and (c). Presence of shrinkage porosity can be easily observed in experimental castings. These experimental results were further employed for superimposing the location of shrinkage porosity in simulated results.

Full factorial design was adopted for performing simulations, and 125 simulations were carried out for each alloy. Detailed methodology adopted for simulations and respective boundary conditions related to simulations are discussed next.

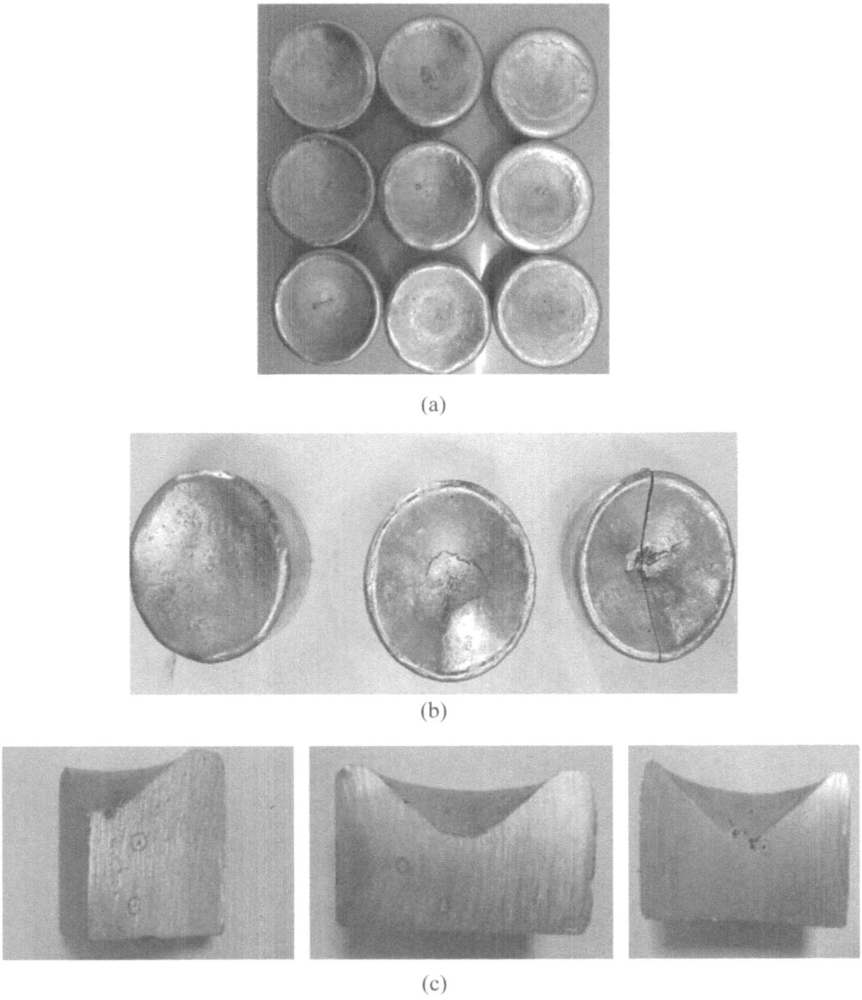

FIGURE 10.5 Representative results of: (a) experimentations, (b) presence of shrinkage porosity and (c) sectional view of castings.

10.3.3 Simulating Benchmark Castings

In total, 125 simulations were conducted for each alloy for achieving temporal information on temperature and thermal gradient for computing the values of criterion function. The simulations were conducted by adopting typical steps (modelling, preprocessing, processing and processing) used in solving complex problems using numerical methods such as FEM. The three-dimensional models were created considering aspect ratio (from 1 to 2 in the steps of 0.25), and used further for preprocessing. The three-dimensional models for mold as well as representative atmospheric (for better results in solutions) conditions were also developed as seen in Figure 10.6.

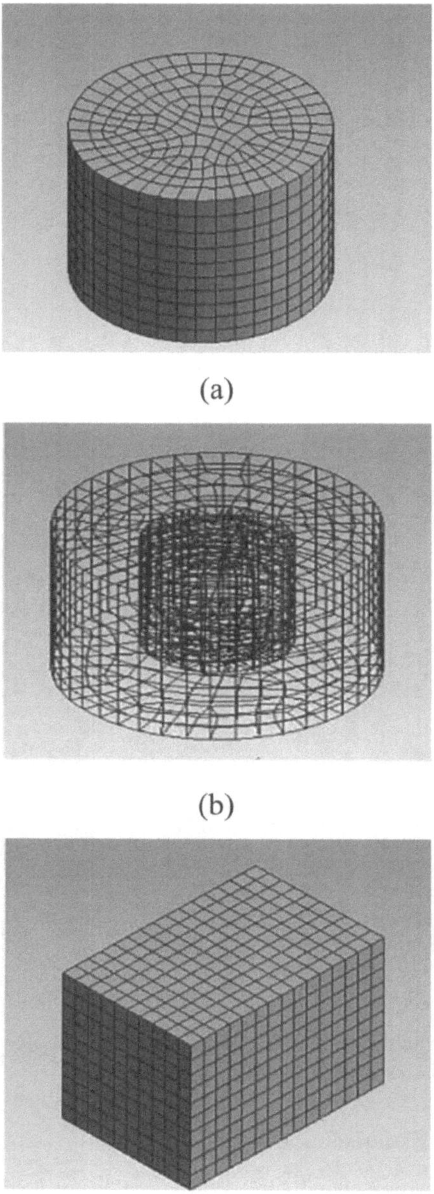

FIGURE 10.6 Three-dimensional model with discretization (a) benchmark shape, (b) mold and (c) representative atmospheric conditions.

TABLE 10.5
Thermal Properties Used for Simulations (Kayal, Behera, and Sutradhar 2011; Divya 2019)

Mold	Initial temperature (°C)	27
(Mild Steel)	Interfacial heat transfer coefficient between mold-casting (LM6 and LM25), W/m²K	800
	Interfacial heat transfer coefficient between mold-air, W/m²K	100
	Density of mold, kg/m³	7900
	Specific heat of mold, J/kg K	640
	Thermal conductivity of mold, W/m K	53
LM6	Density, kg/m³	2660
	Specific heat, kg/m³	960
	Thermal conductivity, W/m K	155
LM25	Density, kg/m³	2670
	Specific heat, J/kg K	900
	Thermal conductivity, W/m K	150

These three-dimensional models were imported, and discretized using *hexahedra* element available from library of elements. Different material models and their specific thermo-mechanical properties were defined as a part of preprocessing for further proceedings. The thermo-mechanical properties imported for analysis are mentioned in Table 10.5.

Five variations in rotation of mold and pouring temperature were taken as seen in Table 10.3. Relevant values for angular velocities have been computed using values of rotation as well radius of castings, and incorporated in FEA solver. Rotation of mold was assumed to be in positive Y direction while rotation in other directions was assumed to be negligible. It is assumed that mold is completely filled up before rotation of mold, and splashing of molten metal during rotation of mold was avoided during analysis. It is also assumed that there is no slip between mold and molten metal during rotation of mold. Transient couple analysis was carried out by incorporating mentioned boundary conditions, and solved for 500 seconds taking time step as 1 second. These simulations were conducted by commercial FEA solver (ANSYS® *workbench*) using fundamentals of thermo-mechanical analysis.

Representative simulation results of temperature profile and thermal gradient for castings of alloys LM6 and LM25 with aspect ratio of 1.5, pouring temperature as 750 °C and rotation of mold as 125 rpm are shown in Figures 10.7 and 10.8. These results will be further used for computing the values of criterion function.

10.3.4 Computing Value of Criterion Function

The *Niyama* criterion function for predicting occurrence of shrinkage porosity is preferred in comparison with other criterion functions for further proceedings as it is an accepted criterion function among many simulation tools available across globe. It is also reliable as well as easy to incorporate with existing simulation tools

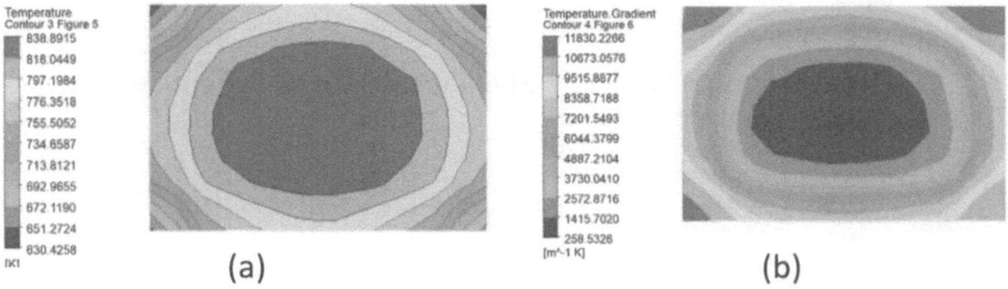

FIGURE 10.7 Simulation results for castings of alloy LM6 (a) temperature profile and (b) thermal gradient.

for predicting occurrence of defects in castings. Computation of values related to criterion function, *Niyama*, requires specific information including thermal gradient and cooling rate from simulation results. The values related to thermal gradient can be directly achieved from the simulation results while cooling rate needs to be computed from initial condition incorporated to achieve simulated results. It is clearly evident that cooling rate is driven by temperature difference and relevant time steps. This in turn requires specific value of temperature and time step at which formation of shrinkage porosity has been started during solidification of alloys. The cooling rate (equation 10.1) was computed for specific temperature (referred as *Niyama temperature*) as mentioned in equation (10.2) (Carlson and Beckermann 2008).

$$Cooling\ rate = R = \frac{dT}{dt} = \frac{T_{\text{pour}} - T_{Niyama}}{t_{\text{pour}} - t_{Niyama}} \tag{10.1}$$

$$T_{Niyama} = T_{\text{solidus}} + 0.1(T_{\text{liquidus}} - T_{\text{solidus}}) \tag{10.2}$$

$$Criterion\ Function,\ Bishop = G = \frac{dT}{dx} \tag{10.3}$$

$$Criterion\ Function,\ Ny = \frac{G}{\sqrt{R}} = \frac{dT/dx}{\sqrt{dT/dt}} \tag{10.4}$$

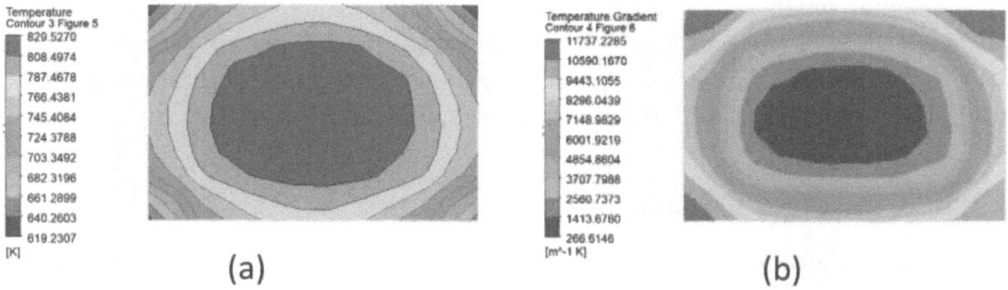

FIGURE 10.8 Simulation results for castings of alloy LM25 (a) temperature profile (b) thermal gradient.

The values of *Niyama temperature* were taken as 566°C and 556.5°C for LM6 and LM25 alloys respectively. These values were used to compute cooling rate for simulated results. These cooling rate and thermal gradient were used in computation for values of criterion function (*Niyama*) at time step (from 500 time steps taken for simulations) where maximum temperature of castings in simulation is in between *Niyama temperature* and solidus temperature of alloy. The twenty-five such representative simulated results related to thermal gradient and cooling rate for the alloy of LM6 and LM25 with aspect ratio of mold as 1 are shown in Tables 10.6 and 10.7. Similarly, 125 simulated results for each alloy were used for computation of values related to criterion function (*Niyama*) similar to values shown in Tables 10.6 and 10.7. These values of criterion function were further used in further proceedings related to superimposing the experimental results on simulated results. It will also be used in modification of existing criterion function.

TABLE 10.6
Twenty-Five Representative Results of LM6 (AR = 1) with Values of Criterion Function

Pouring Temperature, T_{pour} (°C)	Rotation of Mold, N_{mold} (rpm)	Thermal Gradient, (G) (°C/mm)	Cooling Rate (°C/sec), R	Criterion Function, Niyama ($G/\sqrt{R}$)
700	50	2.51	102.8196	0.23
700	75	2.57	98.01	0.24
700	100	2.75	95.8441	0.26
700	125	2.67	91.0116	0.26
700	150	2.84	89.8704	0.28
725	50	2.34	101.6064	0.21
725	75	2.52	98.8036	0.23
725	100	2.53	95.4529	0.24
725	125	2.51	91.7764	0.25
725	150	2.71	91.0116	0.27
750	50	2.27	102.6169	0.21
750	75	2.42	98.6049	0.22
750	100	2.45	94.6729	0.23
750	125	2.50	93.7024	0.24
750	150	2.48	93.1225	0.24
775	50	2.27	110.25	0.21
775	75	2.37	105.2676	0.22
775	100	2.37	98.6049	0.22
775	125	2.30	96.8256	0.22
775	150	2.65	97.8121	0.26
800	50	2.19	111.7249	0.20
800	75	2.33	105.8841	0.22
800	100	2.33	99.6004	0.22
800	125	2.31	98.6049	0.22
800	150	2.37	100.4004	0.23

TABLE 10.7
Twenty-Five Representative Results of LM25 ($AR = 1$) with Values of Criterion Function

Pouring Temperature, T_{pour} (°C)	Rotation of Mold, N_{mold} (rpm)	Thermal Gradient, (G) (°C/mm)	Cooling Rate (°C/sec)	Criterion Function, Niyama ($G/\sqrt{R}$)
700	50	2.51	111.7249	0.23
700	75	2.57	110.25	0.24
700	100	2.75	106.9156	0.26
700	125	2.67	104.6529	0.26
700	150	2.84	99.2016	0.28
725	50	2.34	112.9969	0.21
725	75	2.52	111.0916	0.23
725	100	2.53	106.2961	0.24
725	125	2.51	99.2016	0.25
725	150	2.71	99.4009	0.27
750	50	2.27	114.2761	0.21
750	75	2.42	112.1481	0.22
750	100	2.45	106.9156	0.23
750	125	2.50	100.6009	0.24
750	150	2.48	100.8016	0.24
775	50	2.27	115.1329	0.21
775	75	2.37	112.7844	0.22
775	100	2.37	107.3296	0.22
775	125	2.30	102.2121	0.22
775	150	2.65	102.8196	0.26
800	50	2.19	115.7776	0.20
800	75	2.33	107.7444	0.22
800	100	2.33	108.3681	0.22
800	125	2.31	104.4484	0.22
800	150	2.37	104.4484	0.23

10.3.5 Superimposing Experimental Results on Simulated Results

As discussed in previous section, the values of thermal gradient and cooling rate have been computed at all nodes available in simulated results at *Niyama temperature*. These were further used in computing values of criterion function at all nodes of simulated results for each benchmark castings using equations (10.3) and (10.4). Experimental results with presence of shrinkage porosity were used for superimposing on the results of simulations (identification of nodes was superimposed on simulated results). The superimposed results helped in identifying the locations of nodes (in turn location of shrinkage porosity in castings) in simulated results that has presence of shrinkage porosity in experimental results. The values of criterion function for nodes (that have presence of shrinkage porosity) have been computed using

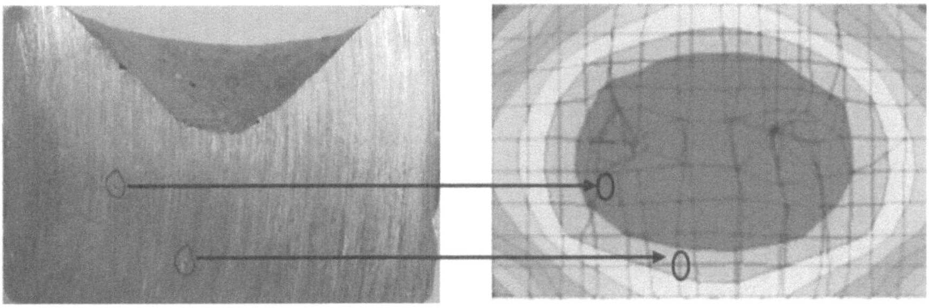

FIGURE 10.9 Superimposing results of experimental result with simulated result.

the values of thermal gradient as well as cooling rate. Some of such results are also shown in Tables 10.6 and 10.7.

The values of criterion function have been computed for all such nodes from simulated results. More than 100 such nodes for each alloy were superimposed with experimental results, and their values of criterion function have been acquired. Representative superimposing of experimental result with simulated result achieved for alloy LM25 with aspect ratio as 1.5, pouring temperature as 700 ^{0}C, as well as rotation of mold as 125 rpm (results of experiment *14* as per Table 10.4) is shown in Figure 10.9. These values are further used in modifying existing criterion function for inclusion of effect of geometry as well as rotation of mold.

10.3.6 Modifying Existing Criterion Function

An existing criterion function (e.g., *Niyama criterion*) mainly considers effect of thermal gradient as well as cooling rate to compute the values related to *threshold values*; however, effect of *aspect ratio* (*AR*), and rotation of mold (N_{mold}) also need to be considered while computing the *threshold values* that are further used for predicting location of shrinkage porosity in castings of LM6 and LM25.

Computed values of criterion function have been employed to develop modified criterion function using multiple regression technique. Multiple regression analysis has been carried out using *Microsoft® Data Analysis* addon. Results of simulations (125 results for each alloy) comprising results of thermal gradient and cooling rate along with aspect ratio (*AR*), as well as rotation of mold (N_{mold}) used for development of statistical model. Statistical information related to multiple regression models developed for each alloy are illustrated in Table 10.8.

It can be observed from the statistical information of developed model that they have exceptional representations related to R Square (statistical measure of how well the developed model is fitted, higher value is better); Adjusted R Square (modified version R square, higher is better); F value (represents the quality of model, higher is better); p value (probability that the independent variable in model has nothing to do with dependent value, lower value is better); t value (represents variations of dependent variable with independent variables, higher is better) (Sata 2010). Separate

TABLE 10.8
Statistical Information of Developed Models

Statistical Information	LM6	LM25
Total Number of Observations	125	125
R Square	0.99	0.99
Adjusted R Square	0.99	0.99
Standard Error	0.01	0.01
F Value	365880.30	395267.57
Intercept	0	0
Coefficient – Thermal Gradient (G)	0.99	1.08
p value – Thermal Gradient (G)	5.1E-83	6.3E-92
t value – Thermal Gradient (G)	50.35	60.05
Coefficient – Cooling Rate ®	–0.50	–0.51
p value – Cooling Rate (R)	1.7E-132	2.12E-135
t value – Cooling Rate (R)	–131.64	–139.19
Coefficient – Aspect Ratio (AR)	0.006	0.002
p value – Aspect Ratio (AR)	0.045	4.8E-05
t value – Aspect Ratio (AR)	2.01	4.21
Coefficient – Rotation of Mold (N_{mold})	–0.006	–0.002
p value – Rotation of Mold (N_{mold})	0.3410	0.42
t value – Rotation of Mold (N_{mold})	–0.95	–0.80

models were developed for both alloys, and their findings in terms of factors are illustrated in Table 10.9.

It can be observed that four factors have been incorporated in the modified criterion function in comparison with two factors available with existing criterion function. It is also observed that the values of factors associated with modified criterion function are sensitive with alloys too. The sensitivity with alloys is represented by material constants. It is clearly evident from the developed models that thermal gradient as well as cooling rate highly affects the formation of

TABLE 10.9
Modified Criterion Function

Existing Criterion Function		Existing Factors		Modified Criterion Function	Modified Factors				
		a	*b*		*Alloy*	*a*	*b*	*c*	*d*
Niyama	$\frac{G^a}{R^b}$	1	0.5	$\frac{(G)^a \cdot (AR)^c}{(R)^b \cdot (N_{\text{mold}})^d}$	LM6	0.99	0.50	0.006	0.006
					LM25	1.08	0.51	0.02	0.002

G: temperature gradient; R: cooling rate; AR: aspect ratio
N_{mold}: rotation of mold; a, b, c, d: material constants

shrinkage porosity. While severity of rotation of mold on formation of shrinkage porosity is slightly higher than the aspect ratio. However, scientific explanation for this phenomenon is yet to be established, and can be considered for future research work.

10.4 CONCLUSIONS

The present work mainly focused on extending an application of *Niyama* criterion to predict shrinkage porosity in castings of LM6 and LM25 alloys manufactured through vertical centrifugal casting. Systematic approach has been adopted to modify existing criterion function that further can be employed for predicting shrinkage porosity in LM6 and LM25 alloys. Benchmark casting has been cast by varying aspect ratio, rotation of the mold and pouring temperature. Casting simulation was carried out by incorporating appropriate boundary conditions and initial conditions to achieve the results related to thermal gradient and cooling rate. Threshold values have been computed by superimposing experimental results to simulated results. It has been observed that threshold values vary between 0.20 and 0.30. These values were further used in the development of modified criterion function that considers the effect of aspect ratio as well as rotation of mold as effect of aspect ratio as well as rotation of mold on criterion function. It can be observed that aspect ratio of mold as well as rotation of mold affect the formation of shrinkage porosity in vertical centrifugal castings of LM6 and LM25 alloys. However, the effect of aspect ratio is relatively less in comparison with effect of rotation of mold on formation of shrinkage porosity. Modified criterion function derived using present work for each alloy can be easily embedded with any existing casting simulation tool, and will be very useful for predicting occurrence of porosity in vertical centrifugal castings of LM6 and LM25. This in turn helps in predicting the quality of vertical centrifugal castings of LM6 and LM25.

ACKNOWLEDGEMENT

The work was supported by the SMART Foundry 2020 model project funded by Department of Science and Technology-Government of India under Advanced Manufacturing Technologies of Technology System Development Program (TSDP).

REFERENCES

Carlson, Kent D., and Christoph Beckermann. 2008. "Use of the Niyama Criterion to Predict Shrinkage-Related Leaks in High-Nickel Steel and Nickel-Based Alloy Castings." In *62nd SFSA Technical and Operating Conference*, 1–18.

Divya, Bhoraniya. 2019. "*Development of Criterion Function for Predicting Defect in Vertical Centrifugal Casting.*" Postgraduate Dissertation, *Gujarat Technological University.*

Divya, Bhoraniya, Pradip Kanzaria, Dhaval Anadkat, andAmit Sata. 2018. "Development of Vertical Centrifugal Casting (VCC) Experimental Setup." In *National Conference on Excellence in Design, Manufacturing & Automation (NCEDMA-2018).*

Jolly, Mark. 2002. "Casting Simulation: How Well Do Reality and Virtual Casting Match? State of the Art Review." *International Journal of Cast Metals Research* 14 (5): 303–13.

Kayal, Sourav, Rabindra Behera, and G. Sutradhar. 2011. "Effect of SiCp on Fluidity of the LM6/SiCp Metal Matrix Composites." *International Journal of Emerging Trends in Engineering and Development* 3 (1): 172–80.

Lewis, R. W., and K. Ravindran. 2000. "Finite Element Simulation of Metal Casting." *International Journal for Numerical Methods in Engineering* 47: 29–59.

Lusk, M., G. Krauss, and H.-J. Jou. 1995. "A Balance Principle Approach for Modeling Phase Transformation Kinetics." *Le Journal de Physique IV* 05 (C8): C8-279–84.

Partheepan, G., D. K. Sehgal, and R. K. Pandey. 2011. "Quasi-Non-Destructive Evaluation of Yield Strength Using Neural Networks." *Advances in Artificial Neural Systems* 2011: 1–8.

Plant, Robert T., and Qing Hu. 1992. "The Development of a Prototype DSS for the Diagnosis of Casting Production Defects." *Computer Industrial Engineering* 22 (2): 133–46.

Rao, C. R. 2005. *Data Mining and Data Visualization. Elsevier B.V.* Vol. 24.

Sahoo, Ajit K., Ming J. Zuo, and M. K. Tiwari. 2012. "A Data Clustering Algorithm for Stratified Data Partitioning in Artificial Neural Network." *Expert Systems with Applications* 39 (8): 7004–14.

Sata, Amit. 2010. "*Shrinkage Porosity Prediction Using Casting Simulation.*" Postgraduate Dissertation, *Indian Institute of Technology Bombay.*

Sata, Amit. 2016. "Investment Casting Defect Prediction Using Neural Network and Multivariate Regression along with Principal Component Analysis." *International Journal of Manufacturing Research* 11 (4): 356–73.

Sata, Amit, and B. Ravi. 2017. "Bayesian Inference-Based Investment-Casting Defect Analysis System for Industrial Application." *International Journal of Advanced Manufacturing Technology* 90: 3301–15.

Sutaria, Mayur. 2015. "*Casting Solidification Feed-Paths: Modeling, Computation and Applications.*" Doctorate Thesis, *Indian Institute of Technology Bombay.*

Totten, G., M. Howes, and T. Inoue. 2002. *Handbook of Residual Stress and Deformation of Steel. ASM International.*

Zhang, Guoqiang, B. Eddy Patuwo, and Michael Y. Hu. 1998. "Forecasting with Artificial Neural Networks: The State of the Art." *International Journal of Forecasting* 14: 35–62.

11 On Type D Fuzzy Cellular Automata-Based MapReduce Model in Industry 4.0

Arnab Mitra

CONTENTS

11.1 INTRODUCTION

Intelligent digitization is considered as the key feature to be adopted by modern-day production houses for an efficient realization of manufacturing line in Industry 4.0 scenarios. We find that, to attain the "threshold of the fourth Industrial Revolution" (Lee, Bagheri and Kao 2015), various cutting-edge Information Technologies (ITs) [for example, Wireless Sensor Networks (WSNs), IoT (Internet of Things), Cloud and Fog Computing, Big Data analytics, Cyber Securities, Block Chain technology-based architectures and new protocols, vertical and, or horizonal scaling of systems, autonomous material handling, Additive Manufacturing Technologies, etc.] are needed to be combined effectively (Lee, Bagheri and Kao 2015; Mitra 2021a). On the other hand, the CPS (Cyber Physical Systems) is referred to as an efficient and transformative technology management in view of the system of interconnected systems. It incorporates the physical assets of CPS to its computational skills (Lee, Bagheri and Kao 2015; Mitra 2021a; Mitra and Banerjee 2021). Hence, an efficient realization of a CPS may result in several benefits in view of Industry 4.0 (refer the attributes of Figure 11.1). A typical CPS architecture with reference to Industry 4.0 of (Lee, Bagheri and Kao 2015; Mitra 2021a; Mitra and Banerjee 2021) is once again produced in Figure 11.1.

DOI: 10.1201/9781003229018-11

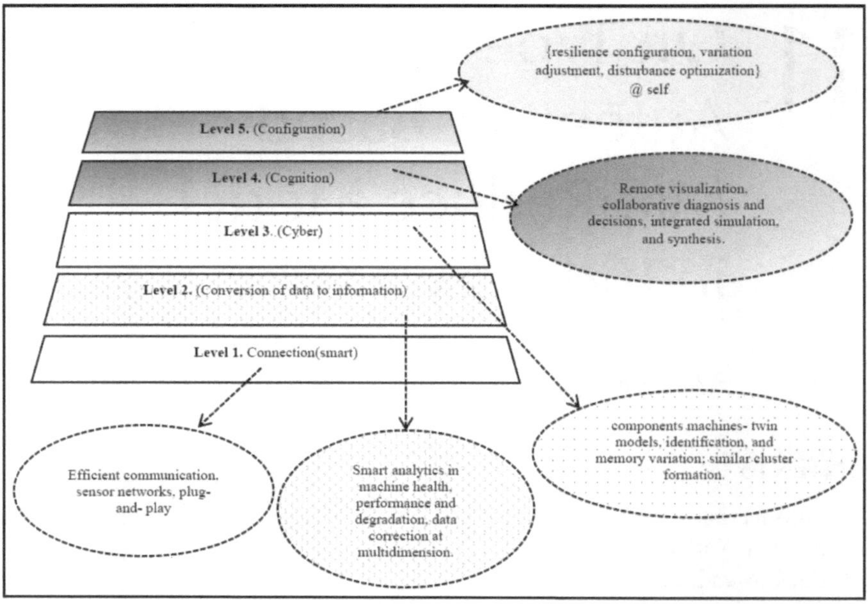

FIGURE 11.1 A typical diagram describing the 5C-based CPS architecture (inspired from Lee, Bagheri and Kao 2015; Mitra 2021a; Mitra and Banerjee 2021).

We learned that the effectiveness of the CPS architecture is primarily reliant on the effective integration of industrial AI (Artificial Intelligence) technologies along with modern cutting-edge IT components and technologies (Lee, Bagheri and Kao 2015; Mitra 2021a). In our studies, we found an interesting 5C-based (i.e., "connection, conversion, cyber, cognition and configuration"; Mitra and Banerjee 2021) CPS architecture (refer Figure 11.1; Lee, Bagheri and Kao 2015; Mitra 2021a; Mitra and Banerjee 2021). An initial CPS architecture was presented in (Lee, Bagheri and Kao 2015) and thus, we followed the same in our presented work. Among several other points (refer Figure 11.1), we understand that "sensor-based networked manufacturing lines are dealing with high volume of data (i.e., Big Data), which in turn are much crucial" (Mitra 2021a) to achieve the manufacturing target (Baheti and Gill 2011; Shi, Wan and Yan et al. 2011; Lee, Lapira and Bagheri et al. 2013; Lee, Bagheri and Kao 2015; Mitra 2021a; Mitra and Banerjee 2021). We observe that scientific communities are at work on numerous challenges to enrich the production-lines in any Industry 4.0 scenario. Though several problem domains in view of Industry 4.0 might be found, we mainly are fascinated towards Big Data processing in production lines. State-of-the-art literatures related to Big Data processing in view of Industry 4.0 explore an emphasis on the model for MapReduce design towards efficient workload processing at smart production-lines (Kumar, Shankar and Choudhary et al. 2016; Wan, Tang and Shu et al. 2016; Lin, Shu and Deng et al. 2017; Manogaran, Thota and Lopez et al. 2017; Mitra, Kundu and Chattopadhyay et al. 2018; Xu and Duan 2019; Mitra 2021a). A short discussion on the different

MapReduce-based workload processing in view of Industry 4.0 scenario is provided in Section 11.2. Among others, we particularly found an interesting effort in (Mitra, Kundu and Chattopadhyay et al. 2018; Mitra 2021a). A Cellular Automata (CA)-based MapReduce model was presented in (Mitra, Kundu and Chattopadhyay et al. 2018; Mitra 2021a) to facilitate energy-efficient and low-cost physical modelling towards Big Data processing in Industry 4.0 scenario. As we are particularly influenced by the said research, a brief discussion on related terminologies as presented in (Mitra, Kundu and Chattopadhyay et al. 2018; Mitra 2021a) is presented next.

CA (Wolfram 1984; Chaudhuri, Roy Chowdhury, and Nandi et al. 1997; Mitra, Kundu and Chattopadhyay et al. 2018; Mitra 2021a; Mitra and Banerjee 2021) is a cost-effective dynamic demonstration tool; CA progresses over both discrete time and space. Elementary CA (ECA) is recognized as extremely simple CA structure containing three cells at one-dimension at three vicinities (right neighbour, left neighbour and self-cell) in null (fixed) or periodic boundary scenario (Chaudhuri, Roy Chowdhury and Nandi et al. 1997; Mitra 2021a). To realize the distinct ECA boundary scenarios, a typical diagram is shown in Figure 11.2 (as inspired from Chaudhuri, Roy Chowdhury and Nandi et al. 1997; Mitra, Kundu and Chattopadhyay et al. 2018; Mitra 2021a; Mitra and Banerjee 2021). The next state of any ECA cell at time t is governed by a mathematical function [refer Equation 11.1 (Chaudhuri, Roy Chowdhury and Nandi et al. 1997; Mitra, Kundu and Chattopadhyay et al. 2018; Mitra 2021a; Mitra and Banerjee 2021)] named as CA rule, also accepted as Wolfram CA rule, total 256 ECA rules (Wolfram 1984; Chaudhuri, Roy Chowdhury and Nandi et al. 1997),

$$x_i^{t+1} = f\left(x_{i-1}^t, x_i^t, x_{i+1}^t\right) \tag{11.1}$$

where at time t, x_{i-1}^t signifies value at left cell, x_i^t signifies value at self-cell, x_{i+1}^t value at right cell and x_i^{t+1} signifies the value of a cell at time $(t+1)$ (Chaudhuri, Roy Chowdhury and Nandi et al. 1997; Mitra 2021a; Mitra and Banerjee 2021). For more insights about ECA and related ECA dynamics, readers are requested to follow (Chaudhuri, Roy Chowdhury and Nandi et al. 1997).

Reasons for the preferences of our CA-based modellings are as follows: (i) cheap physical implementation is possible at the cost of D-flip-flops (Chaudhuri, Roy Chowdhury and Nandi et al. 1997; Mitra and Kundu 2017; Mitra 2021a; Mitra and

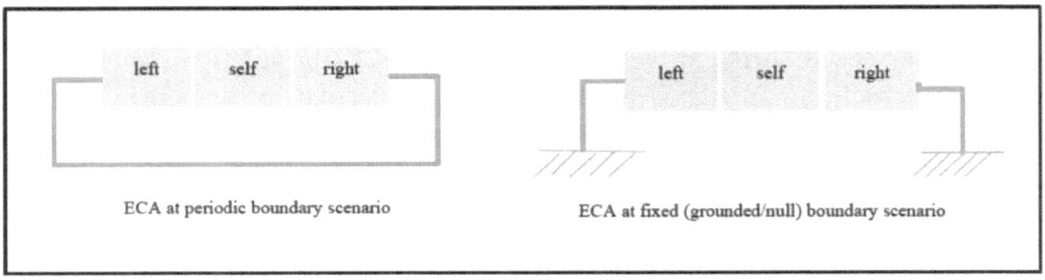

FIGURE 11.2 A typical diagram for ECA boundary scenarios (inspired from Chaudhuri, Roy Chowdhury and Nandi et al. 1997; Mitra, Kundu and Chattopadhyay et al. 2018; Mitra 2021a; Mitra and Banerjee 2021).

Banerjee 2021), (ii) both parallel computing facility and easy incorporation at VLSI (Very Large-Scale Integration) are possible with CA (Chaudhuri, Roy Chowdhury and Nandi et al. 1997; Mitra and Kundu 2017; Mitra 2021a; Mitra and Banerjee 2021) and (iii) very low amount of energy consumption (ranging from a maximum of $1.20E-05$ watt to a minimum of $1.17E-07$ watt at different fabrication technologies) is found with CA-based model (Mitra and Kundu 2017).

Major contributions of this chapter are as follows.

i. It presents Type D fuzzy representation of two CA rules (i.e., rules 102 and 153) towards its uses in CA-based MapReduce design in Industry 4.0 scenarios.
ii. It further examines its dynamics at several fixed boundary scenarios for several higher automata (cell) sizes towards its consideration in CA-based MapReduce design.

Rest of the chapter is organized as follows: related works are in Section 11.2; background work is presented in Section 11.3; Type D Fuzzy CA (FCA) representation for ECA rules targeting MapReduce design in Section 11.4; simulation results are presented in Section 11.5; finally, concluding remarks are presented in Section 11.6.

11.2 RELATED WORKS

Existing works connected with the CPS architecture in Industry 4.0 scenario were briefly presented in Section 11.1. As already concluded, any possible enhancements in Big Data processing in any smart production-line may appreciably improve the production in view of both quality and quantity. Since the past, investigators have concentrated on the developments of different technologies targeting smart production involving Big Data processing with MapReduce data processing in view of non-uniform data access at intermittent patterns (Chen, Alspaugh and Katz 2012). In a different investigation, MapReduce-based learning approach was introduced in (Lin, Shu and Deng et al. 2017) to improve the CBM (Condition Based Maintenance) services. In this regard, uses of software defined Industrial IoTs (IIoTs) were suggested in (Wan, Tang and Shu et al. 2016) to efficiently manage the physical devices in Industry 4.0 scenarios. On the other hand, a MapReduce structure-based fault-diagnosis involving an automatic pattern-recognition was introduced in (Kumar, Shankar and Choudhary et al. 2016) to facilitate the Cloud-based production scenario. In a different approach, an enhanced MapReduce scheduler was presented in (Xu and Duan 2019) towards heterogeneous workload processing in heterogeneous Cloud environment, while CA-based MapReduce design targeting Cloud applications was introduced in (Mitra, Kundu and Chattopadhyay et al. 2018) and was further refined in (Mitra 2021a) towards uses in Industry 4.0. Besides MapReduce data processing, Big Data security issues, etc. were always in focus among researchers. Thus, we found a focus on security issues towards Big Data processing in (Manogaran, Thota and Lopez et al. 2017).

We found sufficient state-of-the-art literatures targeting MapReduce design. We were particularly impressed with the CA-based MapReduce design model as it was well compatible towards a cheap physical implementation at low energy consumption in view of Industry 4.0 (Mitra 2021a). We believe that further investigation on the CA-based MapReduce model of (Mitra, Kundu and Chattopadhyay et al. 2018; Mitra 2021a) might be useful and may lead to an augmented implementation of Big Data processing in Industry 4.0 scenarios.

As CA is rich in dynamics, several researchers presented different CA-based models. We found FCA-based investigation in (Ahmed 1996). Such investigation presented the "elementary FCAs in immunology, using three values of confidence degree, 0, 0.5, 1" (Teodorescu 2015b). In a different research, Type D FCA-based model was investigated and discussed in (Teodorescu 2015b) at "only linear (non-circular 1-D) CAs with fixed boundary cells" (Teodorescu 2015a,b). The confidence degree of an FCA is in the reality that the state value is '0' or '1' and may be represented by $s(j, k, c_1, c_2)$, the state of the cell k-th cell $(k \in \{1, \ldots, N\})$ at t-th time $(k \in \{1, \ldots, T\})$, "where the confidence value for c_1 is 0 and for c_2 is 1 (assuming that $c_1 + c_2 = 1$) . The state of the cell at time moment $j+1$ is derived from the states of the cells in vicinity $V = \{-p_1, \ldots, +p_2\}$ at the current moment, $s(j+1, k, c_1, c_2) = g(s(j, k-p_1, c_1, c_2), \ldots, s(j, k+p_2, c_1, c_2))$ where g is the state transition function of the cells" (Teodorescu 2015b). The confidence degree (c_1, c_2) may be explained with the help of fuzzy logic (may be with the min and max logic), and thus is considered as Type D FCA (Teodorescu 2015b). We further learned in (Teodorescu 2015b) that most commonly triangular m.f.s [refer Equation 11.2 (Teodorescu 2015b) and Equation 11.3 (Teodorescu 2015b)] may be used.

$$\mu_0(x) = \begin{cases} 1 - x & x \in [0,1] \\ 0 & \text{elsewhere} \end{cases} \tag{11.2}$$

$$\mu_1(x) = \begin{cases} x & x \in [0,1] \\ 0 & \text{elsewhere} \end{cases} \tag{11.3}$$

We further observe that "defuzzification corresponds to TSK systems, with the result forced to 0.5 whenever both m.f.s are null" and may be represented by following Equation 11.4 (Teodorescu 2015b).

$$Def = \begin{cases} \dfrac{0.\mu_1 + 1.\mu_2}{\mu_1 + \mu_2} = \dfrac{\mu_2}{\mu_1 + \mu_2}, & \text{for } \mu_1 + \mu_2 \neq 0 \\ 0.5, & \text{else} \end{cases} \tag{11.4}$$

For further discussions related to the fuzzified ECA rules, readers are suggested to go through (Adamatzky 1994; Mraz, Zimic and Lapanja et al. 2000; Maji and Chaudhuri 2005; Teodorescu 2015a,b). Based on the investigations as presented in (Teodorescu 2015a,b), we find that CA-based MapReduce model may further be investigated to explore its true potential towards its possible consideration in Industry 4.0 uses. We found such investigation was not considered previously. For this reason, we continued our chapter.

As this chapter is mostly influenced by the design of CA-based MapReduce model of (Mitra, Kundu and Chattopadhyay et al. 2018; Mitra 2021a), we present a brief discussion of that model in Section 11.3.

11.3 BACKGROUND

The CA-based MapReduce proposal of (Mitra, Kundu and Chattopadhyay et al. 2018) was introduced in a null-boundary scenario involving the rule set (total 36 ECA rules) for a special class of group CAs known as ELCAs (Equal Length CAs). The said design of (Mitra, Kundu and Chattopadhyay et al. 2018) was presented in view of Big Data processing in Cloud environment. The MapReduce design of (Mitra, Kundu and Chattopadhyay et al. 2018) was further investigated in (Mitra and Banerjee 2021) towards its capabilities for possible uses in Industry 4.0 scenarios. In (Mitra and Banerjee 2021), a detailed investigation simplified the existing CA-based MapReduce design with the uses of two additive ECA rules (ECA rules 102 and 153) irrespective of the consideration of the ECA boundary conditions while ensuring quality of shuffle at produced MapReduce block. High data lessening ability (about 87.5% to 93.75%) was concluded in (Mitra, Kundu and Chattopadhyay et al. 2018; Mitra 2021a) for the said CA-based MapReduce model. A typical diagrammatic representation of such model as presented in (Mitra, Kundu and Chattopadhyay et al. 2018; Mitra 2021a) is once again presented in Figure 11.3.

It was found from the design of (Mitra, Kundu and Chattopadhyay et al. 2018; Mitra 2021a) and also from Figure 11.3 that 2^m number of equal length data blocks and size of (2^{n-m}) length for each equal length blocks may be produced from a primary population of total (2^n) independent data and finally may be shrunk to a population of (2^m) size of data block, for $n \geq 1$ and $m = 1,2,3,\ldots,(n-1)$. Scientifically, the relationship among the primary population and definitive MapReduce population as defined in (Mitra, Kundu and Chattopadhyay et al. 2018; Mitra 2021a) is presented again in Equation 11.5.

$$2^n = 2^m \times 2^{n-m} \text{ for } n \geq 1 \text{ and } m = 1, 2, 3, \ldots\ldots(n-1) \tag{11.5}$$

11.4 TYPE D FUZZY CA REPRESENTATION FOR ECA RULES TARGETING MAPREDUCE DESIGN

Discussions on the Type D FCA presentation on two ECA rules 102 and 153 are shown next. We followed the investigation procedure as presented in (Teodorescu 2015b).

An N-cell null boundary CA at three vicinity scenario was considered, where initial confidence values of the central cell was compelled to $(0, 1)$, i.e., $B(1, N/2, 2) = 1$, and $B(1, N/2, 1) = 0$. As the three vicinities at N-cell CA is considered, we received, $c_1 = B(j, k-1, 1)$; $c_2 = B(j, k, 1)$; $c_3 = B(j, k+1, 1)$; $d_1 = B(j, k-1, 2)$; $d_2 = B$ $(J, K, 2)$; $d_3 = B(j, k+1, 2)$; where dummy variables c's indicated to 0 and d's indicated to 1.

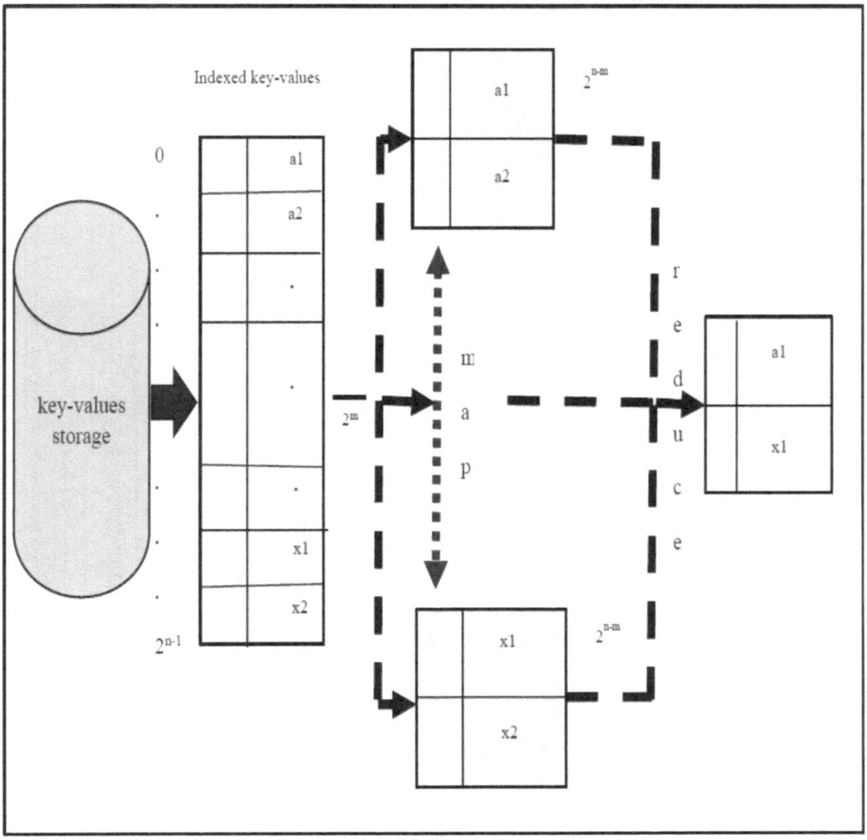

FIGURE 11.3 A typical MapReduce design at null boundary ECA scenario (motivated from the model of Mitra, Kundu and Chattopadhyay et al. 2018; Mitra 2021a).

As discussed in Section 11.2, the min and max logic was applied on ECA rules 102 and 153 towards its Type D representation. Thus, we received the following for said ECA rules.

For ECA rule 102 at its 8-bit representation (i.e., r_0 for $i = 0, \ldots, 7$), the min version with the AND operator is as follows:

$$r_0 = \min\left(\min\left(d_1, d_2\right) d_3\right); \rightarrow \text{result is state 0 when } 0,0,0$$

$$r_1 = \min\left(\min\left(d_1, d_2\right) d_3\right); \rightarrow \text{result is state 1 when } 0,0,1$$

$$r_2 = \min\left(\min\left(d_1, d_2\right) d_3\right); \rightarrow \text{result is state 1 when } 0,1,0$$

$$r_3 = \min\left(\min\left(d_1, d_2\right) d_3\right); \rightarrow \text{result is state 0 when } 0,1,1$$

$$r_4 = \min\left(\min\left(d_1, d_2\right) d_3\right); \rightarrow \text{result is state 0 when } 1,0,0$$

$$r_5 = \min\left(\min\left(d_1, d_2\right) d_3\right); \rightarrow \text{result is state 1 when } 1,0,1$$

$$r_6 = \min\left(\min\left(d_1, d_2\right) d_3\right); \rightarrow \text{result is state 1 when } 1,1,0$$

$$r_7 = \min\left(\min\left(d_1, d_2\right) d_3\right); \rightarrow \text{result is state 0 when } 1,1,1$$

and the max version is with the OR operator:

$$B\left(j+1,\ k,\ 1\right) = \max\left(r_0, \max\left(r_3, \max\left(r_4, r_7\right)\right)\right); \rightarrow \text{for state 1}$$

$$B\left(j+1,\ k,\ 2\right) = \max\left(r_1, \max\left(r_2, \max\left(r_5, r_6\right)\right)\right); \rightarrow \text{for state 0}$$

Similarly, for ECA rule 153 at its 8-bit representation (i.e., r_0 for $i = 0, \ldots, 7$), the min version with the AND operator is as follows:

$$r_0 = \min\left(\min\left(d_1, d_2\right) d_3\right); \rightarrow \text{result is state 1 when } 0,0,0$$

$$r_1 = \min\left(\min\left(d_1, d_2\right) d_3\right); \rightarrow \text{result is state 0 when } 0,0,1$$

$$r_2 = \min\left(\min\left(d_1, d_2\right) d_3\right); \rightarrow \text{result is state 0 when } 0,1,0$$

$$r_3 = \min\left(\min\left(d_1, d_2\right) d_3\right); \rightarrow \text{result is state 1 when } 0,1,1$$

$$r_4 = \min\left(\min\left(d_1, d_2\right) d_3\right); \rightarrow \text{result is state 1 when } 1,0,0$$

$$r_5 = \min\left(\min\left(d_1, d_2\right) d_3\right); \rightarrow \text{result is state 0 when } 1,0,1$$

$$r_6 = \min\left(\min\left(d_1, d_2\right) d_3\right); \rightarrow \text{result is state 0 when } 1,1,0$$

$$r_7 = \min\left(\min\left(d_1, d_2\right) d_3\right); \rightarrow \text{result is state 1 when } 1,1,1$$

and the max version is with the OR operator:

$$B\left(j+1,\ k,\ 1\right) = \max\left(r_1, \max\left(r_2, \max\left(r_5, r_6\right)\right)\right); \rightarrow \text{for state 1}$$

$$B\left(j+1,\ k,\ 2\right) = \max\left(r_0, \max\left(r_3, \max\left(r_4, r_7\right)\right)\right); \rightarrow \text{for state 0}$$

The defuzzification was achieved using the following:

$$Def\left(j,k\right) = \begin{cases} \dfrac{B\left(j,k,2\right)}{B\left(j,k,1\right) + B\left(j,k,2\right)}, & \text{if } B\left(j,k,1\right) + B\left(j,k,2\right) \neq 0 \\ 0.5, & \text{if } B\left(j+1,k,1\right) + B\left(j+1,k,2\right) = 0 \end{cases}$$

As it was a Type D FCA, the defuzzied value indicated a new belief degree.

In our studies we found that the Type D FCA "lay in the middle between B-FCAs and coupled maps, specifically fuzzy coupled maps (FCMs)" (Teodorescu 2015b) and thus, Type D FCAs "combine the simplicity and ease of implementation of CAs and type B FCA with the generalization power of FCMs and of the self-organizing, uncertainty-based networks" (Teodorescu 2015b). For this reason, it may be concluded that same advantages might be present in CA-based MapReduce model.

11.5 SIMULATION RESULTS AND RELATED DISCUSSIONS

A simulation code in 'C' was availed to investigate the dynamics of ECA rules 102 and 153. We used C free IDE version 4.0. It is already presented in Section 11.2 that for binary representation of Cas, we may consider "the confidence value for c_1 is 0 and for c_2 is 1 (assuming that $c_1 + c_2 = 1$) " (Teodorescu 2015b). Thus, we considered the CA dynamics. On the other hand, it was already presented in (Mitra 2021a) that ECA rules 102 and 153 both "do not depend on left side boundary value" (Mitra 2021a) as identical state space diagrams were achieved for the homogeneous (uniform) CA for said ECA rules at low cell (automata) size. Higher cell (automata) sizes were not investigated in (Mitra 2021a). We further investigated the dynamics for said homogeneous CA at different cell (automata) size. Dynamics as achieved in simulation for such homogeneous (uniform) CA for said ECA rules at different automata sizes at different fixed boundary conditions [as described in (Mitra 2021a)] are presented in Figures 11.4 and 11.5. Please note that same initial condition (i.e., all 1's) was applied in all simulations. Achieved state-space diagrams towards the dynamics for said CA at different automata (cell) sizes (e.g., automata size 21, 51 and 101) are presented in Figures 11.4 and 11.5. A sample simulation code (as inspired from Mitra and Teodorescu 2016) to produce the CA dynamics at different fixed boundary scenarios is presented in Annex (Mitra 2021b available at "http://dx.doi.org/10.17632/wkkttw7mgk.1").

It may be observed from Figures 11.4 and 11.5 that identical CA dynamics were present for the homogeneous CA with rule 102 and 153 even at higher automata (cell) size. Thus, we observed those achieved dynamics for homogeneous CA were independent of the left-hand boundary values at different fixed boundary conditions and it further supports the achieved dynamics in (Mitra 2021a). For this reason, said CA rules may have an advantage for lesser requirement of program memory even at higher automata (cell) sizes which is like the result of (Mitra 2021a).

11.6 CONCLUSIONS

Type D fuzzified version of two ECA rules (i.e., rule 102 and 153) were investigated in presented research to explore its true potential towards uses in Industry 4.0 scenarios. Type D Fuzzy-based representation for said ECA rules explored its efficiency, and it was observed that the presented fuzzy model was "close in operation to fuzzy coupled maps" (Teodorescu 2015b). Thus, it ensured "the simplicity and ease of implementation" to deal with uncertainty-based scenarios. Dynamics of said ECA rules at higher automata (cell) sizes (i.e., up to an automata size of 101)

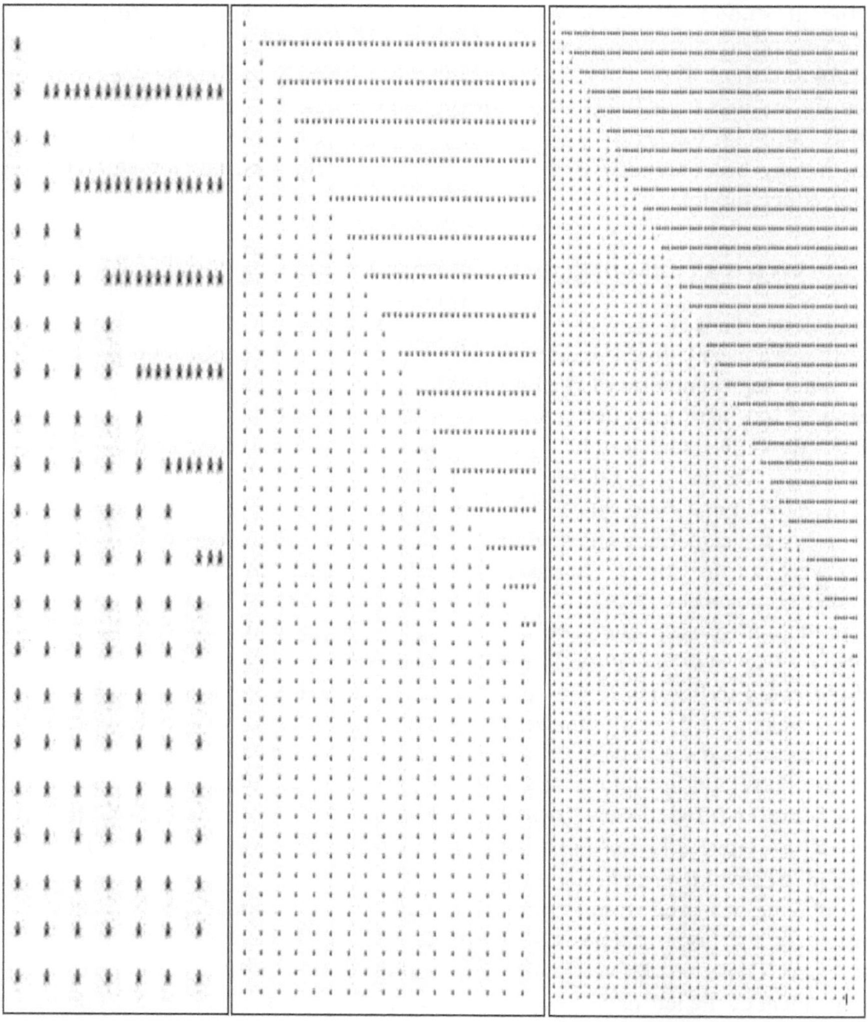

a. Automata size 21 b. Automata size 51 c. Automata size 101

FIGURE 11.4 Dynamics for homogeneous CA with ECA rule 102 at 0…0 fixed boundary condition.

for several fixed-boundary scenarios further ensured that "there is no need for assigning program memory towards its software implementations" (Mitra 2021a) as said ECA rules were found to be left-boundary independent in view of fixed-boundary ECA scenarios. Hence, it may be concluded that the presented Type D fuzzy-based representation for said ECA rules further strengthens the cost-effectiveness at ECA-based MapReduce model towards its easy physical implementation in Industry 4.0 scenarios.

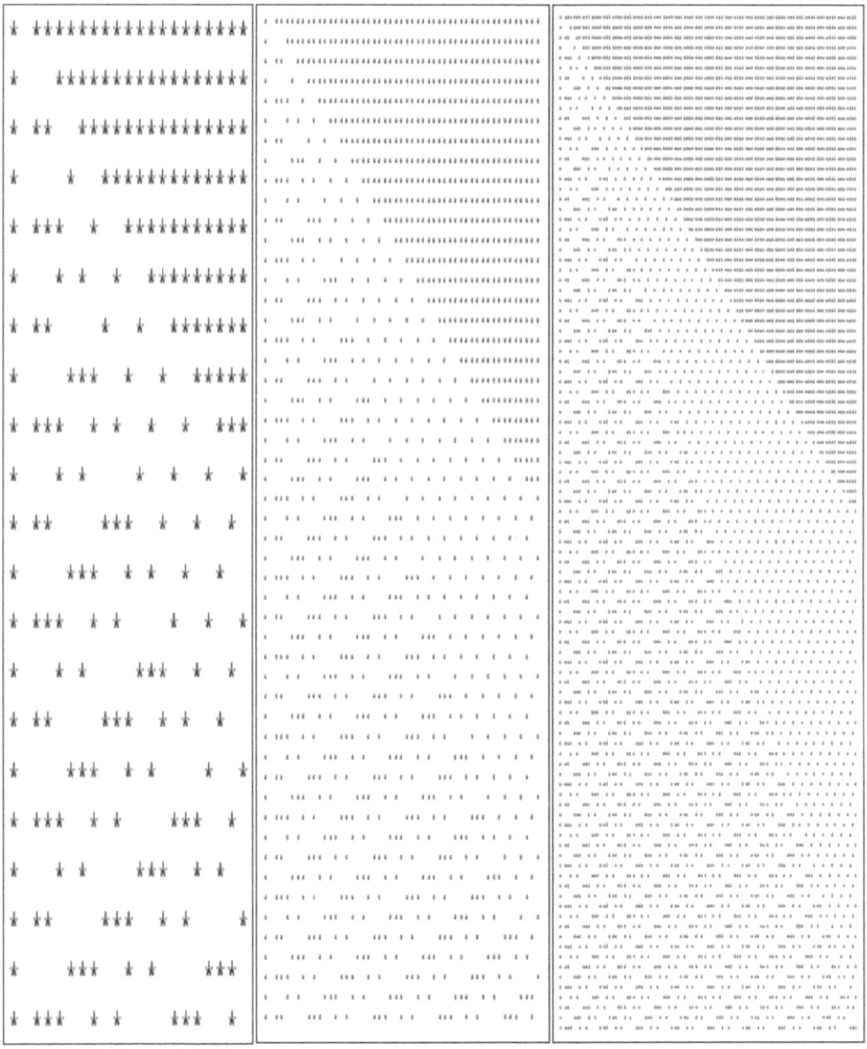

a. Automata size 21 b. Automata size 51 c. Automata size 101

FIGURE 11.5 Dynamics for homogeneous CA with ECA rule 153 at 0...0 and 1...0 fixed boundary conditions.

ACKNOWLEDGEMENTS

Author sincerely acknowledges that he has learned advanced research and related skills from Prof. Horia-Nicolai Teodorescu, "Gheorghe Asachi" Technical University of Iasi, Romania. Author also acknowledges the reviews received from the anonymous reviewers and believes that those reviews have further strengthened the final version of this chapter.

REFERENCES

Adamatzky, Andrew I. "Hierarchy of fuzzy cellular automata." Fuzzy Sets and Systems, vol. 62, no. 2 (1994): 167–174.

Ahmed, E. "Fuzzy cellular automata models in immunology." Journal of Statistical Physics, vol. 85, no. 1 (1996): 291–294.

Baheti, Radhakisan, and Helen Gill. "Cyber-physical systems." The Impact of Control Technology, vol. 12, no. 1 (2011): 161–166.

Chaudhuri, Parimal Pal, Dipanwita Roy Chowdhury, Sukumar Nandi, and Santanu Chattopadhyay. Additive Cellular Automata: Theory and Applications. Los Alamitos, (California): IEEE Computer Society Press, 1997.

Chen, Yanpei, Sara Alspaugh, and Randy Katz. "Interactive analytical processing in big data systems: A cross-industry study of MapReduce workloads." Proceedings of the VLDB Endowment, vol. 5, no. 12 (2012): 1802–1813.

Kumar, Ajay, Ravi Shankar, Alok Choudhary, and Lakshman S. Thakur. "A big data MapReduce framework for fault diagnosis in cloud-based manufacturing." International Journal of Production Research, vol. 54, no. 23 (2016): 7060–7073.

Lee, Jay, Behrad Bagheri, and Hung-An Kao. "A cyber-physical systems architecture for industry 4.0-based manufacturing systems." Manufacturing Letters, vol. 3 (2015): 18–23.

Lee, Jay, Edzel Lapira, Behrad Bagheri, and Hung-An Kao. "Recent advances and trends in predictive manufacturing systems in big data environment." Manufacturing Letters, vol. 1, no. 1 (2013): 38–41.

Lin, Chun-Cheng, Lei Shu, Der-Jiunn Deng, Tzu-Lei Yeh, Yu-Hsiang Chen, and Hsin-Lung Hsieh. "A MapReduce-based ensemble learning method with multiple classifier types and diversity for condition-based maintenance with concept drifts." IEEE Cloud Computing, vol. 4, no. 6 (2017): 38–48.

Maji, Pradipta, and P. Pal Chaudhuri. "Fuzzy cellular automata for modeling pattern classifier." IEICE Transactions on Information and Systems, vol. 88, no. 4 (2005): 691–702.

Manogaran, Gunasekaran, Chandu Thota, Daphne Lopez, and Revathi Sundarasekar. "Big data security intelligence for healthcare industry 4.0." In Lane Thames and Dirk Schaefer (Eds.), Cybersecurity for Industry 4.0, pp. 103–126. Cham: Springer, 2017.

Mitra, Arnab. "On the capabilities of cellular automata-based MapReduce model in industry 4.0." Journal of Industrial Information Integration, vol. 21 (2021a): 100195.

Mitra, Arnab. "Annex to 'on type-D fuzzy cellular automata-based MapReduce model in industry 4.0.'" Mendeley Data, V1 (2021b). Doi: http://dx.doi.org/10.17632/wkkttw7mgk.1

Mitra, Arnab, and Anirban Kundu. "Energy efficient CA based page rank validation model: A green approach in cloud." International Journal of Green Computing (IJGC), vol. 8, no. 2 (2017): 59–76.

Mitra, Arnab, Anirban Kundu, Matangini Chattopadhyay, and Samiran Chattopadhyay. "On the exploration of equal length cellular automata rules targeting a MapReduce design in cloud." International Journal of Cloud Applications and Computing (IJCAC), vol. 8, no. 2 (2018): 1–26.

Mitra, Arnab, and Avishek Banerjee. "On the dynamics of cellular automata-based green modelling towards job processing with group-based industrial wireless sensor networks in industry 4.0." In Om Prakash Jena, Alok Ranjan, and Zdzislaw Polkowski (Eds.), Green Engineering and Technology Innovations, Design, and Architectural Implementation, pp. 207–224. Boca Raton, FL: CRC Press, 2021.

Mitra, Arnab, and Horia-Nicolai Teodorescu. "Detailed analysis of equal length cellular automata with fixed boundaries." Journal of Cellular Automata, vol. 11, no. 5–6 (2016): 425–448.

Mraz, Miha, Nikolaj Zimic, Iztok Lapanja, and I. Bajec. "Fuzzy cellular automata: From theory to applications." In Proceedings 12th IEEE Internationals Conference on Tools with Artificial Intelligence (ICTAI), pp. 320–323. IEEE, 2000.

Shi, Jianhua, Jiafu Wan, Hehua Yan, and Hui Suo. "A survey of cyber-physical systems." In 2011 International Conference on Wireless Communications and Signal Processing (WCSP), pp. 1–6. IEEE, 2011.

Teodorescu, Horia-Nicolai. "On the regularities and randomness of the dynamics of simple and composed CAs with applications." Romanian Journal of Information Science and Technology, Romanian Academy, vol. 18, no. 2 (2015a): 166–181.

Teodorescu, Horia-Nicolai. "Type-D fuzzy CAs for medical and social sciences." In 2015 E-Health and Bioengineering Conference (EHB), pp. 1–4. IEEE, 2015b.

Wan, Jiafu, Shenglong Tang, Zhaogang Shu, Di Li, Shiyong Wang, Muhammad Imran, and Athanasios V. Vasilakos. "Software-defined industrial internet of things in the context of industry 4.0." IEEE Sensors Journal, vol. 16, no. 20 (2016): 7373–7380.

Wolfram, Stephen. "Computation theory of cellular automata." Communications in Mathematical Physics, vol. 96, no. 1 (1984): 15–57.

Xu, Li Da, and Lian Duan. "Big data for cyber physical systems in industry 4.0: A survey." Enterprise Information Systems, vol. 13, no. 2 (2019): 148–169.

12 Fine-Grained Feature Classification of Objects by Learning-Based Feature Selection Using Region-Proposal Convolutional Neural Network (RCNN)

Radhamadhab Dalai and K. K. Senapati

CONTENTS

12.1 INTRODUCTION

Fine-grained visualization techniques [1, 2, 3, 4] have achieved popularity in presenting a regular number of classifications in expanding manner. In earlier works, extensive learning for Bag of Words, Scale Invariant Feature Transform (SIFT)-based optimized feature learning processes are utilized for feature-based recognizable subset and characterization of object's true class. Presently, Convolution Neural Network-based region classification has found a worldwide application because of its exactness and semantic precision. Since it is based on supervised learning process; it is trained and tested on a standard dataset with Region-Proposal Convolutional Neural Network (RCNN) model and performances are also noted [5]. The result has shown significant improvement in precision from 23% to 55.7% after 1000 iterations. But since the model uses semantic-based segmentation approach, an average feature-based model

DOI: 10.1201/9781003229018-12

has been attempted further. On the information side, while observing the features of object in an image, there has been a possibility of advancements for growing the structure of fine-grained regions we have been searching for. A standard dataset has been prepared, which now incorporates random objects such as solid object parts, minerals, rigid metal elements, etc. On the other hand, compared to non-comparative issue for acknowledgment, fine-grained acknowledgment [1, 2] provides helpful more information than learning basic parts of the particles in a region that has helped to adjust objects of a similar class and separate between neighboring classes. In the current approach, the best classification method that comes about is, accordingly, from models that required part explanations as a major feature of a regulated model-based layer preparing process [6]. This has presented an issue of scaling up fine-grained information such as optimized feature set or subset of features to an expanding number of areas. An imperative objective fact has been proposed toward the objective of preparing fine-grained classifiers without part explanations, which also accumulates minute feature point. There are a few drawbacks to a fine-grained class with a high level of shape similarity that can be changed solely through spatial division. Classifications in a fine-grained feature set share comparable form, which take into consideration arrangement to be done simply in light of feature point segregation. We can take in the trademark elements without the finer features of images in this preparation technique. In this chapter, we offer a strategy for creating parts that can recognize novel features in images and determine which parts are important for image registration recognition. To fragment the preparation photos, we in our method apply the late advance in co-division for producing segments and clusters. Next, we precisely change the images that are similar in posture, conducting an arrangement over all the images as the structure of these are more reliable neighborhood arrangements. Furthermore, it can sum up to fine-grained spaces without segment correlations, establishing a new suitable methodology on exact classification to quantify the goodness of materials from the Common Object in Context (COCO) dataset by a large margin. The COCO dataset was utilized as a basis for the experiment.

Deep learning methods, which show promise in region-based segmentation and classification, offer a wealth of research options in computerized feature extraction. RCNN [6, 7], Fast-RCNN [8], Faster-RCNN, and Mask-RCNN [5] are some of these learning algorithms. Using such procedures, more unique characteristics (important feature points) are identified as far as lower-level features [9]. This method can be quite useful for recognizing and determining the general shape of things on the surface, such as building materials and mined metals.

Because additional theoretical representations are regularly created in light of edge or shape-based features of an image, the deep learning methods discussed above encourage dynamic representations. One of the most significant advantages of more conceptual representations is that they can be invariant to transformational changes in data depending on regions. The ability to adapt such invariant features has become an essential goal in design recognition. Such representations can highlight the variables of variation in data in addition to being invariant. In fine-grained categorization models that require multifeatured communications across various tiers of the network, genuine data is used as part of component-based modeling technologies. An image, for example, is made up of well-built qualities like light intensity, defined forms, and object materials. These conceptual representations generated by profound learning computations can isolate various wellsprings of variations when employed in a CNN-based network layer. Additional part-based characteristics will normalize the primary feature set, allowing for substantially faster training.

Deep learning algorithms are basically multi-layered operational designs with several levels [5]. Here, the inputs are subject to nonlinear transformation, which results in a representation of the output. In order to develop a sophisticated and abstract representation of data, a number of CNN-based techniques [10, 11, 12] were used here in fine-grained classification model. These data are presented in lower to higher order by argumentation through multiple transformation layers. The first layer of our network receives data received from different sensors, such as pixels in an image. The output of each layer is then delivered as input to the layer above it.

12.2 DESIGN OF FINE-GRAINED VISUALIZATIONS THROUGH CNN

In order to reduce intra-class variation, fine-grained image classification has focused on performing components detection. If a feature set contains a large number of features, the part-based one-versus-one feature system has been a good choice for classification techniques [2, 13], where duplicate outlier points can be removed. Triplet feature optimization is a good implementation of such classification mechanism, where subsets of features are gradually selected based on triangular-based optimization technique as shown in figure 12.1. The deformable parts-model [3] is another technique that gets features from a set of pre-defined granular components. CNN-based segmentation has been used to implement the part localization method, and parts detection has aided this segmentation process.

Shaoqing et al. [14] recommended that interdependent correspondences of distinct regions be trained using pose-based normalized features extracted from weak semantic annotations from images. This finding focuses that, CNNs are effective for fine-grained image categorization, which is primarily used for part-based localization and region-based detection. Thus, region proposal methods combined with a CNN architecture will localize object parts with better accuracy. With experiments on solid object for feature extraction, it has been shown that a CNN can use normalized features as edges and corners for training of part-based localization and visibility prediction [15]. Part is considered as certain region in an image where distinct feature points are gathered to give a meaningful information about an object or its part. Thus, part-based classification along with fine-grained mechanism has produced a high precision result when applied on the COCO dataset. Although the aforementioned part-based delineation segmentation method is fully automated during testing, it requires manual annotation of a significant number of images in order to train the model. Recent work has provided a number of approaches to do fine-grained categorization with a mixed part annotations network, obviating the requirement for time-consuming manual annotations. Bo Zhao et al. [3] demonstrated that, even without part-based annotations, CNNs can outperform full-feature-based classification algorithms like RCNNs for fine-grained classification tasks. The fine-grained approach yielded cluster-based classification, demonstrating that the data has been partitioned into K non-overlapping sets, as well as an expert

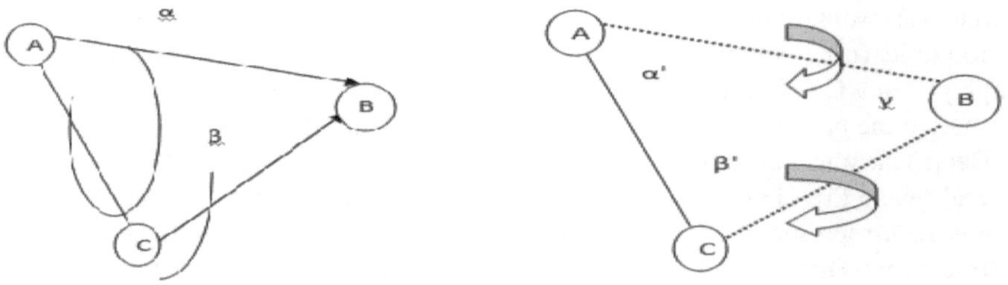

FIGURE 12.1 (a) The triplet feature points initial stage (b) The feature points' gradient transform with relative angular momentum.

feature extraction algorithm that has been trained further for each of the K sets using DCNNs (Delineating CNN).

We proposed a unique hybrid strategy combining Faster-RCNNs [16] to improve fine-grained image classification network performance by partitioning the data into K non-overlapping chunks. A learning-based training has been tried using an expert classifier for each training set. This solution uses a patch-based network with CNN layers, similar to the gated neural network suggested by Shaoqing Ren et al. [14], or to the fine-grained classification problem. In the earlier approach, there was a parameterized feature-based CNN architecture but currently there is a triplet patch-based network producing a reduced feature subset. So the discriminative region proposal approach used in Faster-RCNN plays a significant role in feature description along with a feature-based training for triplet patch network. Figure 12.2 depicts a high-level summary of these two approaches. The triple patch network design [4, 17] also entails learning K expert networks that make judgements regarding a subsequent subset of the data input for the following layer. This simplifies the recognition performance that is being modeled by each component in an image. There are two such approaches which are being able to assign sample-based distribution on the movement of shape features, in an image at the appropriate network. This region may be specified by a set of points, edges, or the angles between them, since the network provides feature values by training a distinct gated neural network which are based

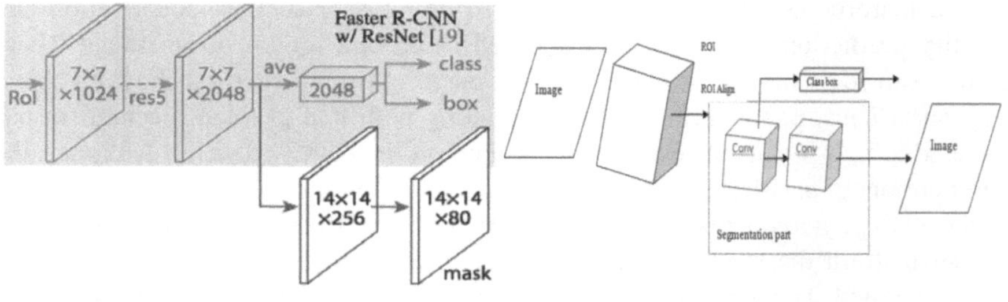

(a) (b)

FIGURE 12.2 (a) Diagram of Faster-RCNN; (b) Diagram for Mask-RCNN.

on region-based localization and segmentation. This generates a moment value specified by a probability distribution function with a value of k, which is derived from a sample from the k-th network. This supervised learning-based model, which is based on a combination of CNN network and probabilistic k-distance model [18, 19], helps in providing a semantic and rational understanding of the data. Such a vector representation corresponds to the extracted representations of data instances, which would provide faster searching and information retrieval. Therefore, an RCNN-based detection along with K-distance feature points storage in a data store [20] is being proposed in this design. Another approach devised by Wang et al. [21]. A triplet path-based feature design, which concerns for overlapping point of three distinct features. In this approach, the rate of change of position vector as the gradient is also taken into consideration to map these distinct key points.

Where $\eta = (\alpha\text{-}\alpha')(\beta\text{-}\beta')/\gamma$

α = angle made by feature vector AB and AC,
γ = angle made by feature vector AB and BC,
β = angle made by feature vector AC and BC

As shown in figure above; since the trained complex data representations contain semantic and relational information, which is depicted as points in triangle. Instead of using merely raw bit data, they can be directly employed for semantic cluster-based classification, with each data point acting as an edge and discriminative point acting as corner. In an image, these feature points are represented by a vector as shown in gradient transform. This allows for a more efficient vector-based processing operation (SIFT) than comparing segmented instances using raw data alone. Data sets with similar vector representations are likely to have similar semantic meanings when represented as triplet triangles. Thus, using vector representations of complex high-level data, findings are indexed as A, B, and C points, marked for the data; which makes semantic indexing feasible. The training and testing of indexes based on knowledge gathered from CNN-based deep learning algorithms [22, 23, 24] have been prioritized at the remainder of this section. However, the fundamental notion of indexing, based on fine-grained data representations, can be expanded to multiple sorts of data instances, allowing for faster information searching and retrieval.

12.3 DATASET DESIGN

The edge part, corner point, maximum length and width of object are considered as discriminative feature points in an image at our network. These points have undergone fine-grained regression with gradient-based SIFT operations on images. Images of solid materials were used as an input and those important spots are estimated for further network input. Individual images from the COCO dataset were used in our experiment and a fine-tuned feature set was created. For this purpose, a scalable 2-node fault tolerant big database has been used to store such a huge number of data points. Since these were one-shot operations, using Big Data environment [15, 25] was suitable for such operations. The view dataset is divided into two groups for this feature extraction process: (1) RAW dataset: file sizes range from 30 to 60 GB. The file is fairly enormous since it contains a lot of relevant points from feature data readings during the feature-based extraction operation from each image. (2) PROCESSED dataset: Here size of data points is of few kilo bytes because only graded and filtered feature points have been selected. The processed dataset was created by taking the original dataset and applying a triplet patch-based function, which is more powerful than a CNN-based technique. It generates well quantized and formative data in the form features. There are roughly 100 features and a few hundred rows in the training dataset. However, the processed dataset generates a large number of resemblances and many datasets, which a well-designed database can handle.

12.4 METHODOLOGY

12.4.1 A Mixture of Faster RCNN and Fine-Grained

For each candidate object that was to be classified, RCNN generated two outputs: a class label and a bounding-box offset. A third branch has been added to this, which outputs the object mask. While Mask-RCNN [5] is a simple and clear strategy for region-based classification, our model's additional mask output is built from the class and box outputs, necessitating the extraction of an object's significantly finer spatial layout. Following that, crucial Faster-RCNN elements such as feature vector, granular module for classification layer, branch modulation, and sigmoid function, as well as pixel-to-pixel alignment, were implemented.

This has become the main advantage feature of Fast/Faster-RCNN. Following the max pool operations, the loss function was utilized, with N denoting the number of training examples, X_i denoting the prediction value of the i_{th} example, and Y_i denoting the ground truth of the i_{th} sample. The loss function which is used in the final activation layer is

$$Loss = \frac{\sum_{i=1}^{N} (X_i - Y_i)^2}{N} \tag{12.1}$$

The distance function used for feature matching is applied as followed in equation (12.2).

$$\text{Distance Function } d = \sqrt{(XCNN_i - XSFT_i)^2 + (YCNN_i - YSFT_i)^2} \tag{12.2}$$

Where ($XCNN_i$, $YCNN_i$) represent feature points before applying SIFT operations, and ($XSFT_i$, $YSFT_i$) represent feature points after applying SIFT operations.

The following convolution operations, fully connected (FC) layer and MaxPool layers for one set of CNN configuration, have been applied.

$$\begin{array}{llll}
\text{Convolution}(32,3,1,0) & \Rightarrow \text{Convolution}(32,3,1,0) & \Rightarrow \text{MaxPool}(2) & \Rightarrow \text{1st Layer} \\
\text{Convolution}(64,\ 3,\ 1,\ 0) & \Rightarrow \text{Convolution}(64,\ 3,\ 1,\ 0) & \Rightarrow \text{MaxPool}(2) & \Rightarrow 2^{\text{nd}} \text{ Layer} \\
\text{FC}(1024) & \Rightarrow \text{FC}(\text{N}) & \Rightarrow \text{Softmax} & \Rightarrow 3^{\text{rd}} \text{ Layer}
\end{array}$$

A weak learner for making predictions [21] and an additive model have been applied as given in algorithm.

Algorithm

Input: A training set represented by a first-order differentiable loss function with M iterations.

Algorithm:

1. Initialize all features set of fine-grained model with constant values.

$$F_0(x) = \arg_\lambda \min \sum_{i=1}^{n} L(y_j, \lambda) \tag{12.3}$$

2. For $m = 1$ to M:
   1. Compute: pseudo-residuals as the loss function minimization from the target vector and training vector set.

$$r_{im} = -[\partial L(y, f(x))/\partial F(x)]_{F_x = F_{m-1}(x)} \tag{12.4}$$

   2. Adjust a base learner as gradient-tree h_m (x) to pseudo-residuals. Later train it using the training set given by $\{x_i, r_{im}\}_{i=1}^{n}$.
   3. Next compute multiplier γ_m factor by solving the following one-dimensional optimization problem as

$$\gamma_m = \arg_\gamma \min \sum_{i=1}^{n} L(y_i, F_{m-1}(x_i) + \gamma h_m(x_i)) \tag{12.5}$$

   4. Update the model as per the equation:

$$F_m(x) = F_{m-1}(x) + \gamma_m h_m(x) \tag{12.6}$$

3. Output F_m (x)

As shown in Figure 12.2, three detailed architectures have been shown for the classification of random object for a particular material type. Faster-RCNN design is shown in Figure 12.2(a), which consists of two 7 × 7 × 2048 convolution layers with a feeder rennet-5 block and a 7 × 7 × 1024 filter block. Following the convolution layer, an average pooling layer with 14 × 14 × 256 block pooling layer and 14 × 14 × 180 mask is applied before moving on to flattened output for classification. On the other hand, in Mask-RCNN model, the two convolution layers with dimension 14 × 14 × 256 follow a fully connected layer (4096) over an Region of Interest [ROI] filter. The segmentation box is having two convolution layers with clear box with mask layer of feature dimension 7 × 7 × 256. Integrating both these layers, fine-grained model produces one ROI layer and one feature layer in two blocks as shown in Figure 12.3. Hence in this model, two image inputs were taken as an input. The reduced image set is for only defining keen feature points (convolution layer, pooling layer, and Rectified Linear

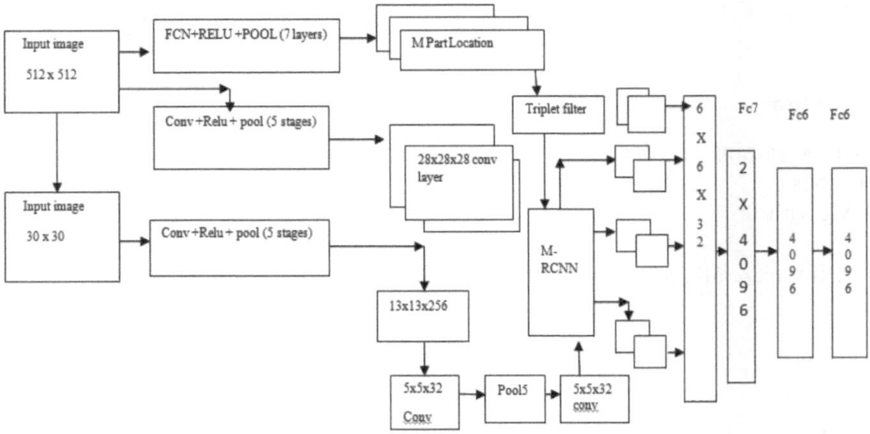

FIGURE 12.3 Diagram for fine-grained hybrid RCNN architecture.

activation Unit [RELU] layer with five stages). [Two convolution layers, Two pooling layers, and one RELU layer]. One of the layers as shown in Figure 12.2(c) has FC layer, convolution layer (28 × 28 × 256), and RELU layer. Then partial regions were found out using pooling layers to find region coherence among distinct features; triplet equality for angular deviation has been used as a filter operation. Mask-RCNN has been used for the intrinsic segmentation properties to classify the image dataset based on features generated from triplet feature and convolution layer. The pooling layer consists of average pooling layer of diminishing feature points of size 7 × 7 × 256.

12.5 EXPERIMENTS AND RESULTS

One more analysis on fine-grained feature has been conducted using bipolar convolution parameters. These parameters confirm that RCNN-based convolutional degradation of image has been balanced with less feature marginalization error. The convolution operation of 64 × 64 block images has been performed using SIFT (64) technique by applying padding and locking operation as shown in Figure 12.4 with solid metal block as shown in Figure 12.5. Padding is done for fitting image of small size and locking is done to fix the image size at each layer. If image size is 64 × 64, then layered image is resized to 64 × 64. In the first stage preprocessing filter has been applied to the image, and this is separated from training process. In the next stage using feature extraction mechanism, the key feature points have been extracted. They are stored in 128 × 128 pixels matrix. The original model was created by training 10,000 examples of bespoke solid objects in two-dimensional photos. The COCO dataset is utilized for the initial training of Faster-RCNN in MATLAB. Big data is used to extract feature sets, and the feature data is saved and available afterwards. This is a one-time operation for database access for feature storing and retrieving. The actual classification is then done using test data, with one experiment using RCNN techniques and the other using a fine-grained RCNN hybrid approach. As shown in Figure 12.6, comparison of performance parameters of various network architectures has been done in which accuracy and precision holds the significant

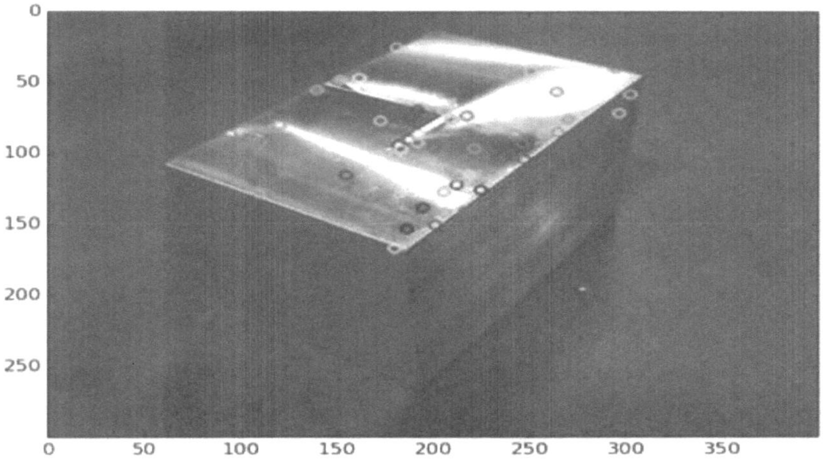

FIGURE 12.4 Key feature points using SIFT algorithm.

FIGURE 12.5 Diagram for a solid metal cuboid block.

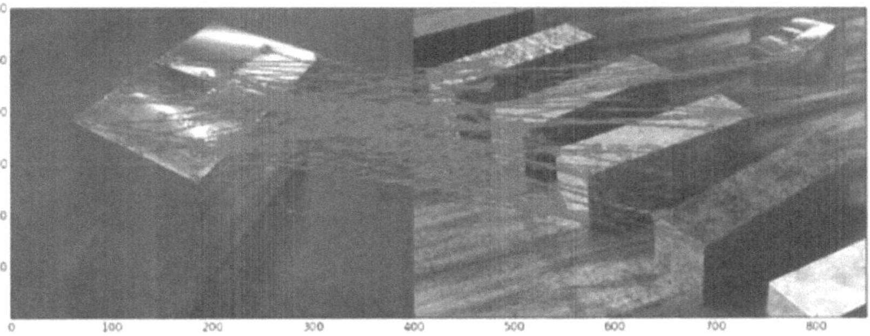

FIGURE 12.6 Diagram showing matching points among two identified objects using RCNN algorithm.

TABLE 12.1
Comparison of Dataset from COCO for Mask-RCNN and FG_RCNN Algorithm for Various Solid Blocks

	Data set	AP [val]	AP	AP 50	Solid block	Block (steel)	Block (Marble)
FGCNN	COCO(Fine)	27.4	29.2	46.3	16.4	28.4	17.4
FRCNN	COCO(Fine)	37.8	28.2	44.5	14.6	30.21	15.65
M-RCNN	COCO(Fine)	43.2	41.4	34,2	21.6	37.2	19.8
FG_CNNT	COCO(Fine)	21.3	28.5	43.8	17.3	31.3	18.7

value displaying the proposed techniques. Here Table 12.1 shows the contrast with the Faster-RCNN fine-grained hybrid technique.

Bounding box correlation factor with coordinates and correlated function given as

$d_i(p)$ where $i \ni \{x, y, w, h\}$

The target parameters are given as

$t_x = (g_x - p_x)/p_w$
$t_y = (g_y - p_y)/p_h$
$t_w = \log(g_w/p_w)$
$t_h = \log(g_h/p_h)$

Where g_x is the filtered value after convolution operation and the transform operation of input x and p_w is triplet filter transform operation of weight matrix w.

So the loss function as defined by given function

$$£ = \Sigma\left(t_i - d_i(p)\right)^2 + \lambda|w|^2 \quad i \ni \{x, y, w, h\} \tag{12.7}$$

Where w is the weight matrix evaluated from training process.

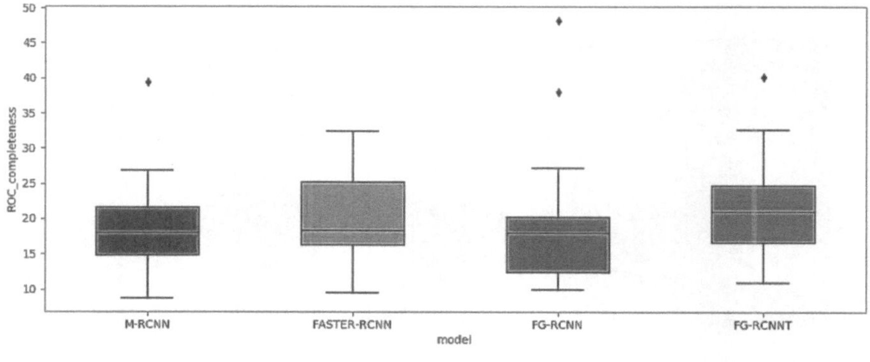

FIGURE 12.7 The diagram shows region of convergence completeness results with respective models.

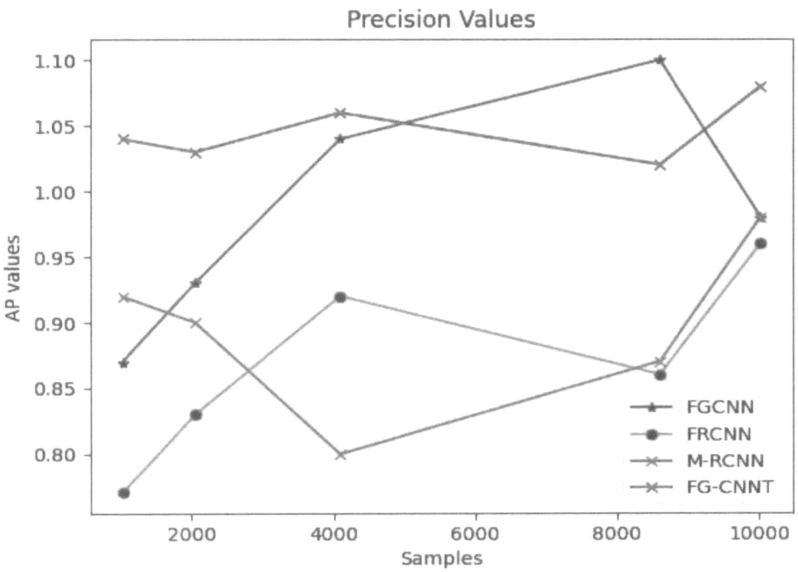

FIGURE 12.8 Precision value graph.

R-CNN is a naive feature-dependent generative model to train; thus, it is called Mask-RCNN. In the case of Faster-RCNN, which runs at 5 frames per second, a little overhead is added. Mask-RCNN is also utilized to make it simple to generalize recognition and region assimilation tasks, such as estimating human poses, inside the same framework. Instance segmentation, bounding-box object detection [26], and person key point detection are among the top three tracks of the COCO suite of tasks.

In any classification model, for the correctness of the model, the parameters which need to be considered are evaluated as mentioned below

1. Accuracy: By definition it provides the correctness of a sample from classification model. It is given as equation as follows.

$$\text{Accuracy} = (\text{True Positive} + \text{True Negative}) / \text{Total Number}$$

2. Precision: True precision means the frequency of occurrence of events if the classification model predicts those events.

$$\text{Precision} = [\text{True Positive}] / [\text{True positive} + \text{False positive}]$$

   Precision takes into consideration all retrieved events, but it can alternatively be evaluated using a cut-off rank, in which case the assessment returns only the topmost results. This statistic is called precision at n.
3. Recall is the percentage of data that is true positive out of all true positive and false positive data in information retrieval.

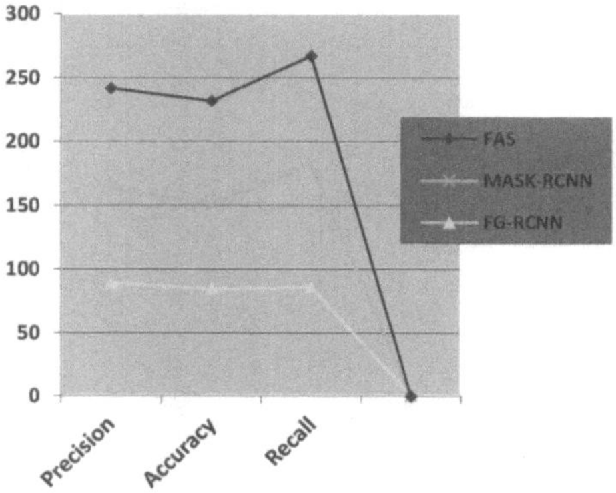

FIGURE 12.7

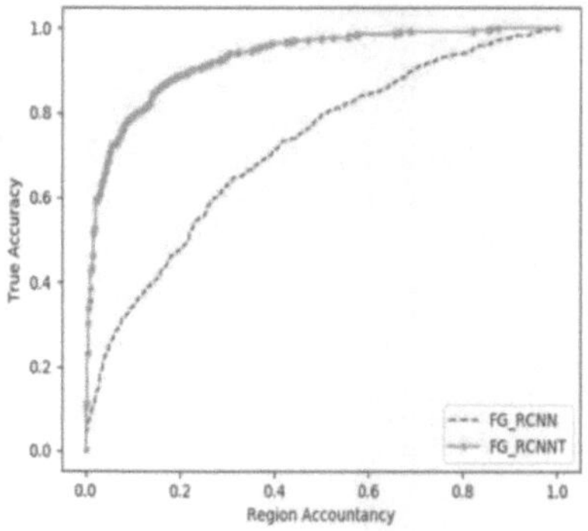

FIGURE 12.8

Graph diagram for comparison of results of performance measures.

Our proposed design performs well for images for solid blocks with similar texture and shape. This may be enhanced to use in feature extraction of dissimilar and high contrasting features such as wooden block or pained polymers. Figure 12.7 and Figure 12.8 shows precision, accuracy, recall values for FG-RCNN and Mask-RCNN which has been provided under training of both real-time objects and COCO data. Figure 12.7 shows the region of convergence values for FG-RCNN and FG-RCNNT (with triple region algorithm).

12.6 CONCLUSIONS

In object categorization and detection, granular convolution and pooling have made significant advances. There are two stages to this outstanding categorization framework using the COCO database. The first step is for the tuning 3–4 convolution layer, such as granular classification and then using SIFT-based feature vectors to extract regions of interest from an image and then feed them to form a deep neural network for classification. Thus, the models – fine-grained RCNN classification and fine-grained classification with triple feature point normalization – show better testing values for the true accuracy and recall values.

REFERENCES

[1] J.T. Turner, Kalyan Gupta, Brendan Morris, David W. Aha, "Keypoint Density-based Region Proposal for Fine-Grained Object Detection and Classification using Regions with Convolutional Neural Network Features", IEEE Applied Imagery Pattern Recognition Workshop (AIPR), 2016.

[2] Tsung-Yu Lin, Aruni Roy Chowdhury, Subhransu Maji, "Bilinear CNN Models for Fine-grained Visual Recognition", ICCV '15 Proceedings of the 2015 IEEE International Conference on Computer Vision (ICCV), Pages 1449–1457, December 07–13, 2015.

[3] Bo Zhao, Jiashi Feng, Xiao Wu Shuicheng Yan, "A Survey on Deep Learning-based Fine-grained Object Classification and Semantic Segmentation", International Journal of Automation and Computing, vol. 14, no. 2, 2017, pp. 119–135.

[4] Jie Fang, Yu Zhou, Yao Yu and Sidan Du, "Fine-Grained Vehicle Model Recognition Using A Coarse-to-Fine Convolutional Neural Network Architecture", IEEE Transactions on Intelligent Transportation Systems, vol. 18, no. 7, 2017, pp. 1782–1792.
[5] Kaiming He, Georgia Gkioxari, Piotr Dollar, Ross Girshick, "Mask R-CNN", Facebook AI Research (FAIR), April 2017.
[6] Jianlong Fu, Heliang Zheng, Tao Mei1, "Look Closer to See Better: Recurrent Attention Convolutional Neural Network for Fine-grained Image Recognition", Microsoft Research, 2015.
[7] C. Chen, M.-Y. Liu, C.O. Tuzel, J. Xiao, "R-CNN for Small Object Detection", TR2016-144, November 2016, Mitsubishi Electric Research Laboratories.
[8] Ross Girshick, "Fast R-CNN Object detection with Caffe", Microsoft Research, ICCV 2015.
[9] Cheng Wang, Ying Wang, Yinhe Han, Lili Song, Zhenyu Quan, Jiajun Li, Xiaowei Li, "CNN-based object Detection Solutions for Embedded Heterogeneous Multicore SoCs", 22nd Asia and South Pacific Design Automation Conference (ASP-DAC), 2017.
[10] Ala Mhalla, Thierry Chateau, Houda Maamatou, Sami Gazzah, Najoua Essoukri, Ben Amara, "SMC Faster R-CNN: Toward a Scene-specialized Multi-object Detector", Computer Vision and Image Understanding, June 2017. Doi: 10.1016/j.cviu.2017.06.008
[11] Subhransu Maji, Alexander C. Berg, Jitendra Malik, "Efficient Classification for Additive Kernel SVMs", Transactions on Pattern Analysis and Machine Intelligence, vol. 39, no. 6, 2017.
[12] Chao-Yung Hsu, Chun-Shien Lu, Soo-Chang Pei, "Image Feature Extraction in Encrypted Domain with Privacy-Preserving SIFT", IEEE Transactions on Image Processing, vol. 21, no. 11, 2012, pp. 4593–4607.
[13] Liang Zhang, Peiyi Shen, Guangming Zhu, Wei, Houbing Song, "A Fast Robot Identification and Mapping Algorithm Based on Kinect Sensor", Sensors, vol. 15, no. 8, 2015, pp. 19937–19967. Doi: 10.3390/s150819937
[14] Shaoqing Ren, Kaiming He, Ross Girshick, Jian Sun, "Faster R-CNN: Towards Real-Time Object Detection with Region Proposal Networks", IEEE Transactions on Pattern Analysis and Machine Intelligence, vol. 39, no. 6, 2017, pp. 1137–1149.
[15] Lu Wang, Min Chen, Lin Wang, Yixue Haoi, Kai Hwang, "Disease Prediction by Machine Learning Over Big Data From Healthcare Communities", IEEE Access, vol. 5, 2017, pp. 8869–8879.
[16] Yuting Zhang, Kihyuk Sohn, Ruben Villegas, Gang Pan, Honglak Lee, "Improving Object Detection with Deep Convolutional Networks via Bayesian Optimization and Structured Prediction", CVPR, 2015.
[17] Hantao Yao, Shiliang Zhang, Yongdong Zhang, Jintao Li, Qi Tian, "Coarse-to-Fine Description for Fine-Grained Visual Categorization", IEEE Transactions on Image Processing, vol. 25, no. 10, 2016, pp. 4858–4872.
[18] Jian Hou, Huijun Gao, Qi Xia, Naiming Qi, "Feature Combination and the kNN Framework in Object Classification", IEEE Transactions on Neural Networks and Learning Systems, vol. 27, no. 6, 2016, pp. 1368–1378.
[19] Jhing-Fa Wang, Han-Jen Hsu, Jyun-Sian Li, "Intelligent Object Extraction Algorithm Based on Foreground/Background Classification", International Conference on Embedded and Ubiquitous Computing, EUC 2005 Workshops, pp. 101–110.
[20] Maryam M. Najafabadi, Flavio Villanustre, Taghi M. Khoshgoftaar, Naeem Seliya, Randall Wald, Edin Muharemagic, "Deep Learning Applications and Challenges in Big Data Analytics", Journal of Big Data, vol. 2, no. 1, 2015.
[21] Yaming Wang, Jonghyun Choi, Vlad Morariu, Larry S. Davis. "Mining Discriminative Triplets of Patches for Fine-grained Classification," Proceedings of the IEEE Conference on Computer Vision and Pattern Recognition, pp. 1163–1172, 2016.

[22] Xiaojiang Peng, Cordelia Schmid, "Multi-region Two-stream R-CNN for Action Detection", European Conference on Computer Vision, October 2016, Amsterdam, Netherlands.

[23] David G. Lowe, "Distinctive Image Features from Scale-Invariant Keypoints", International Journal of Computer Vision, vol. 60, no. 2, 2004, pp. 91–110.

[24] Lei Yu, Zhixin Yu, Yan Gong, "An Improved ORB Algorithm of Extracting and Matching Features", International Journal of Signal Processing, Image Processing and Pattern Recognition, vol. 8, no. 5, 2015, pp. 117–126.

[25] Cheng Wang, Ying Wang, Yinhe Han, Lili Song, Zhenyu Quan, Jiajun Li, Xiaowei Li, "CNN-based Object Detection Solutions for Embedded Heterogeneous Multicore SoCs", 22nd Asia and South Pacific Design Automation Conference (ASP-DAC), 2017.

[26] Liang Zhang, Peiyi Shen, Guangming Zhu, Wei, Houbing Song, "A Fast Robot Identification and Mapping Algorithm Based on Kinect Sensor", Sensors, vol. 15, 2015, pp. 19937–19967. Doi: 10.3390/s150819937

13 An Embedded Implementation of a Traffic Light Detection System for Advanced Driver Assistance Systems

Riadh Ayachi, Mouna Afif, Yahia Said, and Abdessalem Ben Abdelali

CONTENTS

13.1 INTRODUCTION

The recent advances in technology of Industry 4.0 [1] have allowed the development of more reliable techniques in the automotive industry. When adopting new technologies, automotive industry always remained at the forefront. From the first industrial revolution, building the first cars has seen many innovations. Integrating computers in cars has allowed the automation of many processes. In addition to computers, the availability of cellular connectivity, data processing techniques, and high-scale datasets has boosted the performance of the automated processes. The fourth industrial revolution combined physical sensors, digital processing techniques, artificial intelligence, and big data to achieve the desired performance.

Pedestrian and vehicle safety is a significant area that automotive companies are focusing on heavily. Recently, automotive manufacturers have developed many technologies to help prevent accidents. These technologies allowed the automation and enhancement of the vehicular system to assist the driver and ensure its safety. Advanced driver assistance systems (ADAS) [2] are one of the automotive

DOI: 10.1201/9781003229018-13

technologies established by Industry 4.0. ADAS is a combination of smart systems that facilitate the control of the vehicle by the driver and can perform easy and repetitive tasks such as parking, highway driving, and cruise control. ADAS collect different sensor data and use the most recent artificial intelligence techniques.

The advances in data processing techniques have improved the performance of automotive technologies. The recent breakthrough of deep learning techniques has achieved great success in computer vision applications. In particular, convolutional neural networks (CNN) [3] were the most used model for image processing applications such as object detection [4], scene recognition [5], indoor object detention and recognition [6, 7], face identification, and many others. Besides CNN models have been widely used to solve traffic-related problems such as traffic sign detection [8], pedestrian detection [9], and traffic light detection [10]. The achieved success of CNN models for image processing applications comes from the self-learning and automatic methodology in feature selection and processing. Also, the CNN models have a decision-making process similar to the biological nervous system. Combining the mentioned characteristics makes the CNN model the first choice for image processing applications. The main problem of CNN is the high computation complexity. However, many works have proved that a lightweight model can be developed by making an optimized architecture that provides high performance without degrading the accuracy.

To reduce the computation complexity of CNN models, many optimization techniques were proposed. Model level optimization aims to develop a lightweight model by proposing a good architecture that ensures lower computation complexity and high accuracy. The second category is model compression techniques that are used to decrease the computation complexity by eliminating redundant and weak connections without big damage to accuracy. The third category is hardware implementation tools which allow the distribution of the computation on the hardware resources optimally. Focusing on the first category, which is the most important, many works were proposed. An overview of CNN optimization techniques is shown in [11].

The squeezeNet model [12] has proposed to replace convolutional layers with a fire module to reduce the resultant parameters and compressing the size by a big margin compared to other models with the same accuracy. The proposed fire module has two convolution stages, the first compresses the input feature maps and the second stage was used to extract features through the combination of 3×3 and 1×1 convolution filters. The proposed fire module was very useful for parameter optimization. The mobile model [13] proposed to separate the functionalities of the convolution layer. The regular convolution layer filters the features and compresses them to be passed to the next layer. A separable depthwise convolution lock was proposed in mobileNet model to take the place of the convolution layer. The proposed block combines a depthwise convolution layer and a pointwise convolution layer. The depthwise convolution layer filters the input features and conserves the input and output channels. In another words, the output channels are equal to the input channels. The pointwise convolution is a 1×1 convolution layer. It was used to optimize feature maps and channels passed to the next block. Separating the filtering and compress processes all to apply more non-linear layers allow the processing of more complex data with a lower number of hidden layers. Besides, the separation allows processing the data nine times faster while generating a low number of parameters. The efficientNet model [14] has been proposed to scale the parameters of the network to find a balance between them. A neural network has three main parameters. The input resolution is the resolution of the input image. The network depth defines the number of hidden layers. The network-wide is the number of output channels. A compound scaling was proposed by the efficientNet model to scale all parameters based on the relationship between them. The efficientNet model was based on the separable depthwise convolution blocks with some modifications.

In this work, a lightweight CNN model was proposed for traffic light detection that fits in an embedded device and runs in real-time while guaranteeing a high detection rate. For this purpose, we proposed to use the squeezeNet model [12] as a backbone thanks to its light structure and its performance. To detect traffic lights, the single-shot multi-box detection (SSD) [15] framework was used. The SSD was to detect objects at different scales through a pyramid features network. The outputs of different pyramid levels were concatenated to generate the final prediction. The SSD was proposed for traffic light detection to spot traffic lights of small size. The proposed model was named SSD-squeezeNet.

The proposed model in its first version does not fit into the small memory of the embedded device. So, we proposed to apply compression techniques to reduce the model size. First, a pruning technique was applied to reduce the number of parameters by removing weak and redundant connections and channels. The channel pruning technique [16] was applied, then the model was fine-tuned to recover accuracy. Second, a quantization technique [17] was applied by replacing a floating-point with a two-bit fixed-point and a one-bit fixed-point for activations and weights, respectively.

The main advantages of the proposed SSD-squeezeNet model are that it's lightweight and the high detection rate of the small traffic lights. The lightweight model was reached through the architecture of the squeezeNet model, which was originally designed for embedded implementation especially on field programmable gate arrays (FPGA). The high detection rate was achieved using the SSD as a detection framework, which allows the detection of small objects using predefined anchors and the features pyramid network.

The pynq z1 board was proposed for the implementation of the inference of the traffic light detection model. The proposed board is a Xilinx MPSoC board equipped with an advanced RISC machine (ARM) processor and a reconfigurable part mounted on the same chip. The pynq board allows to implement hardware architecture using the high-level Python programming language and support famous deep learning frameworks. These features allow an easy translation from graphical processing unit (GPU) implementation to embedded implementation with minor modifications.

The SSD-squeezeNet model was trained and evaluated on the Drive U traffic light dataset (DTLD) [18]. The DTLD contains more than 230000 annotated images. The images were collected from 11 German cities using a stereo camera with a resolution of 2 mega-pixels mounted on a car. The traffic lights in the dataset provide 5 traffic light states, which are red, yellow, green, red-yellow, and no light. The model evaluation on the DTLD has achieved 94.3% of mean average precision (mAP). The inference model has achieved 16 FPS of inference. The robustness of the proposed approach has been proved.

The major contributions of this work are the following: (1) proposing a traffic light detection system based on CNN model, (2) proposing the use of a lightweight CNN model to fit into embedded devices, (3) the proposed model was compressed to reduce its size and accelerate the inference processing time, and (4) evaluating the performance of the proposed model on the DTLD.

The chapter is organized as follows. Section 13.2 presents related works to traffic light detection on embedded devices. In Section 13.3, the proposed approach is detailed. In Section 13.4, experimental results are presented and discussed. Section 13.5 provides conclusions.

13.2 RELATED WORKS

Traffic light detection is an important system to ensure safety in urban spaces. Modern cars have been equipped with intelligent technology to help in the control process. ADAS is based on embedded devices and the proposed traffic light system must provide high performance, run in real-time, and fit into the small memory of the embedded device. To achieve the desired performance, many works have been proposed.

Wu et al. [19] proposed a traffic light detection system to recognize traffic light state by visually impaired people. The proposed system was based on Adaboost classifier [20]. The Haar-like descriptor [21] was used for feature extraction. The input image was converted to the YCrCb space color from the RGB space color. Multi-layer features were extracted at different channels. Different features were extracted from the Y, Cr, and Cb channels. The integral window was used to extract the features at different channels. The Adaboost classifier was trained using the extracted features. The sliding window technique was used to detect the traffic light in the image, and the Adaboost-based classifier was used to recognize the traffic light state. The proposed system takes input images with a resolution of 640×480 and can process video at 30 FPS.

A CNN was proposed in [22] for traffic light detection. The CNN model has two convolutional layers each followed by a max-pooling layer and non-linear activation layer, then two fully connected layers are deployed. The dropout technique was used to gain more performances. The proposed traffic light system was used to recognize the traffic light state while the detection task was performed using an external mechanism. The Xilinx Zynq Ultrascale MPSoC+ FPGA was used to implement the inference of the proposed system. An accuracy of 98.3% was achieved with a processing speed of 2.5 FPS.

Ouyang et al. [23] have proposed an embedded implementation of a traffic light detection system for autonomous buses. The proposed system was based on a lightweight multi-backbone detection model. The proposed model was composed of three backbones, a normal CNN model, the ResNet model [24], and the DenseNet model [25]. Features at different levels were concatenated and passed to the detection stage. A features pyramid module was used to detect traffic lights at different scales. The channel compression technique was used to reduce the model size. The color space fusion technique was used to enhance classification accuracy. For training and evaluation, a custom dataset was collected. The proposed traffic light detector was implemented on a low-power embedded device, and a processing speed of 14 FPS was achieved with an accuracy of 94%.

In [26], a traffic light detector was proposed based on fully CNN and clustering techniques. First, an encoder-decoder scheme based on fully CNN was used to extract relevant features. Second, the images were grayscaled and binarized to be passed to the next stage. Third, the Density-based spatial clustering of applications with noise (DBSCAN) algorithm [27] was used for image clustering. Finally, the generated clusters were filtered and only clusters with a size of fewer than 30 pixels were maintained and bounding boxes for those clusters were generated. The proposed model was evaluated on the most popular traffic light datasets such as Bosch Small Traffic Lights Dataset [28] and VIVA traffic light detection benchmark [29]. The NVIDIA Jetson TX2 [30], an embedded GPU, was used for the implementation of the inference of the proposed traffic light detector. A testing precision of 43.23% was achieved with a processing time of 50 ms per image.

Most of the existing works are designed for traffic light recognition without any focus on the detection task or achieved low precision or have a slow processing speed. For safety purposes, a traffic light detection system must be accurate and runs in real-time conditions. In the next section, the proposed traffic light detection will be detailed and it will achieve superior performance compared to existing works.

13.3 PROPOSED APPROACH

In this chapter, we proposed a traffic light detection system based on a CNN. The SSD object detection framework [15] with the squeezeNet model [12] as a backbone was used. SSD is an objects detection model at different scales through the use of a pyramid features module. The SSD architecture is presented in Figure 13.1.

The SSD framework proposed to add extra layers at the end of the backbone. The size of the additional layers was decreased progressively to allow the detection at a different scale. These layers form a pyramid feature which is considered for detecting the small object in high-resolution images. For traffic light detection, the size of the traffic light in the image is considered very small. To generate a prediction, a convolution layer with a 3×3 kernel size was used instead of fully connected. The convolution layer can predict either the class or the bounding box of the object. To predict bounding

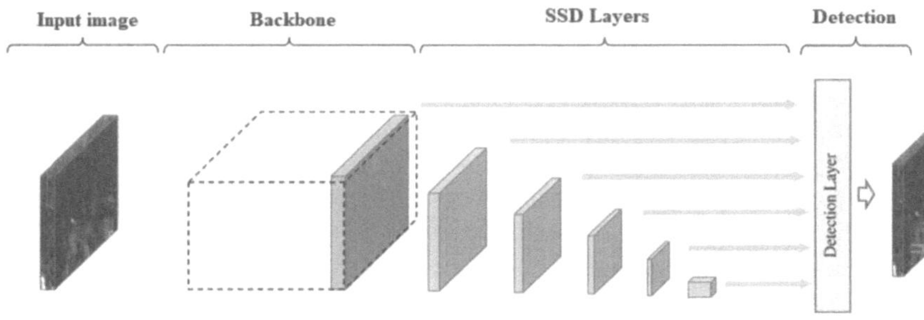

FIGURE 13.1 SSD architecture.

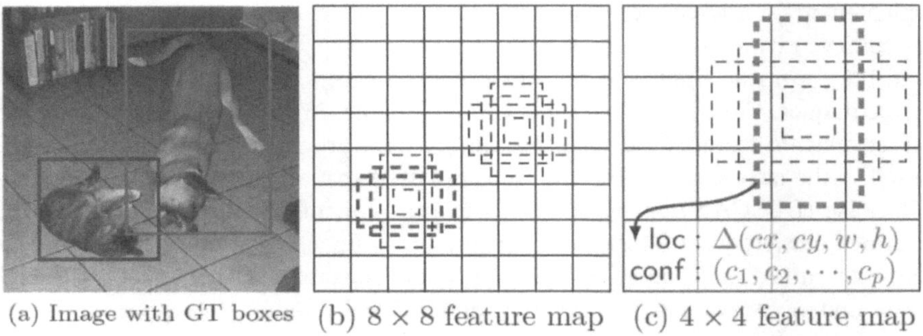

FIGURE 13.2 Object detection using anchors at different feature maps resolution [15].

boxes, the SSD proposed the use of a set of predefined anchors. The proposed anchors were tested at different locations of the feature maps in a convolutional manner. At each cell of the feature map, the offset relative to the tested anchors and the score of object presence was predicted. For each predicted bounding box, c class probabilities were computed in addition to the four offset parameters. The proposed anchors were applied to the feature map at different resolutions to efficiently detect objects at different scales. An example of object detection in different resolution feature maps is presented in Figure 13.2.

To train the SSD, a matching strategy was proposed. This strategy was used to determine the corresponding ground truth detection of each tested anchor. For each ground truth, different anchors, aspect ratios, and scales were selected. The selected anchors were matched to ground truth bounding boxes based on the Jaccard overlap [30]. Any tested anchor that has an overlap higher than 0.5 was matched to the corresponding ground truth, then the network was trained to map the prediction. This technique allows simplifying the training by generating high score predictions for multiple anchors rather than only picking the anchor with the highest overlap.

In order to detect objects at different scales, the SSD proposes a methodology to choose the scales and the aspect ratios of the predefined anchors. The scales and aspect ratios were used to make the anchor responsive to a specific scale of the detected object. After matching the anchors, most of the predictions were negative. This results in an imbalance in training samples. To handle this problem, the negative predictions were sorted according to their score, and the highest score prediction was picked and others were ignored. This technique was very effective to speed up the training and make it more stable.

To extract features, the SSD needs a high-performance backbone. For this purpose, we proposed the use of the squeezeNet model. It is a CNN model designed for embedded implementation especially for FPGA implementation. The model has a lightweight size and high classification accuracy. Many optimizations were proposed by the squeezeNet model to achieve high performance with a lightweight model. A fire module was proposed to replace regular convolution layers. The proposed fire module is composed of two convolution stages, a squeeze stage and an expanding stage. A set of convolution layers with a kernel size of 1 × 1 was used to compress the input feature maps of the model. The expanded layer was a mix of

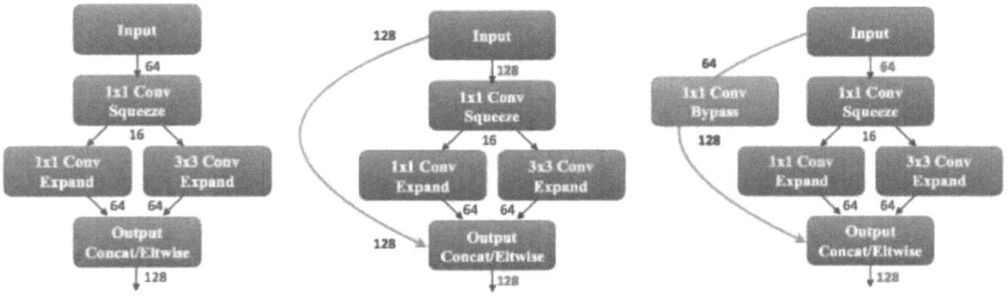

FIGURE 13.3 Proposed structures of the fire module.

3×3 and 1×1 convolution layers. This layer was proposed to extract relevant features with the minimum number of parameters. Three different fire modules were proposed by the squeezeNet model. The proposed structures of the fire module are presented in Figure 13.3.

The normal fire module presents some limitations due to the severe channel compression of the input channels of the fire module. The skip connection allows sharing information around the squeeze layers. The simple bypass is a connection through the identity function. The complex bypass was performed by connecting the input and output of the model through a 1×1 convolution. The residual connection has improved the accuracy but additional parameters were added. The model size with complex bypass is still suitable for possible implementation on embedded devices. The fire module must produce channels in input and output equally. So, only half of the fire modules can have a simple bypass connection. For the rest of the modules, a complex bypass was used to meet the requirement of having the same number of channels at the input and output of the module. The squeezeNet model with simple and complex bypass is presented in Figure 13.4. An alternation between complex and simple bypass was used starting with a complex bypass. For better implementation of the squeezeNet model on the FPGA, we proposed to replace all pooling layers with a strided convolution layer. As proved in [31], using a strided convolution layer is more efficient than max-pooling layers for implementation on hardware. The strided convolution layers generate more parameters but enhance accuracy.

To use the squeezeNet model for traffic light detection, the average pooling and softmax layers were removed and the truncated model was embedded in the SSD

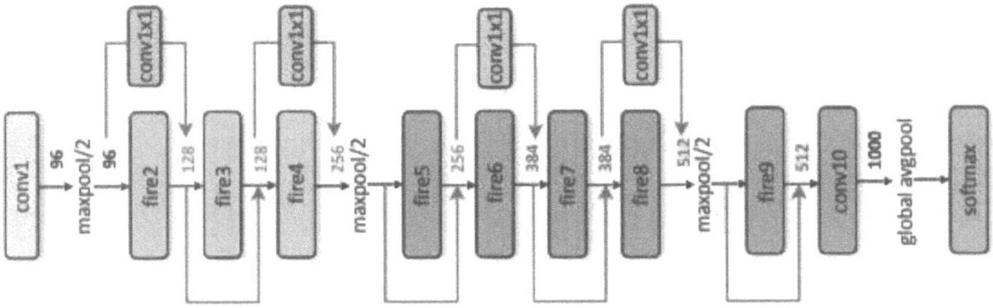

FIGURE 13.4 SqueezeNet model with complex bypass

framework. The SSD-squeezeNet model does not fit into the small on-chip memory of the FPGA. So, we proposed to reduce the size of the model using compression techniques. As a first step, we apply a pruning technique to remove the redundant and weak connection. For this, a channel pruning technique [16] was proposed to filter channels and remove channels that do not degrade the accuracy. The channel pruning technique is an iterative two-steps algorithm. In the first step, the algorithm selects the most relevant channels. The LASSO regression algorithm was used for this step. In this step, features were pruned by removing redundant channels and maintaining relevant features. In the second step, the model was reconstructed with the remaining channels using the least square technique. The proposed steps were applied alternatively on each layer of the model until the completion of all the layers of the model. For more compression, the quantization technique [17] was applied. In this work, we propose to replace the 32-bits floating point with a two-bit fixed-point for activations and one-bit for weights.

13.4 EXPERIMENTS AND RESULTS

In this work, the experiments were conducted on two devices. First, the training and evaluation were performed on a computer. The open cv library was used to load and display images. Second, the pynq z1 board was used for inference. An illustration of the pynq z1 is presented in Figure 13.5. It is an Xilinx MPSoC device designed to enable the implementation of programmable SoCs using the python language. The pynq z1 is equipped with a quad-core ARM cortex A9 CPU with 512 MB of RAM and Artix-7 family programmable logics with 220 digital signal processing (DSP) slices. The xfopen cv was used to load and display images. The xfopen cv was designed by Xilinx for optimized use on Xilinx devices. The xfDNN was used to implement the proposed CNN model and the software-hardware codesign. The xfDNN allows an optimal distribution of the CNN model on the device.

The training and evaluation of the proposed CNN model were performed on the DTLD dataset. The dataset was collected in 11 German cities. The DTLD contains more than 230000 images spotted using two calibrated cameras to get stereo images. The images were labeled manually while considering additional annotations tags such as pictogram, installation orientations, number of light units, and many other tags. The dataset was divided into 3 sets. The training set has 60% of the data, 10% of the data was used as a validation set, and the remaining 30% was used as a testing set.

As an evaluation metric, we proposed to use the mAP. Generally, the precision provides an idea of the performance of the model when the correct prediction is generated. The mAP is the mean of the average precision of all classes on the totality of images in the test set.

The evaluation of the SSD-squeezeNet on the validation set achieved 97.3% of mAP and 95.6% of mAP on the texting set. The compressed model has lost 2.6% of mAP. The compressed model has achieved 93% of mAP.

Table 13.1 summarizes the achieved mAP of the original model and the compressed model and shows the number of parameters and the model size.

As shown in Table 13.1, the compressed model does not degrade the performance while the model size was greatly reduced and can fit into the memory of the embedded device.

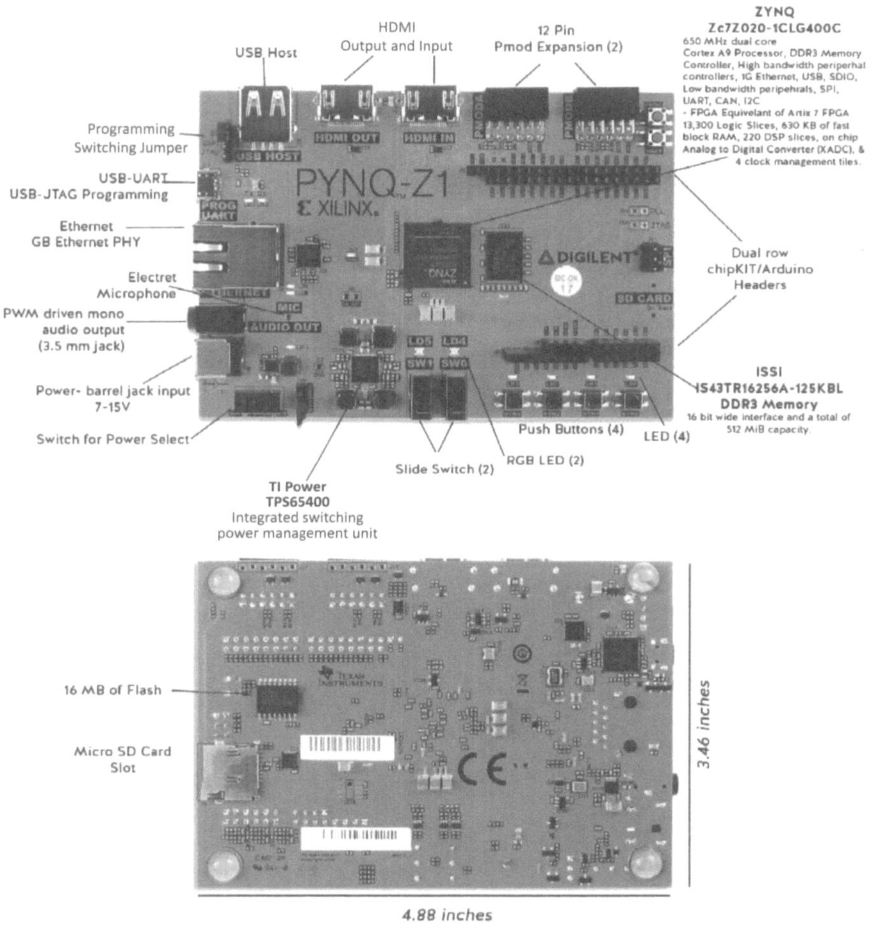

FIGURE 13.5 Illustration of the pynq z1 board.

TABLE 13.1
Summary of the Achieved Results, the Number of Parameters, and the Model Size

Model	mAP (%)	Number of parameters (million)	Model size (MB)
SSD-squeezeNet	95.6	0.976	6.4
SSD-squeezeNet compressed	93	0.454	2.6

The pynq z1 board was used for inference. The proposed model has been distributed on the software-hardware codesign. Convolution layers need a lot of parallel computation. So, the features extraction part and the SSD layers were implemented on the programmable logic. The detection layer needs much storage memory and fewer computations. For this, the detection part was implemented on the CPU. The proposed codesign was optimized using the xfDNN library. An inference speed of 16 FPS was achieved using an input RGB image with a resolution of 128 × 128. The output images were displayed on an external screen through the HDMI out. A sample of the demo output of the traffic light detection system is presented in Figure 13.6. The color of the bounding box was used to define the traffic light state, a red box for the red state, a yellow box for the yellow state, and a green box for the green state.

FIGURE 13.6 Output of the traffic light detection demo.

The obtained results have proved the efficiency of the proposed traffic light detection system. The use of the squeezeNet model was very important to achieve the embedded implementation. The SSD framework has a big role in detecting the small traffic lights in the image. Replacing the max-pooling layers with strided convolution layers has enabled us to build a better hardware engine to process the convolution layers. Thus, the processing speed can be accelerated. The compression techniques allow the reduction of the model size to fit into the limited memory of the pynq z1 board.

13.5 CONCLUSIONS

Traffic light detection is one of the most important technologies for driver assistance. Traffic light detection is a hard task due to many challenges such as similarity to vehicle backlights and the small size of the traffic lights. Building a robust traffic light detection system with high performance that runs in real-time is an obligation to make urban spaces safer. In this chapter, we proposed a traffic light detection system based on the CNN model. The SSD object detection framework was proposed for the detection task. The squeezeNet model was used as a backbone to the SSD. The squeezeNet model was used thanks to its high performance and lightweight size. It was originally designed for FPGA implementation. Combining the SSD and squeezeNet was very effective for the desired task but the obtained size cannot fit into the small memory of the embedded devices. So, we proposed to reduce the model size by applying compression techniques. First, the channel pruning technique was used to remove redundant and weak channels. Second, the quantization technique was applied by replacing the floating-point with a fixed-point for activations and weight. The compression techniques allowed to perfectly reduce the model size. The proposed SSD-squeezeNet was evaluated on the drive U traffic light dataset. The compression version of the proposed model has achieved 93% of mAP and inference of 16 FPS on the pynq board. The robustness of the model is proved by the achieved results. In future works, the proposed system will be tested in a real-world environment under different conditions.

REFERENCES

[1] Lasi, Heiner, Peter Fettke, Hans-Georg Kemper, Thomas Feld, and Michael Hoffmann. "Industry 4.0." Business & Information Systems Engineering 6, no. 4 (2014): 239–242.

[2] Shaout, Adnan, Dominic Colella, and S. S. Awad. "Advanced driver assistance systems-past, present and future." In 2011 Seventh International Computer Engineering Conference (ICENCO'2011), pp. 72–82. IEEE, 2011.

[3] Gu, Jiuxiang, Zhenhua Wang, Jason Kuen, Lianyang Ma, Amir Shahroudy, Bing Shuai, Ting Liu, et al. "Recent advances in convolutional neural networks." Pattern Recognition 77 (2018): 354–377.

[4] Ayachi, Riadh, Yahia Said, and Mohamed Atri. "A convolutional neural network to perform object detection and identification in visual large-scale data." Big Data 9, no. 1 (2021).

[5] Afif, Mouna, Riadh Ayachi, Yahia Said, and Mohamed Atri. "Deep learning based application for indoor scene recognition." Neural Processing Letters 51, no. 1–2 (2020): 1–11.

[6] Afif, Mouna, Riadh Ayachi, Yahia Said, Edwige Pissaloux, and Mohamed Atri. "An evaluation of RetinaNet on indoor object detection for blind and visually impaired persons assistance navigation." Neural Processing Letters 51 (2020): 2265–2279.
[7] Afif, Mouna, Riadh Ayachi, Edwige Pissaloux, Yahia Said, and Mohamed Atri. "Indoor objects detection and recognition for an ICT mobility assistance of visually impaired people." Multimedia Tools and Applications 79, no. 41 (2020): 31645–31662.
[8] Ayachi, Riadh, Mouna Afif, Yahia Said, and Mohamed Atri. "Traffic signs detection for real-world application of an advanced driving assisting system using deep learning." Neural Processing Letters 51, no. 1 (2020): 837–851.
[9] Ayachi, Riadh, Yahia Said, and Abdessalem Ben Abdelaali. "Pedestrian detection based on light-weighted separable convolution for advanced driver assistance systems." Neural Processing Letters 52, no. 3 (2020): 2655–2668.
[10] Vitas, Dijana, Martina Tomic, and Matko Burul. "Traffic light detection in autonomous driving systems." IEEE Consumer Electronics Magazine 9, no. 4 (2020): 90–96.
[11] Ayachi, Riadh, Yahia Said, and Abdessalem Ben Abdelali. "Optimizing neural networks for efficient FPGA implementation: A survey." Archives of Computational Methods in Engineering 28, no. 7 (2021): 1–11.
[12] Iandola, Forrest N., Song Han, Matthew W. Moskewicz, Khalid Ashraf, William J. Dally, and Kurt Keutzer. "SqueezeNet: AlexNet-level accuracy with 50x fewer parameters and < 0.5 MB model size." arXiv preprint arXiv:1602.07360 (2016).
[13] Howard, Andrew G., Menglong Zhu, Bo Chen, Dmitry Kalenichenko, Weijun Wang, Tobias Weyand, Marco Andreetto, and Hartwig Adam. "Mobilenets: Efficient convolutional neural networks for mobile vision applications." arXiv preprint arXiv:1704.04861 (2017).
[14] Tan, Mingxing, and Quoc V. Le. "Efficientnet: Rethinking model scaling for convolutional neural networks." ICML (2019): 6105–6114.
[15] Liu, Wei, Dragomir Anguelov, Dumitru Erhan, Christian Szegedy, Scott Reed, Cheng-Yang Fu, and Alexander C. Berg. "SSD: Single shot multibox detector." In European Conference on Computer Vision, pp. 21–37. Springer, Cham, 2016.
[16] Zhuang, Zhuangwei, Mingkui Tan, Bohan Zhuang, Jing Liu, Yong Guo, Qingyao Wu, Junzhou Huang, and Jinhui Zhu. "Discrimination-aware channel pruning for deep neural networks." In Advances in Neural Information Processing Systems, pp. 875–886. 2018.
[17] Jacob, Benoit, Skirmantas Kligys, Bo Chen, Menglong Zhu, Matthew Tang, Andrew Howard, Hartwig Adam, and Dmitry Kalenichenko. "Quantization and training of neural networks for efficient integer-arithmetic-only inference." In Proceedings of the IEEE Conference on Computer Vision and Pattern Recognition, pp. 2704–2713. 2018.
[18] Fregin, Andreas, Julian Müller, Ulrich Kreβel, and Klaus Dietmayer. "The DriveU traffic light dataset: Introduction and comparison with existing datasets." In 2018 IEEE International Conference on Robotics and Automation (ICRA), pp. 3376–3383. IEEE, 2018.
[19] Wu, Xue-Hua, Renjie Hu, and Yu-Qing Bao. "Parallelism optimized architecture on FPGA for real-time traffic light detection." IEEE Access 7 (2019): 178167–178176.
[20] Schapire, Robert E. "Explaining Adaboost." In Robert E. Schapire (Ed.), Empirical Inference, pp. 37–52. Springer, Berlin, Heidelberg, 2013.
[21] Park, Ki-Yeong, and Sun-Young Hwang. "An improved Haar-like feature for efficient object detection." Pattern Recognition Letters 42 (2014): 148–153.
[22] Fukuchi, Tomohide, Mark Ogbodo Ikechukwu, and Abderazek Ben Abdallah. "Design and Optimization of a Deep Neural Network Architecture for Traffic Light Detection." In SHS Web of Conferences, vol. 77, p. 01002. EDP Sciences, 2020.
[23] Ouyang, Zhenchao, Jianwei Niu, Tao Ren, Yanqi Li, Jiahe Cui, and Jiyan Wu. "MBBNet: An edge IoT computing-based traffic light detection solution for autonomous bus." Journal of Systems Architecture 109 (2020): 101835.

[24] He, Kaiming, Xiangyu Zhang, Shaoqing Ren, and Jian Sun. "Deep residual learning for image recognition." In Proceedings of the IEEE Conference on Computer Vision and Pattern Recognition, pp. 770–778. 2016.
[25] Huang, Gao, Zhuang Liu, Laurens Van Der Maaten, and Kilian Q. Weinberger. "Densely connected convolutional networks." In Proceedings of the IEEE Conference on Computer Vision and Pattern Recognition, pp. 4700–4708. 2017.
[26] Yudin, Dmitry, and Dmitry Slavioglo. "Usage of fully convolutional network with clustering for traffic light detection." In 2018 7th Mediterranean Conference on Embedded Computing (MECO), pp. 1–6. IEEE, 2018.
[27] Ester, Martin, Hans-Peter Kriegel, Jörg Sander, and Xiaowei Xu. "A density-based algorithm for discovering clusters in large spatial databases with noise." KDD 96, no. 34 (1996): 226–231.
[28] "Heidelberg Collaboratory for Image Processing. Bosch Small Traffic Lights Dataset", URL: https://hci.iwr.uni-heidelberg.de/node/6132
[29] "Laboratory for Intelligent and Safe Automobiles, UC San Diego. Vision for Intelligent Vehicles and Applications (VIVA) Challenge", URL: http://cvrr.ucsd.edu/vivachallenge, 2015.
[30] Jaccard, Paul. "The distribution of the flora in the alpine zone. 1." New Phytologist 11, no. 2 (1912): 37–50.
[31] Ayachi, Riadh, Mouna Afif, Yahia Said, and Mohamed Atri. "Strided convolution instead of max pooling for memory efficiency of convolutional neural networks." In International Conference on the Sciences of Electronics, Technologies of Information and Telecommunications, pp. 234–243. Springer, Cham, 2018.

14 CO2 Emissions, Financial Development, and Renewable Energy Consumption (REC)

A Metadata Analysis

Van Chien Nguyen and Thu Thuy Nguyen

CONTENTS

DOI: 10.1201/9781003229018-14

14.1 INTRODUCTION

One of the most important factors related to the socio-economic development of every country all over the world is definitely energy. Currently, fossil fuel sources, for example, gas, oil and coal are accommodating the energy needs of countries. Nevertheless, there is a limitation of fossil energy sources, therefore researching and using green and renewable energies, for instance, solar, wind power, geothermal and biomass energy, is an essential requirement. Besides, the energy consumption dramatically reflects the quality of life (Tran et al., 2020). At the initial step of socio-economic development, the importance of fossil fuels, for example, coal, gas and oil, has been greatly attributed not only for socio-economic development but also caused environmental pollution. These resources are being disapproved as they are some of the major causes of damage to green sustainable development through rocketing CO2 emission, and other greenhouses gases, namely methane and dinitrogen oxide (Nguyen et al., 2020a; Tran et al., 2020; Zeppini and Bergh, 2020).

In recent years, climate change, that is, mainly demonstrated by global warming and sea level rise, has created extreme weather phenomena. This is exactly one of the greatest challenges which humanity is facing in the 21st century since whole ecosystems as well as human lives are obviously affected by climate change. Also, it is discussed that it is time for humans, the government, policymakers and researchers to widely support the transition from a high-carbon to lower-carbon energy to recover from environmental degradation (Zeppini and Bergh, 2020). The wide range of literature reviews on the environmental quality and renewable energy consumption (REC) has been referred in empirical studies (Sugiawan & Managi, 1990; Sadorsky, 2009; Doytch and Narayan, 2016; Khan et al., 2017, Bamati and Raoofi, 2020; Gozgor et al., 2020; Zeppini and Bergh, 2020). In addition to microeconomics, fossil fuels and renewable energy (RE) are exactly substitute goods. In the context of increasing or decreasing fossil fuels, consumers will prefer using RE. This suggestion will support researchers that changes in fossil fuel price, such as gas price or oil price, could be determined to investigate this relationship (Sadorsky, 2009; Reboredo, 2015; Shah et al., 2018).

The major finding in this literature derived from econometric approach could help government, policymakers and researchers to find out the environmental policies to achieve the sustainability goal. The study's aim is to fill the actual literature gap by discovering the relationship among rise in CO2 emissions, financial development, fossil fuel energy consumption and REC in an emerging region of Asia. This chapter aims to examine how prime factors affected REC throughout 1990–2015. This chapter is to discuss panel data based on the Pooled Ordinary Least Square (POLS), Fixed Effect Model (FEM) and Random Effect Model (REM) approaches.

The layout of this study is constructed as follows: Section 14.2 presents the literature review, while the data, methodology and research model are described in Section 14.3. Next, Section 14.4 is devoted to the empirical analysis. Sections 14.5 and 14.6 cover the discussions and implications regarding REC in Asian countries from the present to the future, followed by conclusions in Section 14.7.

14.2 LITERATURE REVIEW

A wide range of previous studies have primarily investigated the determinants of REC, especially in emerging economies (Doytch and Narayan, 2016; Bamati and Raoofi, 2020; Gozgor et al., 2020; Zeppini and Bergh, 2020). However, studies conducted for situation in Asian economies are still restricted. Some discussions on environmental issues have been conducted in some countries in Asia such as China, Japan and South Korea, in which problems in the global environment and climate change have been raised and the direction to create equilibration between economic growth and environmental protection have been proposed. This is exactly the issue that many policymakers, government and the government in relevant economies are interested in. There is a contradictory result in some empirical studies in accordance with the relationship among CO2 emissions, population growth, private credit by deposit money banks to GDP, fossil fuel and REC (Tianyu et al., 2014; Muhammad et al., 2020; Mehdi et al., 2020; Razmjoo et al., 2021). The findings indicate that eliminating fossil energy and subsidies for RE are a must in the transition trend all over the world. Considering the case of G7 economies, including the USA, the UK, Japan, Germany, Italy, France and Canada, Sadorsky (2009), it is indicated that energy security and especially global warming is indeed the most important reason that affect REC basing on panel cointegration estimation. The results of empirical research reveal that increases in per capita income and environmental pollution could be found to be main operators behind REC in the long run.

In some studies such as Razmjoo et al. (2021) and Mehdi et al. (2020), the impact of REC on CO2 emissions has been found. On the opposite side, the influence of CO2 emissions on REC is expected to fill the gap in the situation of Asian economies. Similarly, Sherbinin et al. (2007) gave evidence that population is one of the key factors that have influence on the environment; further, fast population growth aggravates other externalities such as wars, governance quality and polluting technologies. Thanks to the study by CBI (2018), it is shown that in order to halve the global emissions by 2050, it will approximately cost $46 trillion, or in other words, $1 trillion yearly. In that situation, there appears a new capital attraction channel, namely green bonds, which is an effectual solution in mobilizing hundreds of billions USD each year to develop a sustainable and green world economy.

Zeppini and Bergh (2020) studied the relationship between the competitive dynamics of fossil fuels, for instance, oil/gas and coal, and RE such as solar and wind. The results suggest that peak oil can cause a transposition to using more coal rather than RE and will probably worsen environmental quality. In the case of eliminating subsidies for fossil fuels and supporting REs, the climate policies for renewables subsidies such as implementation for carbon taxes, market adoption, and R&D activity subsidies which can be focused, empirical evidence indicates that a potential transition trend from a high-carbon to lower-carbon energy will exist. Also, climate policies can be discussed on two features and will be the channel to solve the climate problem. It concludes that the government should not overpass critical carbon budget and uncertainty of energy sources' market shares. In a recent study, Nguyen et al. (2020b) showed that many developed and developing energy systems rely on fossil fuel energy. European economies are pioneers in promoting RE sources (see World Energy Council, 2013). However, Martins et al. (2018) revealed that fossil fuels still

TABLE 14.1
Variable Measurements

	Abbreviation	Source
Dependent variables		
Renewable energy consumption (% of total final energy consumption)	REC	WDI
Independent variables		
CO_2 emissions (10^6 kilo tons)	CO2	WDI
Population growth (annual %)	POP_ growth	WDI
Private credit by deposit money banks to GDP (%) – Financial development	PC	WDI
Fossil fuel energy consumption (% of total)	FOSSIL	WDI

Source: Authors, WDI (2020a), WDI (2020b), WDI (2020c), WDI (2020d), WDI (2020e).

play an essential role in energy system. It is time for the Asian region to catch up with the modern trend of green sustainable development.

14.3 RESEARCH DATA AND METHODOLOGY

14.3.1 Research Data

This study used the dataset which is extracted from the dataset of World Development Indicator (WDI) in Asian countries. The data covered 13 countries including China, Bangladesh, Indonesia, Japan, India, Cambodia, South Korea, Myanmar, Philippines, Malaysia, Pakistan, Thailand and Vietnam. The dataset covered the period of 1990–2015 and was analyzed by Stata 15 software. Table 14.1 presents the measurement of variables chosen in the study.

14.3.2 Research Model

Influence of CO2 emissions, population growth and fossil fuel energy consumption have been studied by some previous studies of Volkan et al. (2007), Don (2014), Katsuya (2017), Lanouar and Montassar (2019) and Muhammad et al., (2020). However, it will be improved compared to previous studies; we will analyze not only traditional impacts of CO2 emissions, population growth and fossil fuel energy consumption on REC but also modern impact of private credit by deposit money banks to GDP thanks to theoretical consideration of using POLS, FEM and REM. For visualization, Figure 14.1 shows the model selection in this study. The procedure of the model analysis can be seen in Figure 14.1.

In Figure 14.1, note that POLS, FEM, REM will be applied. To find out this relationship, we propose the following model:

$$\text{REC}_{\text{it}} = \textit{function}\left(\text{CO2}_{\text{it}}, \text{POP\_growth}_{\text{it}}, \text{PC}_{\text{it}}, \text{FOSSIL}_{\text{it}}\right) \tag{14.1}$$

The equation for the FEM is as follows:

$$\text{REC}_{\text{it}} = \alpha_i + \beta_1\text{CO2}_{\text{it}} + \beta_2\text{POP\_growth}_{\text{it}} + \beta_3\text{PC}_{\text{it}} + \beta_4\text{FOSSIL}_{\text{it}} + u_{\text{it}} \tag{14.2}$$

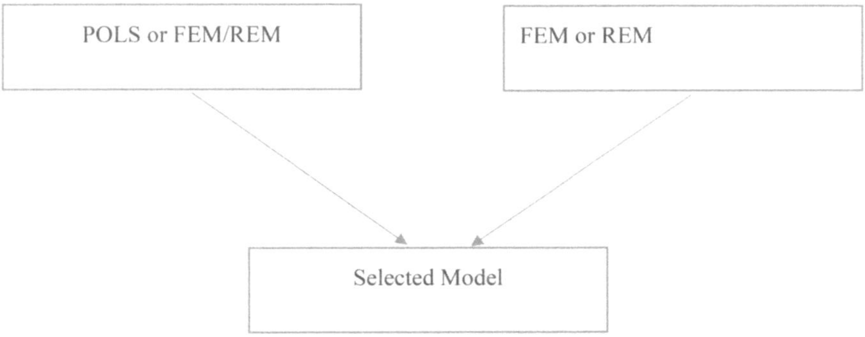

FIGURE 14.1 Model selection.

Source: Authors.

Where:

- α_i ($i = 1....n$) is the estimated intercept for each economy,
- Y_{it} is the dependent variable; i indicates the country and t indicates the time. In this model, Y_{it} is REC_{it},
- X_{it} denotes an independent variable. They include CO2, POP_growth, PC and FOSSIL,
- β_i is the estimated coefficient of independent variables,
- u_{it} is the term of error.

The equation for the REM is as follows:

$$REC_{it} = \alpha_i + \beta_1 CO2_{it} + \beta_2 POP\_growth_{it} + \beta_3 PC_{it} + \beta_4 FOSSIL_{it} + u_{it} + \varepsilon_{it} \quad (14.3)$$

Where:

- α_i (i=1....n) is the estimated intercept for each economy,
- Y_{it} is the dependent variable; i indicates the country and t indicates the time. In this model, Y_{it} is REC_{it},
- X_{it} denotes an independent variable. They include CO2, POP_growth, PC and FOSSIL,
- β_i is the estimated coefficient of independent variables,
- u_{it} is the term of error between entity error,
- ε_{it} is the term of error within entity error.

14.4 RESEARCH DATA AND METHODOLOGY

14.4.1 Statistic Descriptions

For the beginning, data descriptions will supply with an overview on variables being investigated as seen in Tables 14.2 and 14.3.

TABLE 14.2
Statistic Descriptions

Country	Variable	REC	CO2	POP_growth	PC	FOSSIL
Bangladesh	Mean	52.4	0.040	1.63	62.72	24.54
	Min	37.63	0.016	1.12	49.46	12.26
	Max	7.056	0.074	2.17	73.77	37.76
	Std. Dev.	10.06	0.018	0.43	7.31	8.20
	Number of observations	22	22	22	22	22
Cambodia	Mean	75.79	0.0032	1.96	24.72	15.77
	Min	68.00	0.0015	1.48	17.92	3.09
	Max	83.02	0.0067	3.23	32.97	55.40
	Std. Dev.	6.28	0.0016	0.57	5.51	14.69
	Number of observations	20	20	20	20	20
China	Mean	23.32	5.374648	0.79	82.40	99.69
	Min	11.70	2.442919	0.48	74.83	75.39
	Max	34.08	10.300000	1.47	88.90	131.57
	Std. Dev.	8.37	2.729869	0.31	4.90	16.88
	Number of observations	25	25	25	25	25
India	Mean	48.93	1.218570	1.66	64.47	32.29
	Min	36.65	0.619154	1.15	53.76	21.03
	Max	58.65	2.232730	2.08	73.58	49.79
	Std. Dev.	6.89	0.474080	0.28	5.81	10.84
	Number of observations	25	25	25	25	25
Indonesia	Mean	45.20	0.31533	1.44	62.14	28.18
	Min	37.45	0.149565	1.31	53.43	15.85
	Max	58.60	0.529632	1.78	67.15	44.30
	Std. Dev.	6.78	0.102400	0.14	3.66	9.36
	Number of observations	25	25	25	25	25
Japan	Mean	4.26	1.189878	0.12	84.19	134.00
	Min	3.57	1.092396	−0.19	79.40	94.47
	Max	6.30	1.262394	0.38	94.63	192.10
	Std. Dev.	0.63	0.049947	0.17	4.78	38.92
	Number of observations	26	26	26	26	26
South Korea	Mean	1.10	0.450255	0.71	83.70	72.77
	Min	0.44	0.246943	0.21	80.55	43.25
	Max	2.84	0.597354	1.04	87.73	128.49
	Std. Dev.	0.61	0.104435	0.23	2017	24.84
	Number of observations	26	26	26	26	26
Malaysia	Mean	6.62	0.151857	2013	95.03	104.30
	Min	3.82	0.056592	1.34	89.91	62.86
	Max	11.98	0.242671	2082	96.94	144.57
	Std. Dev.	2.44	0.055203	0.43	1.99	17.92
	Number of observations	25	25	25	25	25

(Continued)

TABLE 14.2
Statistic Descriptions (*Continued*)

Country	Variable	REC	CO2	POP_growth	PC	FOSSIL
Myanmar	Mean	82.98	0.009645	1.00	24.67	5.95
	Min	66.13	0.004180	0.62	13.81	2.75
	Max	91.12	0.016347	1.47	44.29	13.40
	Std. Dev.	5.48	0.003358	0.25	6.52	2.54
	Number of observations	25	25	25	25	25
Pakistan	Mean	50.12	0.122057	2.50	58.11	19.53
	Min	44.28	0.068242	2.09	52.31	15.06
	Max	58.09	0.182414	2.96	62.48	27.10
	Std. Dev.	3.90	0.037864	0.30	2.82	3.38
	Number of observations	25	25	25	25	25
Philippines	Mean	35.52	0.070730	2.04	55.22	29.10
	Min	28.58	0.041763	1.65	43.08	14.97
	Max	51.96	0.104516	2.54	62.43	45.66
	Std. Dev.	7.09	0.015569	0.31	4.83	8.14
	Number of observations	25	25	25	25	25
Thailand	Mean	23.31	0.196506	0.82	77.79	103.57
	Min	20.02	0.090721	0.43	63.84	69.78
	Max	33.64	0.287217	1.40	82.06	163.21
	Std. Dev.	3.89	0.057164	0.29	5.30	23.65
	Number of observations	25	25	25	25	25
Vietnam	Mean	51.79	0.080337	1.22	52.25	50.13
	Min	34.80	0.021708	0.92	28.11	13.65
	Max	74.70	0.152169	2.06	70.33	99.98
	Std. Dev.	12.40	0.044490	0.37	13.52	30.96
	Number of observations	22	22	22	22	22

Source: Authors.

TABLE 14.3
Statistic Descriptions for All Countries

Country	Variable	Mean	Min	Max	Standard Deviation	Number of Observations
All Asian Countries	REC	37.49	0.44	91.12	25.77	316
	CO2	0.733678	0.001540	10.300000	1.614475	316
	POP_growth	1.37	–0.19	3.23	0.74	316
	PC	64.51	13.81	96.94	21.37	316
	FOSSIL	56.67	2.75	192.10	45.26	316

Source: Authors.

Tables 14.2 and 14.3 present data descriptions, including 316 observations of 13 Asian economies covering 26 years from 1990 to 2015. In fact, for each country, the study collects some main indicators such as number of observations, mean, standard deviation, minimum and maximum used in the study. Specifically, Table 14.3 shows that REC has remarkably contributed a rate of 37.49% of total EC in Asian economies from 1990 to 2015, with the minimum value of 0.44% and maximum value of 91.12% of REC on energy consumption, respectively. It indicates that in the early stages, a few Asian countries had not consumed REC because of the extremely cheap price of energy from hydrocarbons. In fact, in the context of the world's energy, coal, crude oil and natural gas are truly the fossil fuel and extracted from the ground, therefore this energy comes from primary fuel. In the environmental field, scientists demonstrate that coal, oil and natural gas are composed mainly of hydrocarbons (carbon and hydrogen compounds) with other trace elements. Table 14.3 also shows that REC has slowly improved in recent years. Also, based on the announcement of United Nations (UN), the global population is the total number of humans currently living and was approximated at about 2.6 billion people in 1950, reaching around 5 billion people in 1987 and 6 billion in the year 1999. Further, the global population was considered to be nearly 7 billion in October 2011, and to have reached 7.8 billion people as of March 2020. Regarding global population growth, it reached a peak in the 1960s with a growth rate of 2.2% per year, but since then, it has declined to 1.05% per year in recent years. In the case of Asian countries, Table 14.2 indicates that population growth in this area reached a peak of 3.23% and a minimum value of –0.19% in recent years. It means that population growth has increasingly declined. According to Sherbinin et al. (2007), population is one of several factors that impact the environment; further, quick population growth aggravates other externalities such as wars, governance quality and polluting technologies.

Regarding CO2 consumption in the region, China, India and Japan are the leading CO2 consumers with average values of 5374648 kt, 1218670 kt and 1189878 kt, respectively. Especially, the maximum value of CO2 consumption of China was 10.3×10^6 kt. Reasonably, China and India are the two countries whose populations are the largest. After 2007, polluting industries in China continued to receive resources such as water, land, oil, electricity, banking, while some market-oriented measures, for example, fuel surcharges and coal were not reviewed by Chinese government even though they had been shown to be successful in some other economies. However, the Chinese government's effort in improving the environment situation is helping China to reduce CO2 emissions these days. India is one of those countries which is most damaged by climate change. The main reason for CO2 emission in India is from people's daily life. Kitchen smoke, which includes black carbon, is a leading driver of environmental pollution. The fact is that India continuously depends on burning coal to generate electricity for millions of poor households in the country. An enormous amount of household waste is repeatedly burned overnight. Besides, dust generated from projects of industrial construction spread throughout the cities is also out of control. Increasing production and consumption without controlling emissions treatment and exploitation of natural resources (coal, iron ore…) are also contributing to climate change in India. Japan, one of the major economies in Asia and the world, is extremely lacking in natural resources. As the fifth largest carbon emitter in the world, Japan has set up a target of 26% reduction in greenhouse gas emissions by 2030 compared to 2013 levels, and, Japan ratified the Paris Agreement 2015 to prevent climate change. Japan is the leading country in the region in consuming fossil fuel energy. Japan significantly lacks domestic reserves of fossil fuels except for coal, and has to import large quantities of natural gas, crude oil and other energy sources, including uranium. Population growth is another factor related to the environment issue. Some under-developed Asian countries such as Pakistan, Philippines, Cambodia and Bangladesh are countries with high population growth. In general, with a large population and a rapidly aging rate, Asia is facing many problems with economic growth, labor shortage and difficulties in social welfare. Although Asia has rich resources and biodiversity, rapid population growth leads to a shortage of resources and destruction of existing resources. Many countries with high rates of population growth are low-income countries, which are less able to handle resources and resource pressures. In general, the majority of the working-age population in Asia work in conditions of scarce resources, outdated technology and unsafe environments. Among studied countries, Malaysia has the highest PC at 95.05% on average, followed by Japan (84.19%) and South Korea (83.7%).

14.4.2 Test of Multi-Collinearity

According to Nguyen et al. (2020b), and Gujarati (2004), the existence of multi-collinearity among independent variables could be obviously recognized if their correlation coefficient is at least 0.8. Table 14.4 presents the correlation matrix of related variables. Visually, all correlation coefficients are less than 0.8. It means that there is no multi-collinearity.

Variance inflation factor (VIF) is another test for the multi-collinearity. The results of VIF in Table 14.5 reveals that there does not exist multi-collinearity problem since the VIF of all independent variables is under 10.

TABLE 14.4
Correlation Coefficents between Studied Variables

Items	REC	CO2	POP_growth	PC	FOSSIL
REC	1				
CO2	–0.2822 (0.0000)	1			
POP_growth	0.4097 (0.0000)	–0.3509 (0.0000)	1		
PC	–0.9372 (0.0000)	0.3424 (0.0000)	–0.329 (0.0000)	1	
FOSSIL	–0.7953 (0.0031)	0.3761 (0.0000)	–0.5195 (0.0000)	0.7589 (0.0000)	1

Source: Authors.

TABLE 14.5
VIF Coefficients of Independent Variables

Independent Variable	VIF	1/VIF
CO2	1.23	0.813790
POP_growth	1.45	0.688068
PC	2.43	0.411915
FOSSIL	2.95	0.339343
Mean VIF	2.01	

Source: Authors.

14.4.3 Discussion

Some econometric models such as POLS, FEM and REM have been used. In order to choose POLS or REM/FEM, Breusch and Pagan Lagrange multiplier test is used. The null hypothesis is that the variance for residual is constant. The statistics of Breusch and Pagan Lagrange multiplier test are presented in Table 14.6.

The corresponding probability value of the Breusch and Pagan Lagrange multiplier test is very close to zero. Therefore, the preferred model is REM or FEM at the significance of 5%. In the next step, one should make a choice of FEM or REM. It can be made thanks to Hausman test. Consider the null hypothesis "REM is preferred". Table 14.7 presents the results of Hausman test.

Hausman test shows probability value of 0.0387, which is less than 0.05, hence FEM model is more appropriate to use than REM. In the final step, FEM is the most appropriate in this study, with the independent variable REC.

14.4.4 Results of Economic Modeling

Thanks to the procedure of the model analysis shown in Figure 14.1, estimated models are presented in Table 14.8.

Table 14.8 shows all three models proposed, POLS, REM and FEM. However, results of Hausman test in Table 14.7 indicate that FEM is the most appropriate

TABLE 14.6
Breusch and Pagan Lagrange for Homoscedasticity

Items	Value
Chibar2 (01)	2653.03
Prob > Chibar2	0.0000
H0: Constant variance	

Source: Authors.

TABLE 14.7
Hausman Test Results

Items	Value
Chi2 (3)	10.11
Prob > Chi2	0.0387

Source: Authors.

choice. Notice that the sign * implies significance level of 1%. Therefore, in the following subsections, backtestings for FEM are carried out sequentially.

14.4.4.1 Autocorrelation Test

Wooldridge test is applied for diagnosing autocorrelation in panel data, where the null hypothesis is that there does not exist first-order autocorrelation. Thanks to the result in Table 14.9, the corresponding probability value of the FEM is very close to zero. The null hypothesis is therefore rejected. This indicates that there exists an autocorrelation among variables.

14.4.4.2 Heteroscedasticity Test

Modified Wald test is performed to find out whether heteroscedasticity of the FEM model exists or not. In Table 14.10, the result shows that there is heteroscedasticity problem in the model.

In general, there exist problems of heteroscedasticity and autocorrelation in the FEM model. Robust method is suggested to be carried out to correct them. The final estimation results are shown in Table 14.11. As mentioned earlier in this chapter, the sign * implies significance level of 1%.

TABLE 14.8
Results of Estimation Model

Variables	POLS	REM	FEM
CO2	1.428	–1.191	–1.219
	(0.000)*	(0.000)*	(0.000)*
POP_growth	3.029	2.617	2.712
	(0.000)*	(0.000)*	(0.000)*
PC	–0.980	–0.827	–0.816
	(0.000)*	(0.000)*	(0.000)*
FOSSIL	–0.095	–0.040	–0.0393
	(0.000)*	(0.000)*	(0.000)*
C	100.919	90.628	89.554
	(0.000)*	(0.000)*	(0.000)*

Source: Authors.

TABLE 14.9
Wooldridge Test Results for Autocorrelation

Items	FEM
F(1,12)	60.890
Prob > F	0.000

Source: Authors.

TABLE 14.10
Modified Wald test

Items	FEM
Chi2 (13)	1178.53
Prob > Chi2	0.000

Source: Authors.

14.5 DISCUSSIONS

The discussion in this research paper is presented in line with the results obtained in Table 14.11. FEM is the most appropriate model. This result is also supported by the back tests thanks to the results in Tables 14.9 and 14.10. According to the results, the influence of CO2 emissions, population growth, PC and fossil fuel energy consumption on REC in the case of Asian economies is carefully analyzed.

TABLE 14.11
Results of Correction Econometric Modeling

Variables	FEM*
CO2	–1.219
	(0.000)*
POP_growth	2.712
	(0.000)*
PC	–0.816
	(0.000)*
FOSSIL	–0.0393
	(0.000)*
C	89.554
	(0.000)*

Source: Authors.

14.5.1 CO2 Emissions

The estimated coefficient of this variable is negative and statistically significant. These findings indicate higher CO2 emission will significantly reduce REC in Asian economies.

As mentioned in the microeconomics, oil, gas and RE are substitute goods. As suggested by Tianyu et al. (2014), for China situation in the period between 2010 and 2020, REC reduced the cumulative CO2 emission by 1.8%. Besides, RE is a crucial solution covering all sectors to reduce CO2 at the lowest cost over a long term. Additionally, Muhammad et al. (2020) also found a similar result in the case of South Asian Association for Regional Cooperation countries. The country-level nexus of energy, ISO 14001, agriculture and CO2 emissions were explored. Some other studies have found the influence of REC on CO2 emissions, such as Razmjoo et al. (2021) and Mehdi et al. (2020), while in this study, we found a significant negative impact of CO2 emissions on REC, i.e., in a converse direction.

14.5.2 Financial Development (Proxied by Private Credit by Deposit Money Banks to GDP)

The findings of the study indicate PC has a significant and negative impact on REC. Issuing green bonds is consequently a global direction with the participation of substantial international financial institutions such as International Monetary Fund (IMF), World Bank (WB) and Asian Development Bank (ADB). Green bonds are also regarded as essential capital mobilization to meet the capital needs for sustainable growth. In each country, the capital demand for development of green projects, especially RE projects, is huge. To meet this demand, promoting green financial products, including green bonds, is indispensable. In 2018, there were eight new green bond markets, Iceland, Lebanon, Indonesia, Namibia, Seychelles, Portugal, Thailand and Uruguay, with an issuance value of 3.3 billion USD, and 204 companies were issuing green bonds for the first time with the total issuance value of 61.2 billion USD, accounting for 37% of the total market value (in 2017, there were 161 new issuers with a market share of 35%). In terms of growth, Asia Pacific is the region with the highest growth rate of 35% and the second largest issuance volume after Europe. In the ASEAN region, the total outstanding value of green bonds and sustainable bonds is 1.6 billion USD, of which, construction is the most financed by green bonds (43% by market volume), followed by energy at 32% (see CBI, 2018). The US, China and France are the three biggest green bond issuers all over the world, with 46% of the total international bond issuance value in 2018. The top three green bond issuers in 2018 were Fannie Mae, USA (USD billion 20.1); Industrial Bank of China ($9.6 billion); and French Republic ($6 billion) (see top 5 ranking of green bond issuance in 2018 in Table 14.12).

14.5.3 Population Growth

In theory, the relationship between population growth and the demand for REC can be found. The empirical investigations in this analysis are carried out by identifying the relationship from FEM estimates; results show that regression coefficient is positive

TABLE 14.12
Top 5 Ranking of Green Bond Issuance in 2018

Country	Value (Billion USD)	Market Share (%)	No. of Publishers
USA	34	20	63
China	31	18	69
France	14	8	12
Germany	7.6	5	14
Netherlands	7.4	4	6

Source: CBI, Green bond the state of the market 2018, WDI (2020a, 2020b, 2020c, 2020d, 2020e).

and statistically significant. This finding can be easily explained that the target of economic development should be consistent with the expansion of the shares of renewables. The rapidly growing population has led to increasing demand for electricity. In parallel, there should be a sustainable development strategy, so using RE is indispensable. Khan et al. (2017) further explained that REs, for example, solar and wind, have been increasingly more competitive in cost than fossil fuel energies, especially in the long run. In the long run, countries have to compete with each other by investing in more technology to reduce its adverse impacts: environmental degradation and climate change, global temperature rise and sea-level rise. It is evident to confirm that the regression coefficient of population growth is statistically significant and positive. It can be included that higher growth of population in Asian countries can drive the use of REC. The POP_growth in Asian countries significantly declined in recent years, especially in high and upper-middle income economies. For example, POP_growth of China, South Korea, Japan and Singapore steadily declined from 1.47%, 1.04%, 0.38% and 5.32% in the 1990s to 0.46%, 0.21%, –0.20% and –1.47%, respectively, while REC in these countries gradually decreased from 1990 to the present. It means that a large volume of energy use has been contributed from fossil energy instead of REC. In contrast, the demand for REC in middle-income economies has increased in this period because, in the 1990s, the supply of REC in these countries was somewhat low. The case of Malaysia, it had a large share of ASEAN's fossil fuel resources in the past, led by policies to switch to fossil energy in response to reducing domestic gas production, the proportion of gas in the power mix declined from 67% in 2005 to 47% in 2015 and increasingly declined over time. Further, Malaysia also set a target of 20% REC by 2025. This finding in Asian countries could be supported by studies in the past. As shown by Sherbinin et al. (2007) in a study, population is one of several factors that impact the environment; further, fast population growth aggravates other externalities such as wars, governance quality and polluting technologies.

14.5.4 Fossil Fuel Energy Consumption

Investigating RE is a popular trend worldwide to substitute fossil energy sources which are continuously diminishing and, moreover, to minimize climate change.

In spite of being invested in and developing quite speedily in current years with impressive accomplishments in a number of countries, on the whole, RE is actually very modest in the complete global energy picture. However, RE in the future will dominate and contribute significantly to the global electricity network. Thanks to estimation results, the coefficient of FOSSIL is significantly negative, which coincides with the current trend. Nguyen et al. (2020b) showed that many energy systems, including both developed and developing countries, rely on fossil fuel energy. However, this traditional manner is a disadvantage to green development. According to World Energy Council (2013), European economies have been making changes and RE sources will be a critical factor in the new energy policy. However, Martins et al. (2018) announced that fossil fuels are still crucial in the energy system.

14.6 POTENTIAL AND FUTURE TRENDS OF RENEWABLE ENERGY CONSUMPTION IN ASIAN COUNTRIES

REC has been an inevitable trend of modern times. The use of non-RE resources in the context of a rapidly growing world population and ever-depleting natural resources, leading to rising world oil prices, will lead to habitat destruction and adversely affect world sustainable growth. Xiong et al. (2017) have concluded that use of more biochar as a catalyst and using chemical composite, bio-diesel production from biomass, and pollutant should be considered.

Figure 14.2 describes the trends in REC in Asia between 1990 and 2015. As suggested in WDI, REC is accounted for the proportion of REC to total final energy consumption (EC). Over the years, the proportion of REC to total final EC in most economies has increasingly declined from approximately 34.08% in 1990 to 12.41% in 2015 for China, 58.60% in 1990 to 36.88% in 2015 for Indonesia, 11.98% in 1990 to 5.19% in 2015 for

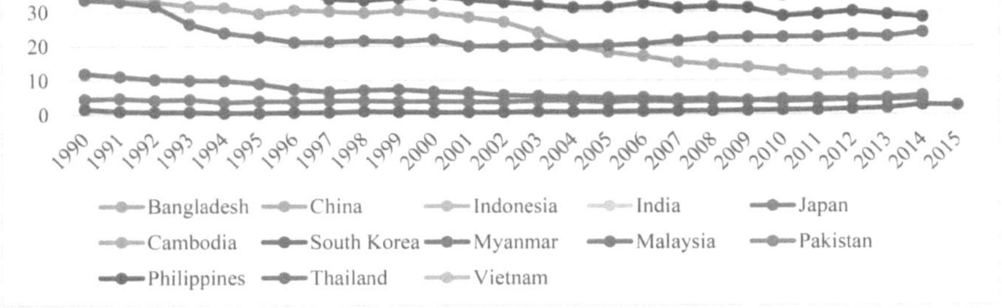

FIGURE 14.2 Trends in REC in Asia from 1990 to 2015.

Source: WDI (2020e).

Malaysia, 51.96% in 1990 to 27.45% in 2015 for the Philippines, 33.63% in 1990 to 22.86% in 2015 for Thailand and 76.08% in 1990 to 34.99% in 2015 for Vietnam.

In the case of South Korea, and Japan with a long winter, the contribution of REC is limited and approximately 0.40% and 1.63% for South Korea and 3.56% and 4.55% for Japan. Singapore, which is a supreme island city-state and a small-scale country with approximately 721.5 km^2 area, has not had many opportunities to invest in RE; therefore, the contribution of REC is around 0.19% and 0.71%.

Climate change mitigation is the main reason for 100% RE targets. However, CO2 reduction benefits for developing RE. In many countries, including Asian ones, decreasing air pollution and reducing health problems due to air pollution is the key driver. Although the total global investment for RE in 2017 was nearly twice as much as fossil fuel, investment for the installation of new RE systems decreased 23% compared to that in the year 2015. In developing and emerging economies, investment in RE decreased by 30% to $116.6 billion, while in developed countries, it decreased by 14% to 125 billion USD. The main reason is due to decrease in the markets of China, Japan and other emerging economies, especially India (mainly due to delays in RE auctions). China still led with the highest investment (32% of total RE financing in the world). In January 2017, the Chinese government announced they would spend $360 billion by 2020, bringing the country to the leading position in the world investing in RE. In Japan, RE has been promoted after the 2011 nuclear disaster in Fukushima. However, the policy change from subsidized electricity tariff financial assistance (FIT) to bidding mechanism led to the decline of nearly 70% of the amount invested in renewable electricity capacity in 2016. Figure 14.3 further suggests that renewables in the global market have predominantly decreased in both 2018 and 2019. Therefore, more investment in renewables worldwide has seriously become a target in order to sustain sustainable development; especially the world's electricity

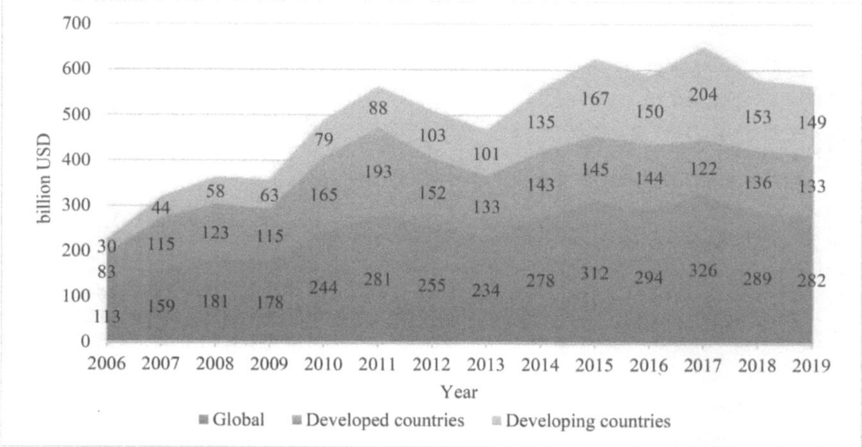

FIGURE 14.3 New investment in global renewable energy in developed, developing and emerging economies 2006–2016.

[*Source:* Bloomberg New Energy Finance (2019), WDI (2020e), IEA (2019), Nguyen (2019).]

consumption needs to be contributed from a greater proportion of renewables. As shown by The International Energy Agency (IEA) (2019), renewables sources still contribute 26% of the world's total electricity and it is likely to reach 30% by 2024. Thereby, it is evident that the continuous efforts of each country, especially the Asian region in developing RE, for sustainable development target are maintained.

14.7 CONCLUSIONS

In the trend of potential transition from polluted and less polluted energy, in particular, use of more RE, CO2 emissions, population growth, PC, fossil fuel energy consumption and REC have been investigated in the case of Asian region covering the period of 1990–2015. The empirical study demonstrates that the proportion of REC to total final EC in most Asian economies has increasingly declined, except in a few economies such as China, South Korea and Japan. Further, changes in CO2 emissions, PC and FOSSIL have statistically significant negative impact on REC. In terms of population growth, it has a significant and positive effect on REC. Additionally, climate change and global warming have predominantly hurt ecology, health and the lives of human beings; the trends of transition from fossil fuels to renewables will occur worldwide.

REFERENCES

Bamati, N. and Raoofi, A. 2020. Development level and the impact of technological factor on renewable energy production. *Renewable Energy*, vol. 151 (May): 946–955. https://doi.org/10.1016/j.renene.2019.11.098

Bloomberg New Energy Finance. 2019. New Energy Outlook (NEO). https://about.bnef.com/new-energy-outlook/ (accessed March 12, 2021).

CBI. 2018. Green bond the state of the market 2018. https://www.climatebonds.net/resources/reports/green-bonds-state-market-2018 (accessed March 20, 2021).

Don, K. 2014. Population growth, energy use, and environmental impact: Comparing the Canadian and Swedish records on CO2 emissions. *Canadian Studies in Population*, vol. 41, no. 1–2 (spring/summer): 120–143. https://www.researchgate.net/publication/287579713_Population_growth_energy_use_and_environmental_impact_Comparing_the_Canadian_and_Swedish_records_on_CO2_emissions

Doytch, N. and Narayan, S. 2016. Does FDI influence renewable energy consumption? An analysis of sectoral FDI impact on renewable and non-renewable industrial energy consumption. *Energy Economics*, vol. 54 (Feb): 291–301. https://doi.org/10.1016/j.eneco.2015.12.010

Gozgor, G., Mahalik, M.K., Demir, E. and Padhan, H. 2020. The impact of economic globalization on renewable energy in the OECD countries. *Energy Policy*, vol. 139 (Feb): 111365. https://www.researchgate.net/publication/339539871_The_impact_of_economic_globalization_on_renewable_energy_in_the_OECD_countries

Gujarati, D. 2004. *Basic Econometrics*. 4th Edition, McGraw-Hill Companies, New York.

Katsuya, I. 2017. CO2 emissions, renewable and non-renewable energy consumption, and economic growth: Evidence from panel data for developing countries. *International Economics*, vol. 151 (10): 1–6. https://ideas.repec.org/a/eee/inteco/v151y2017icp1-6.html

Khan, M.I., Yasmeen, T., Shakoor, A., Khan, N.B. and Muhamad, R. 2017. 2014 oil plunge: Causes and impacts on renewable energy. *Renewable and Sustainable Energy Reviews*, vol. 68, Part 1 (Feb): 609–622. https://doi.org/10.1016/j.rser.2016.10.026

Lanouar, C. and Montassar, K. 2019. Impact of renewable energy consumption and financial development on CO2 emissions and economic growth in the MENA region: A panel vector autoregressive (PVAR) analysis. *Renewable Energy*, vol. 139 (Aug): 198–213. https://doi.org/10.1016/j.renene.2019.01.010

Martins, F., Felgueiras, C. and Smitková, M. 2018. Fossil fuel energy consumption in European countries. *Energy Procedia*, vol. 153 (Oct): 107–111. https://doi.org/10.1016/j.egypro.2018.10.050

Mehdi, B.J., Sahbi, F. and Khaled G. 2020. Renewable energy, CO2 emissions and value added: Empirical evidence from countries with different income levels. *Structural Change and Economic Dynamics*, vol. 53 (Jun): 402–410. https://doi.org/10.1016/j.strueco.2019.12.009

Muhammad, I., Qingyu, Z., Robert, S. and Syed, Z.A.S. 2020. Towards a sustainable environment: The nexus between ISO 14001, renewable energy consumption, access to electricity, agriculture and CO2 emissions in SAARC countries. *Sustainable Production and Consumption*, vol. 22 (Apr): 218–230. https://www.semanticscholar.org/paper/Towards-a-sustainable-environment%3A-The-nexus-ISO-to-Ikram-hang/a300e8332372799bd74916f24c76044a422c247c

Nguyen, H.T. 2019. Trends in investment to renewables nowadays. *Financial and Monetary Review*, vol. 19: 35–39.

Nguyen, T.T., Nguyen, V.C., Tran, T.N. 2020a. Oil price shocks against stock return of oil and gas-related firms in the economic depression: A new evidence from a copula approach. *Cogent Economics & Finance*, vol. 8 (1): 1799908. https://doi.org/10.1080/23322039.2020.1799908

Nguyen, V.C., Thanh, H.P. and Nguyen, T.T. 2020b. Do electricity consumption and economic growth lead to environmental pollution? Empirical evidence from Association of Southeast Asian Nations countries. *International Journal of Energy Economics and Policy*, vol. 10 (5): 297–304. https://www.econjournals.com/index.php/ijeep/article/view/9753/5337

Razmjoo, A., Kaigutha, G.L., Rad, M.A.V., Marzband, M., Davarpanah, A. and Denai, M. 2021. A technical analysis investigating energy sustainability utilizing reliable renewable energy sources to reduce CO2 emissions in a high potential area. *Renewable Energy*, vol. 164 (Feb): 46–57. https://doi.org/10.1016/j.renene.2020.09.042

Reboredo, J.C. 2015. Is there dependence and systemic risk between oil and renewable energy stock prices?. *Energy Economics*, vol. 48 (Mar): 32–45. https://doi.org/10.1016/j.eneco.2014.12.009

Sadorsky, P. 2009. Renewable energy consumption, CO2 emissions and oil prices in the G7 countries. *Energy Economics*, vol. 31, 3 (May): 456–462. https://doi.org/10.1016/j.eneco.2008.12.010

Shah, I.H., Hiles, C. and Morley, B. 2018. How do oil prices, macroeconomic factors and policies affect the market for renewable energy?. *Applied Energy*, vol. 215, 1 (Apr): 87–97. https://doi.org/10.1016/j.apenergy.2018.01.084

Sherbinin, A., Carr, D., Cassels, S. and Jiang, L. 2007. Population and environment. *Annual Review of Environment and Resources*, vol. 32: 345–373. https://www.ncbi.nlm.nih.gov/pmc/articles/PMC2792934/

Sugiawan, Y. and Managi, S. 1990. New evidence of energy-growth nexus from inclusive wealth. *Renewable and Sustainable Energy Reviews*, vol. 103 (Apr): 40–48. https://doi.org/10.1016/j.rser.2018.12.044

The International Energy Agency (IEA). 2019. Renewable 2019 Report. https://www.iea.org/reports/renewables-2019# (accessed February 10, 2021).

Tianyu, Q., Xiliang, Z. and Valerie, J.K. 2014. The energy and CO2 emissions impact of renewable energy development in China. *Energy Policy*, vol. 68 (May): 60–69. https://globalchange.mit.edu/sites/default/files/MITJPSPGC_Reprint_14-9.pdf

Tran, T.N., Nguyen, T.T., Nguyen, V.C. and Vu, T.T.H. 2020. Energy consumption, economic growth and trade balance in East Asia: A panel data approach. *International Journal of Energy Economics and Policy*, vol. 10 (4): 443–449. https://www.econjournals.com/index.php/ijeep/article/view/9401/5184

Volkan, Ş.E., Enes, H., Neşen, S.A. and Hüseyin, T. 2007. Fossil fuel sustainability index: An application of resource management, *Energy Policy*, vol. 35, 5 (May): 2969–2977. https://doi.org/10.1016/j.enpol.2006.10.011

World Development Indicator (WDI) (2020a). CO2 emissions (metric tons per capita). Available at https://data.worldbank.org/indicator/EN.ATM.CO2E.PC

World Development Indicator (WDI) (2020b). Population growth. Available at https://data.worldbank.org/indicator/SP.POP.GROW

World Development Indicator (WDI) (2020c). Private credit by deposit money banks to GDP (%). Available at https://databank.banquemondiale.org/id/c89e17b8?Report_Name=add-data

World Development Indicator (WDI) (2020d). Fossil fuel energy consumption (% of total). Available at https://data.worldbank.org/indicator/EG.USE.COMM.FO.ZS

World Development Indicator (WDI) (2020e). Renewable energy consumption (% of total final energy consumption). Available at https://data.worldbank.org/indicator/EG.FEC.RNEW.ZS

World Energy Council. 2013. *World Energy Scenarios: Composing Energy Futures to 2050.* World Energy Council, London, UK.

Xiong, X., Yu, I.K.M., Cao, L., Tsang, T.C.W., Zhang, S. and Ok, Y.S. 2017. A review of biochar-based catalysts for chemical synthesis, biofuel production, and pollution control. *Bioresource Technology*, vol. 246 (Dec): 254–270. https://doi.org/10.1016/j.biortech.2017.06.163

Zeppini, P. and Bergh, J.C.J.M. 2020. Global competition dynamics of fossil fuels and renewable energy under climate policies and peak oil: A behavioral model. *Energy Policy*, vol. 136 (Jan): 110907. https://doi.org/10.1016/j.enpol.2019.110907

15 A Machine Learning Approach for Translating Weather Information into Actionable Advisory for Farmers

Santosh Kumar Behera and
Darshan Vishwasrao Medhane

CONTENTS

15.1 INTRODUCTION

Agriculture and climate are related to each other to the very depth; factors like rainfall, temperature, humidity, and a lot more affect crops. Humans practice various agricultural methods in various parts of the world depending on climate and terrain.

Machine learning (ML) approaches are used in many fields, ranging from behavioral patterns to geography or weather. ML is also extensively being used in precision agriculture, and many applications of it have already been addressed. Crop prediction, disease detection in plants, fruit quality check, and a lot more are the applications in which ML is used. A large part of our developing country relies to a large extent on agriculture, and many of the farmers remain unaware of recent technological advances in farming. Thus, some organizations advise farmers depending upon the region and weather.

DOI: 10.1201/9781003229018-15

Our purpose is to let computers handle the operation by learning from data generated over the years and make good advisory themselves. We put forward the concept of using the data created over the years by various agricultural and meteorological organizations to help computers learn.

The key contributions of this work are as follows:

- Describes a method for utilization of available unorganized non-uniform data of agricultural advisory by introducing a specific format for the organization of data.
- Explores which architecture of the neural network to be used and the reason for the selection.
- Highlights the possible advancement in the field of agro-meteorology and encourages further work.
- Describes the way to utilize such models for society and agriculture works.

Computational agriculture has not too much to give right now in the field of meteorology, mostly because of the unpredictability of weather and lack of organized open datasets. But there lies a possibility of advancement, which we bring forth in the chapter. While it makes advancement toward agriculture, this chapter also describes the way it can be utilized for the benefit of the community and society.

The chapter is organized into four sections excluding the Introduction. Under the section of Literature Review (15.2), some of the notable works in the field of agro-meteorology and computational agriculture that are relevant to this chapter are described. Next, the Methodology (15.4) section is divided into more subsections over which the work is described: data collection, pre-processing, input cases for recurrent neural network (RNN) and the RNN model with mathematical formulas and definition. Input cases are defined as worst-case and best-case input. In the next section (15.5), the result of the work is briefed, followed by the conclusion of the work and its contribution to agriculture.

15.2 LITERATURE REVIEW

In this section, we are going to cover some of the works related to the field of agricultural meteorology. ML techniques have been long in use in agriculture, and these are used to predict crops, fertilizers, quality of fruit/vegetable, and a lot more. Let's see some of the works and how significant they are to our work.

The first work studied based on crop yield prediction using ML [1] provides the details about researches related to agro-meteorology and how these researches helped. The work in [1] does a systematic literature review of various works done over the years in the field of crop yield prediction. It briefs the features and number of times they are used, like rainfall, temperature, soil type, and a lot more. A variety of algorithms used and their results are also described in a sequential manner, which gives a very good overview.

Helping the Ineloquent Farmers: Finding Experts for Questions with Limited Text in Agricultural QA Communities [2] is also a work worth looking at. It studies the

Farm-Doctor dataset of questions and answers and proposes a solution for limited textual data from farmers. The model learns the limited text questions and suggests the farmer of a probable user who may answer the question. The model is based on a graph neural network and is trained over a Farm-Doctor dataset containing over 690 thousand questions and over 3 million answers.

Analysing Impact of Weather Forecasting through Deep Learning in Agricultural Crop Model Predictions [3] is the work on determining crop generation in season progressively. The significance lies for agricultural producers to make informed crop management and financial decisions.

An IoT Based Smart Agricultural System for Pests Detection [4] proposes a solution for pest detection using neural networks and long-short term memory (LSTM).

15.3 MATHEMATICAL BACKGROUND

Let's look at the concept of LSTM and why we will be using this concept. An LSTM cell has the following basic components – forget gate, input gate, output gate, and cell state.

Forget gate decides what information to discard or keep from the previous step. It uses one sigmoid layer.

$$f_t = \sigma(W_f \cdot [h_{t-1}, x_t] + b_f)$$

Equation 15.1: Forget Layer

Next, the input gate contains two layers to decide which new information is going to be stored in the cell. One sigmoid cell to decide which values to update and another tanh layer to create a vector of new data.

$$\begin{aligned} i_t &= \sigma(W_i \cdot [h_{t-1}, x_t] + b_i) \\ \tilde{C}_t &= \tanh(W_C \cdot [h_{t-1}, x_t] + b_C) \end{aligned}$$

Equation 15.2: Input Gate Layer

The cell state layer acts as the memory of the LSTM.

$$C_t = f_t * C_{t-1} + i_t * \tilde{C}_t$$

Equation 15.3: Cell State Layer

In the output layer, the cell state obtained is passed through a hyperbolic function (tanh) to filter the cell state values between –1 and 1.

$$\begin{aligned} o_t &= \sigma(W_o[h_{t-1}, x_t] + b_o) \\ h_t &= o_t * \tanh(C_t) \end{aligned}$$

Equation 15.4: Output layer equation

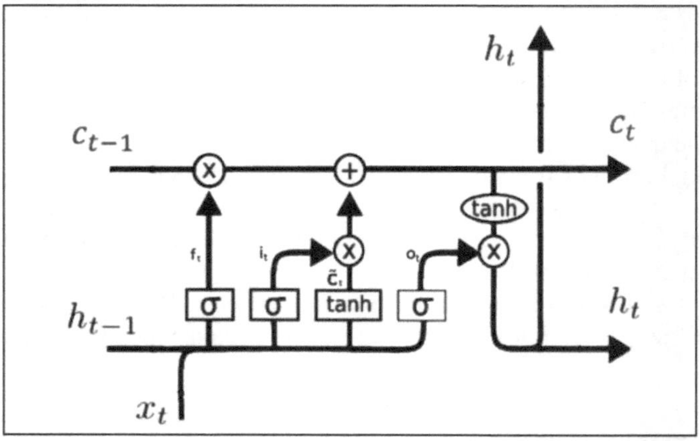

FIGURE 15.1 Representation of a single LSTM cell.

Figure 15.1 shows a single LSTM cell. Now, we will see the use of LSTM architecture in our model. RNN is different than traditional neural networks because it allows information to persist within the network. But RNN having a single layer of tanh in repeating modules also poses a problem of long-term dependencies. LSTM is a special RNN architecture that eliminates the problem of long-term dependency and is better at persisting information. This will help our model to keep information based on previous results or advice as we loop through different months.

15.4 METHODOLOGY

In this section, we will discuss the proposed methodology of work, which will include subsections to go through data collection, data processing, and probable model.

15.4.1 Data Collection

Agrimet has various sub-centers in every state ranging from 1 to 10 in each state. These centers provide local weather forecasts, soil data, agricultural observation data, and advisory for local farmers. These centers help give advisory to the lowest block levels for the region, which can be found on imdagrimet.gov.in and also on the Meghdoot app for Android.

This advisory data, containing forecasts for weather and prescribed advice, is very significant to this work. Below is a given sample of data from their website.

Tables 15.1 and 15.2 show the forecast data provided on each report of Agrimet for each district. For a total of 5 days, the local weather forecast is given, which consists of the following features – rainfall, maximum temperature, minimum temperature, cloud cover, wind speed, wind direction, and maximum and minimum relative humidity. As these reports are provided by the local bodies, they have a varied style of forecast table.

TABLE 15.1
Weather Forecast for the Upcoming Days, Given at the Top of the Advisory Report [5]

Weather Parameters	Weather Forecast				
Date	21	22	23	24	25
Rainfall (mm)	0	0	0	0	0
Max. Temp.	37	38	39	40	40
Min. Temp.	21	21	22	23	23
Cloud Cover	2	5	7	7	3
Max. RH (%)	46	39	45	44	46
Min. RH (%)	28	30	35	36	35
Wind Speed (km/hr)	13	12	10	9	8
Wind direction (deg)	284	291	291	280	285

We can notice that there is a variation in how the data is presented, by every other local body. Note that the color or dimensions of the table don't affect our data collection. If we look at the following features, there are variations:

- Rainfall – The units in Table 15.1 are written once, while in Table 15.2 units are written with each number. This makes the variation that the rainfall record will be treated as an integer in Table 15.1 but as a string in Table 15.2.
- Cloud Cover – The records are written as integers on the scale of 10 in Table 15.1, while in Table 15.2, it is written as human interpreted words.
- Wind Direction – The unit of wind direction is degrees in Table 15.1 and given as NSWE in Table 15.2.

TABLE 15.2
Weather Forecast for the Upcoming Days, Given at the Top of the Advisory Report [6]

Date	24-Feb	25-Feb	26-Feb	27-Feb	28-Feb
Weather Indices					
Rainfall	0 mm	0 mm	0 mm	0 mm	0 mm
Maximum Temperature	33	34	37	36	36
Minimum Temperature	15	16	16	18	18
Cloud Cover	Intermittent C louds	Sunny	Sunny	Sunny	Sunny
Maximum Relative Humidity	33	29	26	27	32
Minimum Relative Humidity	14	12	9	10	11
Wind Speed (km/hr)*	2	4	4	3	5
Wind Direction	WN-ly	W-ly	WN-ly	W-ly	SW-ly

General Advisory: Recommends to adopted as prophylactic measures to protect against anticipated upside-down in temperature, dry weather and deficiet rainfall as follows :. Use pesticides and fungicides - Vegetables-Spray 2% Urea after light irrigation to protect from cold wind. Provide straw mulch and go for intercultural operations to avoid moisture losses from soil. For leaf webbers in cabbage, spray NSKE solution @ 5 per cent. For the management of purple blotch in onion crop, spray Mancozeb @ 2.5g or Ridomil-MZ @1g /lt water. For the management of mite attack in chili, apply carbofuran 3G @ 1 g/10 lt water. Due to cloudy weather condition, there is a chance of leaf blight disease in potato crop, spray Mancozeb 75% WP @600 g/acre or Carbendazim 12% + Mancozeb 63% WP @ 400g/acre. Pulses- Present weather condition favours for Collar rot and root rot disease in chickpea crop. It is advised to rogue out the affected plants and drench the affected patches of the field with Thiram or Captan @ 2g/lt water. For the management of pod borer in chickpea crop, spray Emamectin benzoate @ 1 g per 3 lt water. In case of powdery mildew incidence in pulses, spray NSKE 5 per cent or Neem oil 3 af curl disease in papaya, it is advised to uproot and destroy the infected papaya plant.

FIGURE 15.2 Example of raw advisory from IMD Agrimet [7].

There can be more-less variation depending upon the state or local body on the forecast. Now, we will see another important part, that is, advice given by local authorities which will later be our labels. This advisory data contains advice for a variety of regional and time-specific crops as well as livestock and poultry. Figure 15.2 is an example of a raw advisory for the report.

As raw advisory is not much useful, we need to filter out specific advice from the advisory, like the ones highlighted.

The data collection can be done in various ways; some are given below.

- Manual Data Collection – Browse through PDF reports of many districts for every month and collect data one by one.
- Semi-Automated Collection – Collect all the PDF reports for a district, and feed them to any PDF to Excel converter, this will extract all the forecast tables and advice into separate sheets which can be easily converted into proper tables using Excel macros.
- Fully Automated Collection – This will require a whole different ML model to be made which will retrieve data from PDFs and filter the useful data only.

15.4.2 Data Pre-Processing

Before being able to feed data into models or for use in our programs, we need to filter and process the data we have gathered. The columns that are present in our data are - state, district, month, rainfall, maximum temperature, minimum temperature, cloud cover, wind speed, wind direction, and maximum and minimum relative humidity.

Let us first have a look at the numerical and categorical data we have collected. Categorical data includes state, district, and month, and numerical data includes weather forecast parameters (Figure 15.3).

Categorical data have a definite set of unique values, and string data may get spelling mistakes when having actual predictions or at the time of result testing. We can

Categorical Data

state	district	month
Orissa	Kendrapara	2
Jharkhand	Saraikela Kharsawan	1
...	...	...

String data type needed to be converted into encoded numbers

String data type needed to be converted into encoded numbers

Encoded numerical data, no processing required

FIGURE 15.3 Categorical data example [7].

tokenize the categorical data like state and district. Note, that the month field is already tokenized. Next, we look at an example for numerical data (Figure 15.4) which consists of rainfall, temperature and more.

Numerical features require some mathematical processing for us to be able to use them. Inputs may have a different unit, which means the variables have different scales, which may increase the difficulty of the problem being modeled. So, it is better to normalize them to a scale of 0–1. Normalization is a rescaling of the data from the original range so that all values are within the range of 0–1. Numerical data can be normalized using the following formula:

$$y = \frac{x - \min(\text{col})}{\max(\text{col}) - \min(\text{col})}$$

Equation 15.5: Normalization equation

Where, x in actual value, col is the column to which it belongs and y is the normalized value.

Now coming to our labels, which also need processing before they can be used. Figure 15.5 shows an example of how our labels look like after the data collection process.

These pieces of text are of unequal length and the computer does not understand your words. So, we need to encode or vectorize the text and also pad them so

Numerical Data

rainfall	max_temp	min_temp	max_rh	min_rh	wind_speed
0	34.6	19.4	88.2	29.6	12
0	32.8	18	62.6	28.4	10.2
...	...	...	...	...	...

FIGURE 15.4 Numerical data example [7].

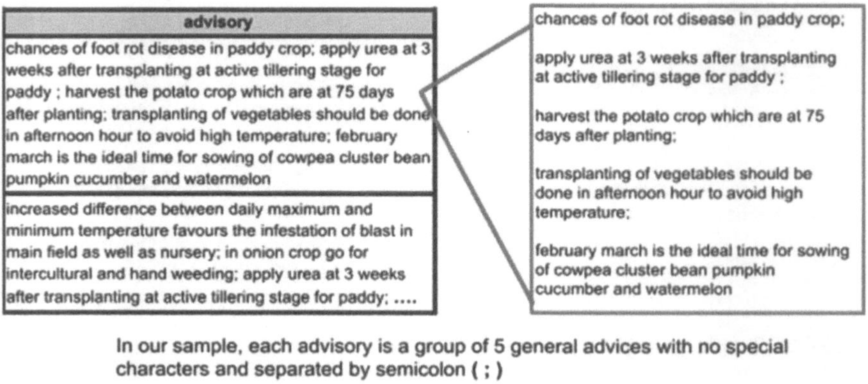

FIGURE 15.5 Depiction of arrangement from raw data.

computers understand the sequences of text. Here we are going to use Keras inbuilt text pre-processing Tokenizer. The padding of the sequence should not be too long, so we check the maximum length of single advice and pad it to the nearest greater multiple of 10 (Figure 15.6).

Note that in the case of short sentences, there are too many zeros but that does not matter too much.

With this, our data pre-processing phase is complete and we can use the data further.

15.4.3 RNN Model

Now that data has been gathered and pre-processed, we can use this data to train a model.

RNN and CNN are the two pivotal deep neural network architectures used. CNNs employ filters within the convolutional layers to transform data. RNN tends to reuse

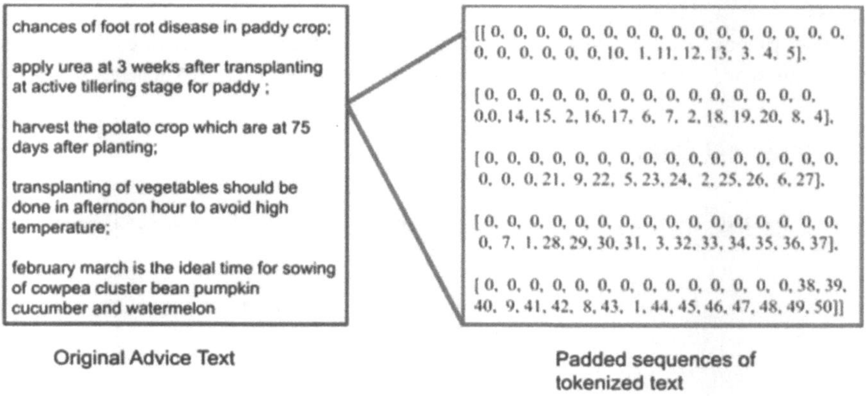

FIGURE 15.6 Filtered data on being converted into a padded sequence.

activation functions from other points in the sequence to generate the next output. Here, we are going to use RNN as a result of previous data point matters to properly generate the next output.

15.4.3.1 Input Cases

Here we will discuss the best-case inputs, average-case inputs, and worst-case inputs. Given some inputs, whether they are best-case or average-case or worst-case depends on several factors: quantity of specific category of input, label's uniqueness, the difference between the values of the same month.

15.4.3.1.1 Best-Case Input

The best-case inputs (Figure 15.7) will be having the following properties:

- Specific states and districts have at least one data for each month.
- The variety of words, in each advice out of 5, is more.
- Less duplication of words in individual advice.
- Less number of repeated advice for a district for every unique month.

The labels column or the advice also needs to have a few conditions to act as a best-case input. Figure 15.8 is an example of the same.

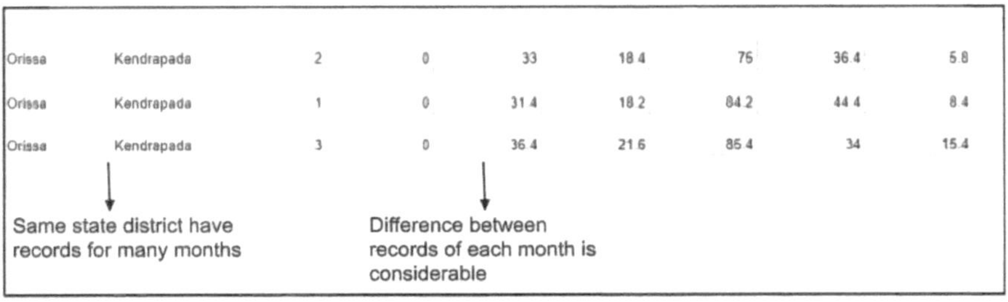

Orissa	Kendrapada	2	0	33	18.4	76	36.4	5.8
Orissa	Kendrapada	1	0	31.4	18.2	84.2	44.4	8.4
Orissa	Kendrapada	3	0	36.4	21.6	85.4	34	15.4

FIGURE 15.7 Example of best-case input [7].

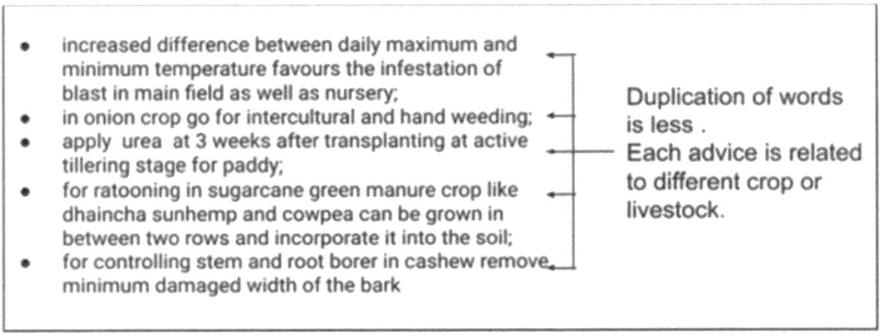

FIGURE 15.8 Example of labels in a best-case input [7].

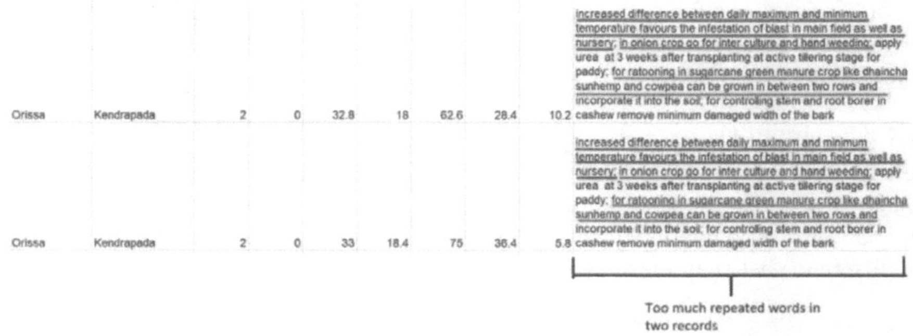

Orissa	Kendrapada	2	0	32.8	18	62.6	28.4	10.2	increased difference between daily maximum and minimum temperature favours the infestation of blast in main field as well as nursery; in onion crop go for inter culture and hand weeding; apply urea at 3 weeks after transplanting at active tillering stage for paddy; for ratooning in sugarcane green manure crop like dhaincha sunhemp and cowpea can be grown in between two rows and incorporate it into the soil; for controlling stem and root borer in cashew remove minimum damaged width of the bark
Orissa	Kendrapada	2	0	33	18.4	75	36.4	5.8	increased difference between daily maximum and minimum temperature favours the infestation of blast in main field as well as nursery; in onion crop go for inter culture and hand weeding; apply urea at 3 weeks after transplanting at active tillering stage for paddy; for ratooning in sugarcane green manure crop like dhaincha sunhemp and cowpea can be grown in between two rows and incorporate it into the soil; for controlling stem and root borer in cashew remove minimum damaged width of the bark

FIGURE 15.9 Example of worst-case input [7].

15.4.3.1.2 Worst-Case Input

The worst-case inputs (Figure 15.9) will be having the following properties:

- The variety of words, in each advice out of 5, is not too much.
- Duplication of words in each advice.
- Repeated advice for a district for every unique month.

Figure 15.10 is the representation of a single label in a worst-case input.

15.4.3.2 Model

We will use the LSTM architecture of RNN, which can be used to remember the sequence of words in advice. This will help to generate close to human-understandable advice. A recurrent neural network built using an LSTM architecture can be given supervised training on a set of text sequences of advice. Also, the features are trained on Dense layers, which are the most basic neural layers. Training a complex Comma Separated Value (CSV) with textual output needs some workaround. The textual data

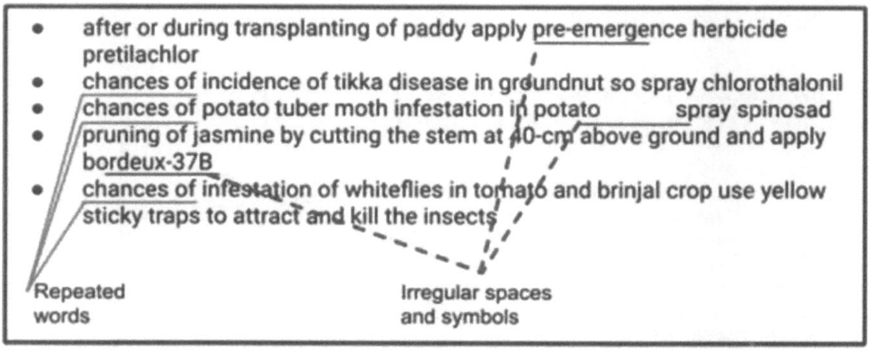

FIGURE 15.10 Example of labels in worst-case input.

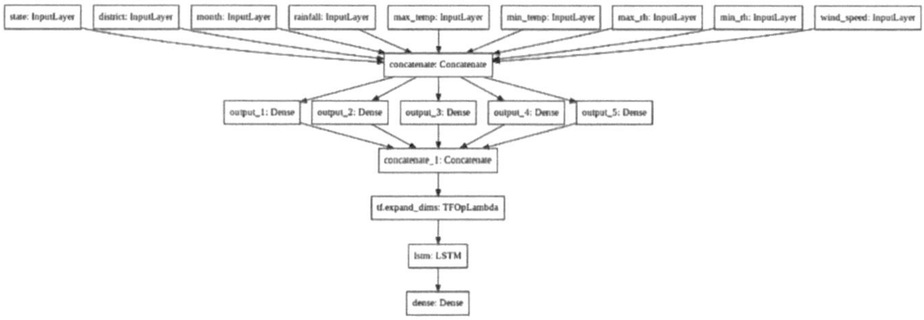

FIGURE 15.11 A proposed model representation.

is needed to be fitted through LSTM, while the input features are processed through Dense layers.

Figure 15.11 depicts an RNN model built with LSTM architecture. It achieves only a maximum accuracy of 56%.

Input features are fed through various input neurons and output is fed through dense neurons.

There are a total of nine features and one output feature of shape [5, 30]. The label's shape is converted [1, 30] and fed to five neurons, instead of feeding [5, 30] to a single neuron.

The output obtained in the shape of [150] is then divided into five to give a shape of [5, 30] output. These [5, 30] output contains many zeros and may contain float point if activation is not set softmax. Our output didn't contain any float as softmax is applied.

15.5 RESULTS AND CONCLUSION

All experimental studies were conducted in the Google Colab environment with the allocated memory. All the codes were written with the Keras framework. Keras is an open-source deep neural network library written in Python language.

The RNN model we have built can generate an output in the shape of [150]. The accuracy of the output suffers due to the shortage of data available to train. Figure 15.12 shows the performance, accuracy, and loss graph after 500 epochs.

We tested on manually created data to see if it matches real advice. Among the points generated, a few of them are near human-understandable. Here, we have separated the [150] shaped output to [5, 30] shaped sequence. All the zeros/no-words are removed from the list to make them look clear, and in each of the five lists, the words are concatenated to form sequences (Figure 15.13).

From the data collected manually, we notice that there are very minute changes in data over adjacent months. The RNN model learns from the data and the results thus produced are not very refined. The sentences produced contain random meaning and only align a little to what the actual result should have been.

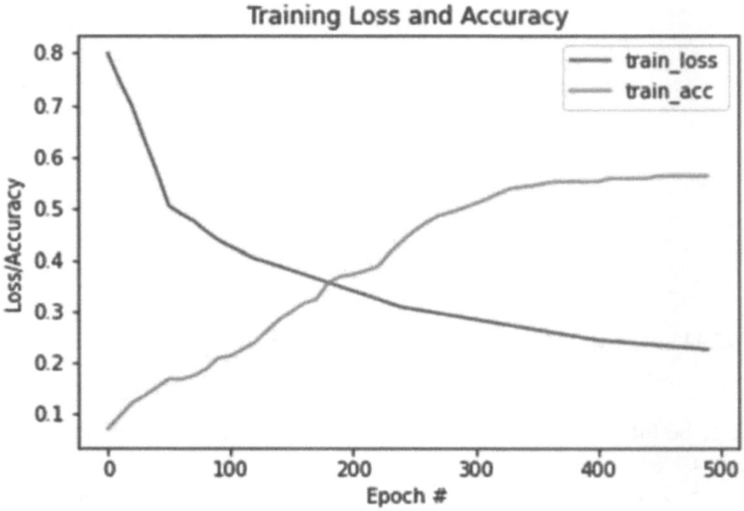

FIGURE 15.12 Graph for loss and accuracy.

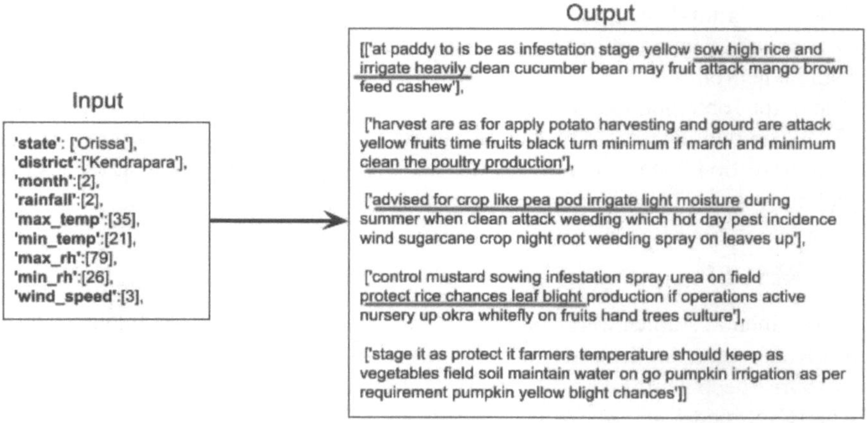

FIGURE 15.13 Actual input and output tested.

15.6 FUTURE SCOPE

In contrast to previous works on agro-meteorology, work related to meteorological advisory generation is rarely found. This newer field and newer approach may give rise to many possibilities from the method proposed in this work. Firstly, the amount of data was limited during the proposed methodology; future works could include large amounts of data that will help computers to give better advice. The accuracy of learning and generating text can be improved. The model generated through such a large amount of data can be fed into android applications that could generate advice without requiring to connect to the internet. The application would need the internet for getting the weather forecast.

REFERENCES

1. Van Klompenburg, Thomas, Ayalew Kassahun, and Cagatay Catal. "Crop Yield Prediction Using Machine Learning: A Systematic Literature Review." Computers and Electronics in Agriculture 177 (2020): 105709.
2. Shen, Xiaoxue, Adele Lu Jia, Siqi Shen, and Yong Dou. "Helping the Ineloquent Farmers: Finding Experts for Questions with Limited Text in Agricultural Q&A Communities." IEEE Access 8 (2020): 62238–62247.
3. L., Naveen and Mohan H.S., "Analyzing Impact of Weather Forecasting Through Deep Learning in Agricultural Crop Model Predictions." International Journal of Applied Engineering Research 14, 23 (2019): 4379–4386.
4. Chen, Ching-Ju, Ya-Yu Huang, Yuan-Shuo Li, Chuan-Yu Chang, and Yueh-Min Huang. "An AIoT Based Smart Agricultural System for Pests Detection." IEEE Access 8 (2020): 180750–180761.
5. "IMD Agrimet", State-Maharashtra, District-Pune, 21-02-2021, https://www.imdagrimet.gov.in/AGDistrictBulletin.php
6. "IMD Agrimet", State-Jharkhand, District-Saraikela Kharsawan.
7. "IMD Agrimet", State-Orissa, District- Kendarapada.

16 Modeling of Fast Charging Electric Vehicles Using Different Controllers

Priyaratnam, Anjali Jain, and Neelam Verma

CONTENTS

16.1 INTRODUCTION

In today's era, circumstances in the developing countries and even developed countries have led to air pollution becoming a big threat to the Earth and its environment. The rate of air pollution is increasing substantially. This rapid increase in pollution is harming the life of humans. Air pollution is causing health issues such as redness in the eyes, headache, and respiratory infection, may lead to respiratory disease, heart disease, and lung cancer, and can even damage the nervous system. The major reason for the increase in air pollution to this extent is the growth in the use of conventional vehicles. We can overcome this situation by switching to electric vehicles (EVs) from conventional vehicles. Even the government is motivating people to use EVs by providing subsidies on the exchange of conventional vehicles for EVs. Switching to EVs

DOI: 10.1201/9781003229018-16

causes a problem of charging these EVs, and to tackle this problem, the government is installing fast-charging stations (CS) for EVs [1]. Nowadays, many research projects are going on to reduce the charging time of EVs. Here, we use a solar photovoltaic (SPV) array [2] as the main source for charging the EV.

In this chapter, two kinds of battery chargers are presented: uni-directional and bi-directional. In a uni-directional battery charger, the power will flow from the source to the battery, and besides circuital simplicity, a uni-directional battery charger reduces source integration and lowers battery degradation. In bi-directional battery chargers, power flows in both directions, and they can perform the ancillary operation.

Authors in [3–7] discussed the modeling of a fast-charging EV using proportional-integral-derivative (PID), fuzzy logic controller (FLC), and maximum power point tracking (MPPT) techniques, and in this chapter, SPV array is used as a charging source. Authors have proposed different kinds of controllers to control SPV array.

Prior to the discussion on the working and modeling of EV charging platform, the following is the juxtaposition between DC fast charging and AC general charging systems [8].

- **AC Charging**

Generally, CS that we use in the house have AC power. Two major types of CS are:

Level 1: Charging the vehicle from a household outlet, and it takes approximately 22 hours to charge the vehicle's battery.
Level 2: In this type, the charging time of the EV's battery is less as compared to Level 1. This two-level CS is known as a generic CS.

- **DC Fast Charging**

There are different kinds of fast CS, namely SAE (The Society for Automotive Engineers) combo and CHAdeMO chargers, which can give a car an 80% charge in just 15–30 minutes.

We classified the different EV charging modes based on maximum charging power and maximum current rating as shown in, Table 16.1:

TABLE 16.1
Different EV Charging Modes [9]

Modes	Maximum Current with Different Phases	Maximum Power
Mode I	12 A-16 A, AC, 1-Phase	3.7 kW
Mode II	32 A, AC, 3-Phase	22 kW
Mode III	63 A, AC, 3-Phase	44 kW
Mode IV	400 A, DC	Appx. 200 kW

In this chapter, MATLAB simulation and modeling of DC fast-charging EVs are proposed and formulated with three different controllers. The authors describe the whole model and MATLAB simulation of the DC fast CS in the following sections. Section 16.2 discusses all the components that are used in DC fast CS for EV. Section 16.3 compares the different controllers that we use to control the system. Section 16.4 provides the design of the MATLAB model and the results, and Section 16.5 concludes the work.

16.2 COMPONENTS OF DC FAST CHARGING STATION

The essential components of EV fast CS are SPV array as a source, battery charger, boost converter, and different controllers [3, 10].

Figure 16.1 shows the flowchart of the EV DC fast CS. In this, we essentially consider the source of power like solar, which converts solar energy (SE) to electric energy. It can control by different kinds of controllers then it controls the boost converter then it goes to the storage of energy system then storage energy charges the vehicle.

16.2.1 Solar Photovoltaic Array

SPV converts SE to electrical energy. When the solar radiation excites the diode junction, then the PN junction generates electricity. This value of generation mainly relies on the value of irradiation and temperature; as the time changes the irradiation changes with constant temperature [4, 11].

The following equation is according to the electrical equivalent of the SPV array model in Figure 16.2 [12]

$$V_{\text{cell}} = \frac{AkT_C}{e}\left(\ln\frac{I_{ph}+I_{D1}+I_{D2}-I_C}{I_{D1}+I_{D2}}\right)-I_C\left(\frac{R_s\times R_p}{R_s+R_p}\right) \tag{16.1}$$

V_{cell}: output voltage of SPV cell

In this chapter, we consider the SPV with variable irradiation with a constant temperature.

$$C_{TV} = 1+\beta(T_a-T_x) \tag{16.2}$$

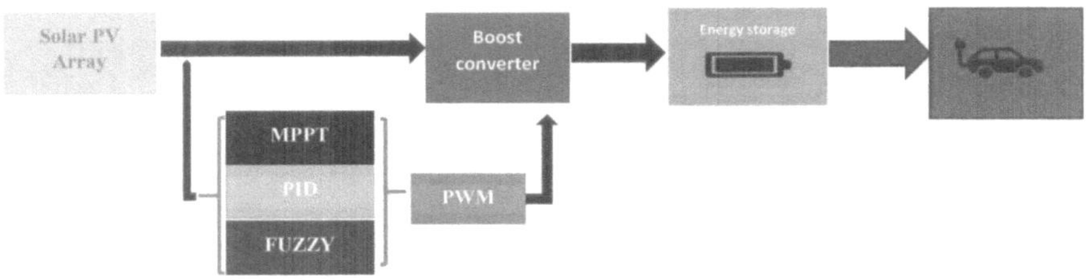

FIGURE 16.1 Flowchart of EV DC fast CS.

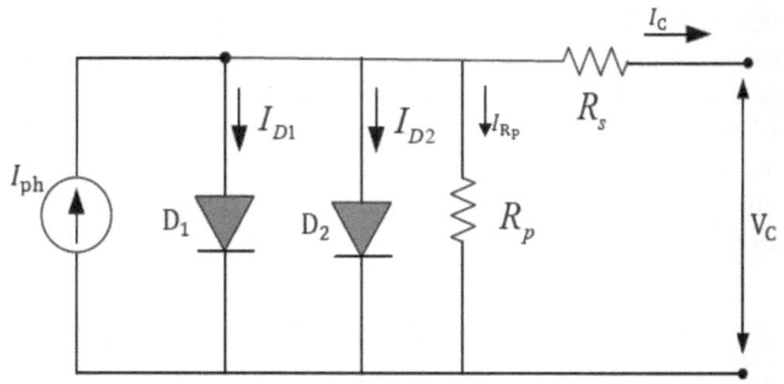

FIGURE 16.2 The electrical equivalent of solar photovoltaic.

$$C_{TI} = 1 + \frac{\gamma T}{Sc}(T_x - T_a) \tag{16.3}$$

Where
C_{TV}: cell voltage
C_{TI}: cell photocurrent

As the voltage is affected by the change in the value of irradiation level (S_x) and PV system current can be denoted by the constants C_{SV} and C_{SI}, these two are the improvement or correction factors for changes in cell voltage and photocurrent I_{ph}. These are represented in Equations (16.4) and (16.5) as,

$$C_{SV} = 1 + \beta_T \alpha_S (S_x - S_c) \tag{16.4}$$

$$C_{SI} = 1 + \frac{1}{S_{cT}}(S_x - S_c) \tag{16.5}$$

S_c = reference solar irradiation

16.2.2 Battery Charger

For an EV CS, the charger should have a bi-directional DC to DC converter with the assistance of buck-boost operation with V2G capability [13]. It serves as a link between solar PV and EV battery systems.. The configuration of the battery charger is shown in Figure 16.3. It carries two insulated gate bipolar transistor (IGBT) switches; these are operated by the control signal.

16.2.2.1 Buck Mode

When we consider S_{buck} (upper switch), then the converter behaves as the buck converter and the input voltage (V_{in}) steps down to battery charging voltage. When we consider the switch is ON, the current is flowing through the inductor and switch to the batteries. In this charging operation, the power is flowing from the G2V. When it is in an OFF state, the current changes its path. It starts flowing in the reverse direction through the inductor and diode of the lower switch and completes the circuit [14].

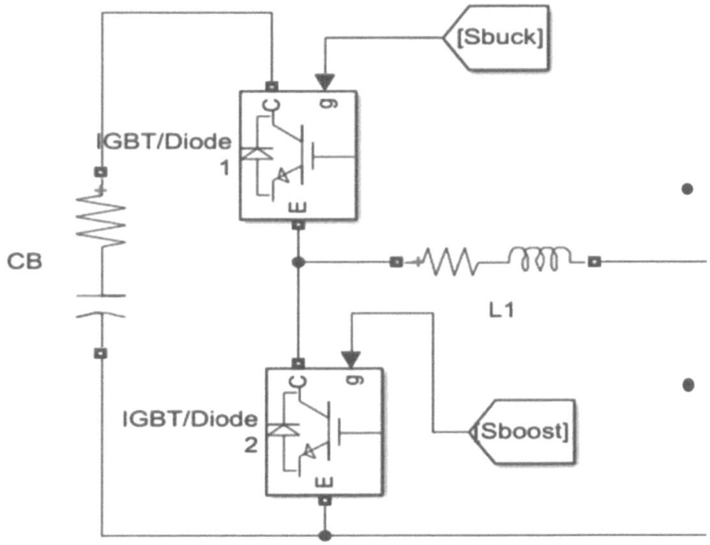

FIGURE 16.3 Battery charger

16.2.2.2 Boost Mode

When the battery voltage is increased to the Direct Current B_{Bus} voltage, the S_{boost} considers the converter to be a boost converter. Similar to the buck converter, we consider both ON and OFF states of boost converter; the current continuously flows via the inductor and completes its whole circuit through the anti-parallel diode of the upper switch and also the capacitor. The batteries operate on the discharge mode and the power starts flowing; this case is known as V2G [14].

16.2.3 Battery Model

It has two types of operations: charging type and discharging type. When the circuit of the battery is positive, it is charging mode, and when the current entering the battery is negative, it is discharging mode. There are many kinds of battery models as shown in Figure 16.4 but for this chapter, we use a lithium-ion battery. The battery model has

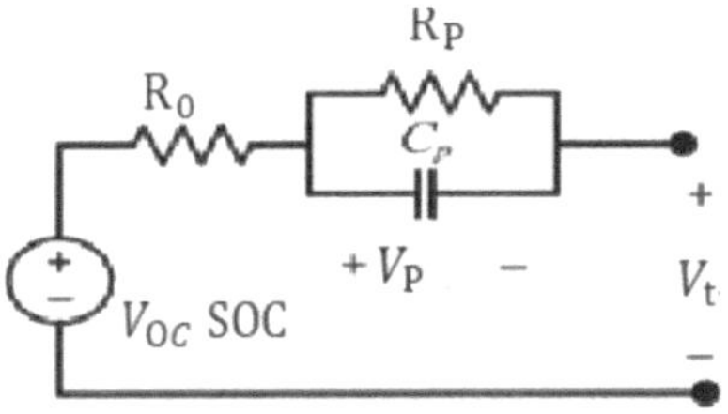

FIGURE 16.4 Battery equivalent circuit [17].

TABLE 16.2
Battery Specification

Nominal voltage	24 V
Rated capacity	50 Ah
State of charging	45%
Battery type	Lithium-ion

different parameters (Table 16.2) like state of charge (SOC), the number of 2-V cells in series, the health of the battery, massic energy, power to weight ratio, energy density, charge/discharge cycles of battery, environmental factors, etc. [15, 16].

16.3 CONTROLLERS OF SYSTEM

16.3.1 Battery Controller

In this chapter, we consider two controllers for EV battery charging, i.e., 'current constant' and 'voltage constant'. First, we use the constant current [Figure 16.5(a)] strategy and then switch to voltage strategy [Figure 16.5(b)] to charge the EV battery.

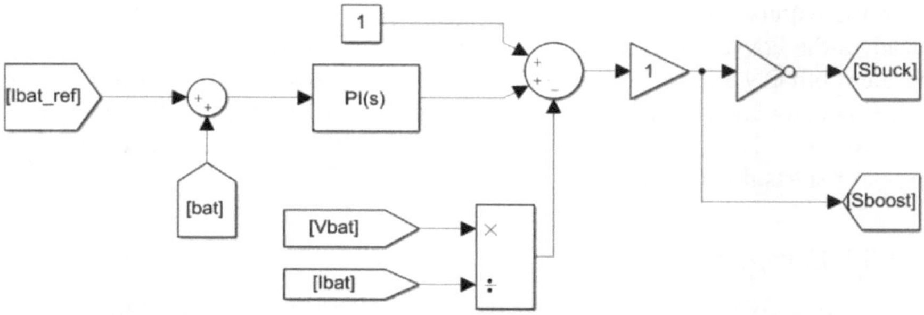

FIGURE 16.5 (a) Constant current.

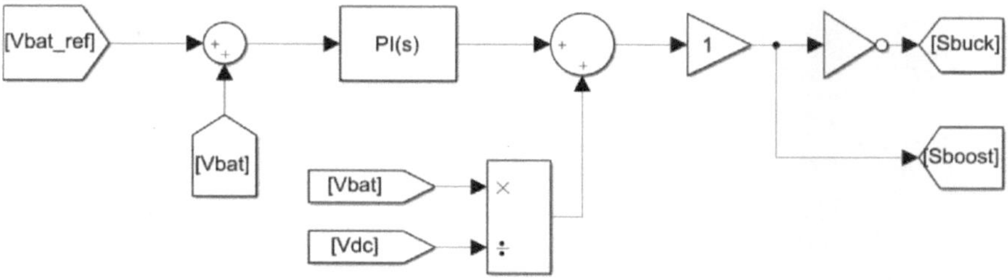

FIGURE 16.5 (b) Constant voltage.

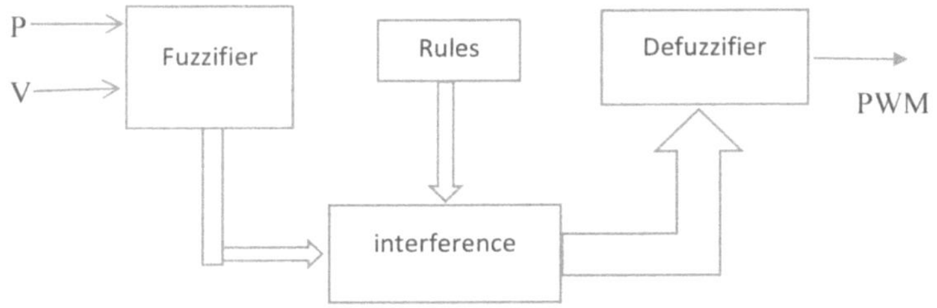

FIGURE 16.6 Block diagram of fuzzy logic controller.

16.3.2 Fuzzy Logic Controller

A Fuzzy logic controller (FLC) was investigating the controller's design during the assessment. FLCs, as we know, are model-free and have an input-output or black box. In Figure 16.6, we use this controller to control the boost converter.

- Fuzzy rule base; where fuzzy sets and fuzzy logic (FL) are considered as tools for representing a different form of knowledge [18].
- Fuzzification; converting a crisp input value to a fuzzy value, i.e., performed by the information in the knowledge base [19].
- Inferences; it's the procedure of devising the design from a given input value to an output using fuzzy logic [20].
- Defuzzification; process of producing a specific result in crisp logic [21]

To introduce FLC, it comprises one output with two input variables. The inputs that we considered in the fuzzy controller are power (P) and voltage (V) and the output is pulse width modulation (PWM) [22]. To make it simple, we considered triangular membership function (Mf) for the fuzzification. The Mf of the fuzzy input-output is shown in Figures 16.7(a), 16.7(b), 16.7(c). Table 16.3 shows the rule base used in FLCs.

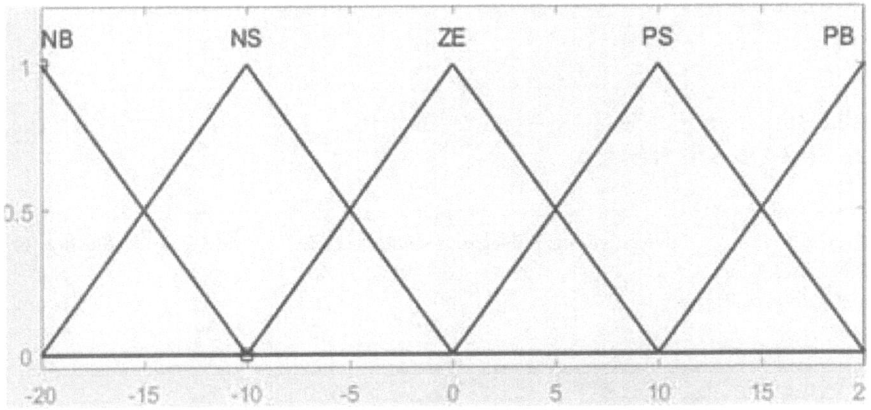

FIGURE 16.7 (a) Mfs of the input voltage.

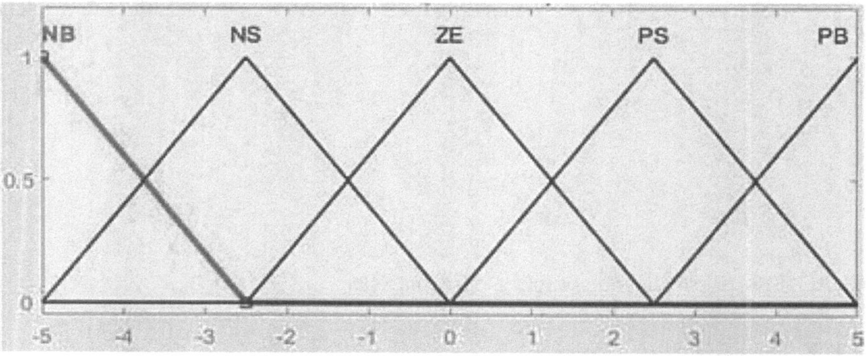

FIGURE 16.7 (b) Mfs of the input power.

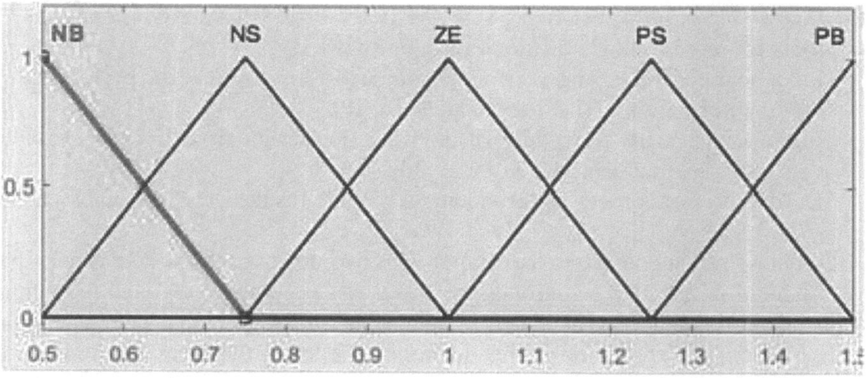

FIGURE 16.7 (c) Mfs of the output (PWM)

TABLE 16.3
Rule Base Used in FLCs

Power (P) / Voltage (V)	NB (Negative Big)	NS (Negative small)	ZE (Zero)	PS (Positive small)	PB (Positive big)
NB (Negative Big)	PB	PS	ZE	NS	NS
NS (Negative small)	PS	PS	ZE	NS	NS
ZE (Zero)	NS	NS	NS	PB	PB
PS (Positive small)	NS	PB	PS	NB	PB
PB (Positive Big)	NB	NB	PB	PS	PB

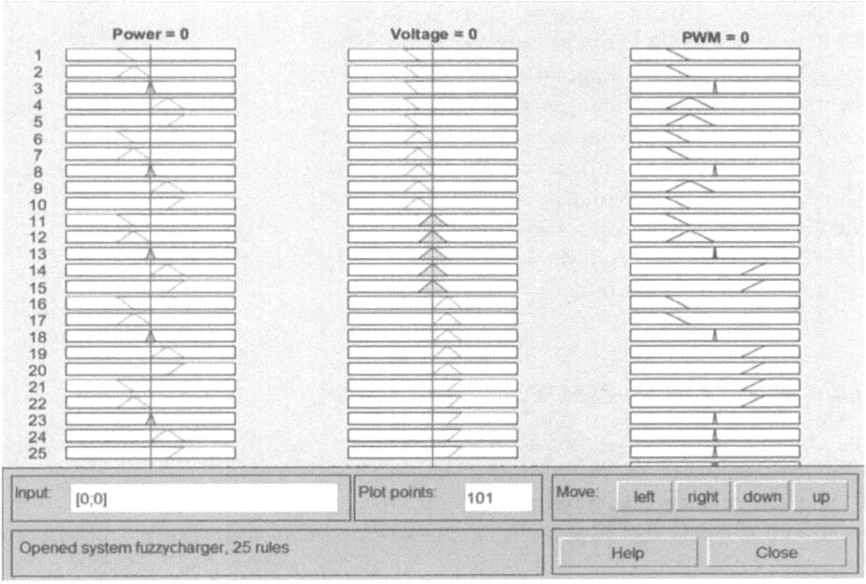

FIGURE 16.8 Rule viewer.

- **Construction of rules and rules viewer**

Here we define 25 'if and then' rules for variables of output as shown below and the rule viewer is also shown in Figure 16.8.

Rules defined for FL:

i. (Power==NB) & (Voltage==NB) (PWM=NB) (1)
ii. (Power==NS) & (Voltage==NB) (PWM=NB) (1)
iii. (Power==Z) & (Voltage==NB) (PWM=Z) (1)
iv. (Power= =PS) & (Voltage==NB) (PWM=NS) (1)
v. (Power==PB) & (Voltage==NB) (PWM=NS) (1)
vi. (Power==NB) & (Voltage==NS) (PWM=NB) (1)
vii. (Power==NS) & (Voltage==NS) (PVVM=NB) (1)
viii. (Power==Z) & (Voltage==NS) (PWM=Z) (1)
ix. (Power==PS) & (Voltage==NS) (PWM=NS) (1)
x. (Power==PB) & (Voltage== NS) (PWM=NB) (1)
xi. (Power==NB) & (Voltage==Z) (PWM=NB) (1)
xii. (Power==NS) & (Voltage==Z) (PWM=NS) (1)
xiii. (Power==Z) & (Voltage==Z) (PWM=Z) (1)
xiv. (Power==PS) & (Voltage==Z) (PWM=PB) (1)
xv. (Power==PB) & (Voltage==Z) (PWM=PB) (1)

xvi. (Power==NB) & (Voltage==PS) (PWM=NB) (1)
xvii. (Power==NS) & (Voltage==PS) (PWM=NB) (1)
xviii. (Power==Z) & (Voltage==PS) (PWM=Z) (1)
xix. (Power==PS) & (Voltage==PS) (PWM=PB) (1)
xx. (Power==PB) & (Voltage==PS) (PWM=PB) (1)
xxi. (Power==NB) & (Voltage==PB) (PWM=PB) (1)
xxii. (Power==NS) & (Voltage==PB) (PWM=PB) (1)
xxiii. (Power==Z) & (Voltage==PB) (PWM=Z) (1)
xxiv. (Power==PS) & (Voltage==PB) (PWM=Z) (1)
xxv. (Power==PB) & (Voltage==PB) (PWM=Z) (1)

16.4 SIMULATION RESULTS

This chapter gives a different kind of idea about the Simulink implementation and gives the results on the Simulink model. The Simulink diagram of the EV DC fast CS is shown in Figure 16.9, in which we consider SPV array as the source and it shifts the power to the energy storage system of the CS via boost converter. Boost converter is applied to boost the power of source that comes from SPV. Boost converter is controlled by three different controllers, PID, MPPT, and FLC, and energy storage is considered for the direct current charging of EVs.

The simulation outputs are shown in Figures 16.12–16.21; charging of the batteries is observed by increasing SOC and minimum SOC is kept at 45%, which is increasing as shown in the output of simulation in Figures 16.16–16.21 with MPPT, FLC, and PID controllers.

16.4.1 MPPT Controller

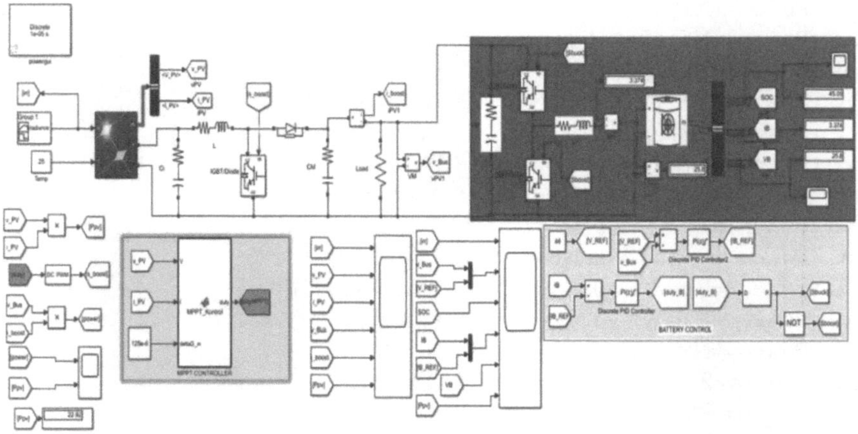

FIGURE 16.9 Simulink diagram of a charging station with MPPT controller.

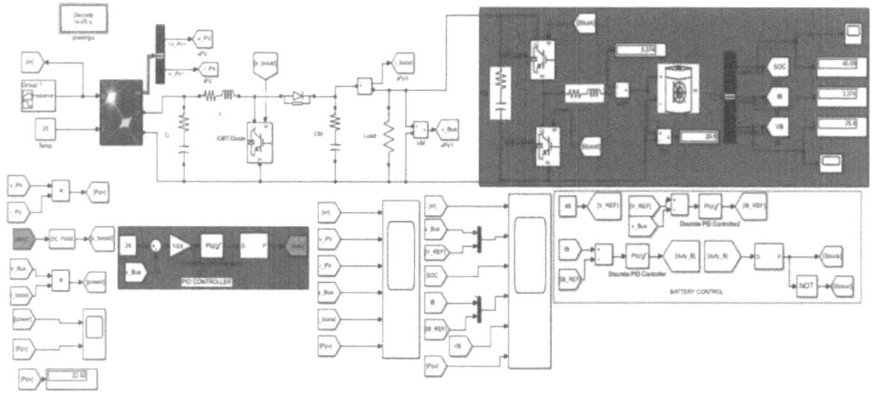

FIGURE 16.10 (a) MATLAB/Simulink model using PID controller

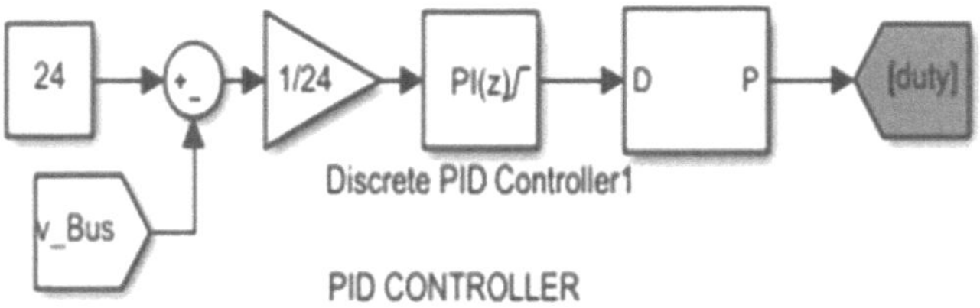

FIGURE 16.10 (b) MATLAB block diagram of PID controller.

16.4.2 PID Controller

Figure 16.10 shows the MATLAB/Simulink of the system using a PID controller, and there we apply a constant value.

16.4.3 Fuzzy Logic Controller

As we showed in Figure 16.11, the FLC is much smarter and faster than the PID controller. And one more thing: FLC takes less time than PID to achieve the steady state, so we considered FLC instead of PID controller.

Following are Simulink results for the three different controllers:

Figure 16.12 shows the SPV input value, i.e., for alterable irradiance and fixed value of temperature. For the value of irradiance, we assumed that as time changes, the value of irradiance changes.

Figures 16.13, 16.14, and 16.15 show the output value of the boost converter with MPPT, FLC, and PID controllers. By looking at the figures, we can say that there is a lot of disturbance in the PID controller as compared to the other two controllers.

Figures 16.16, 16.18, and 16.20 show the SOC of EV, and Figures 16.17, 16.19, and 16.21 show the EV battery power using all three controllers and comparing which

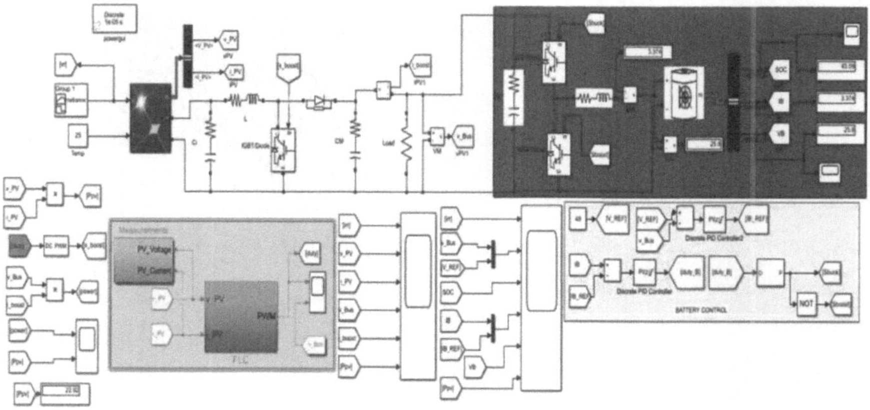

FIGURE 16.11 (a) MATLAB/Simulink model of the system using FLC

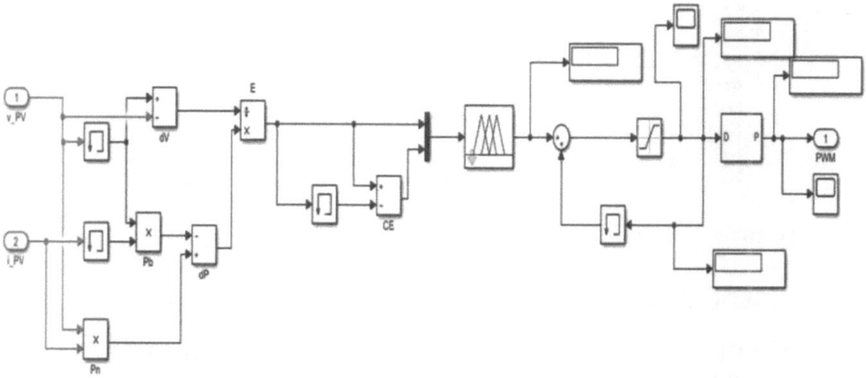

FIGURE 16.11 (b) MATLAB block diagram of FLC

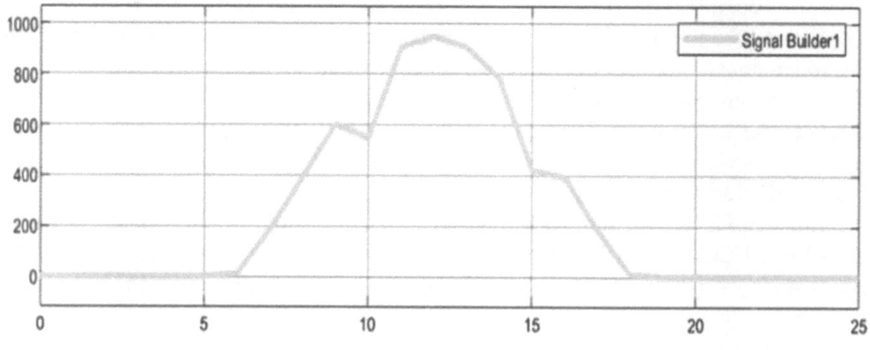

FIGURE 16.12 Solar input variable irradiance.

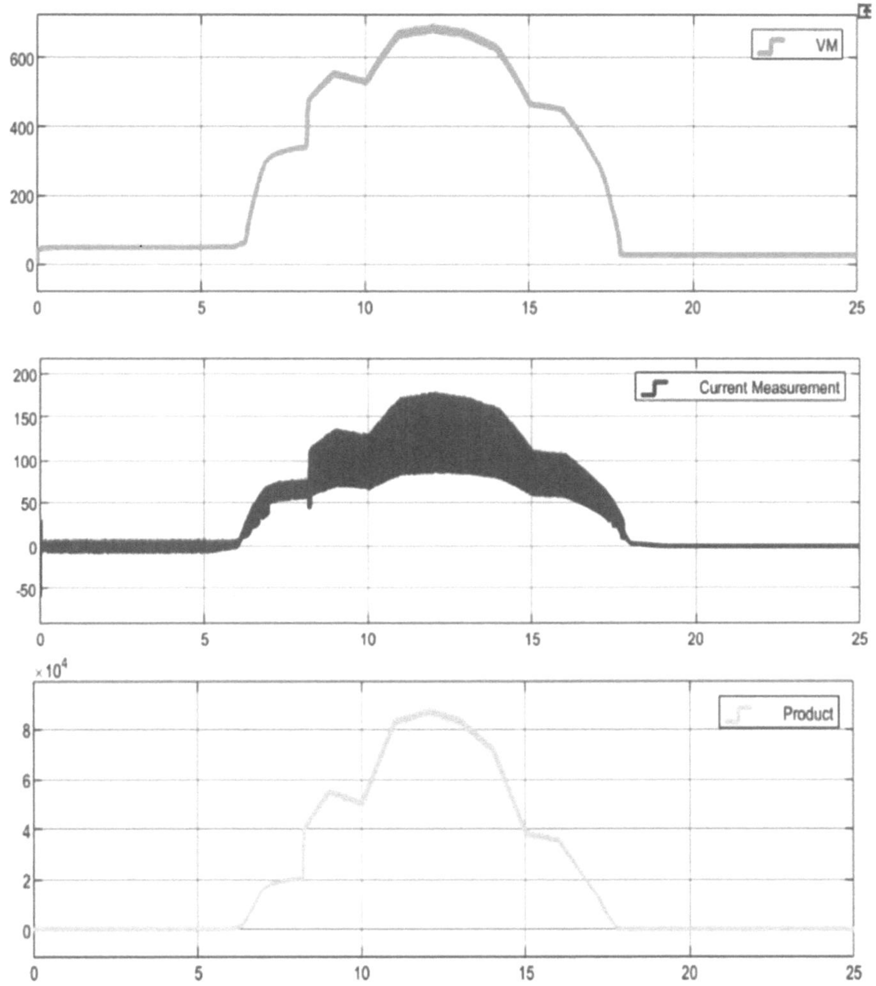

FIGURE 16.13 Boost converter output using MPPT.

controller has more accuracy. By comparing all the controllers, we concluded that FLC has better accuracy than the others. According to the figure, we can say that first, the battery starts discharging as the time changes, and then it starts charging as the time changes.

16.5 CONCLUSION

This project shows modeling and designing of a V2G and G2V in SPV array using EV DC CS. An EV DC CS takes the power through an SPV source, i.e., connected to a boost converter to boost the power that flows from the source. The converter has

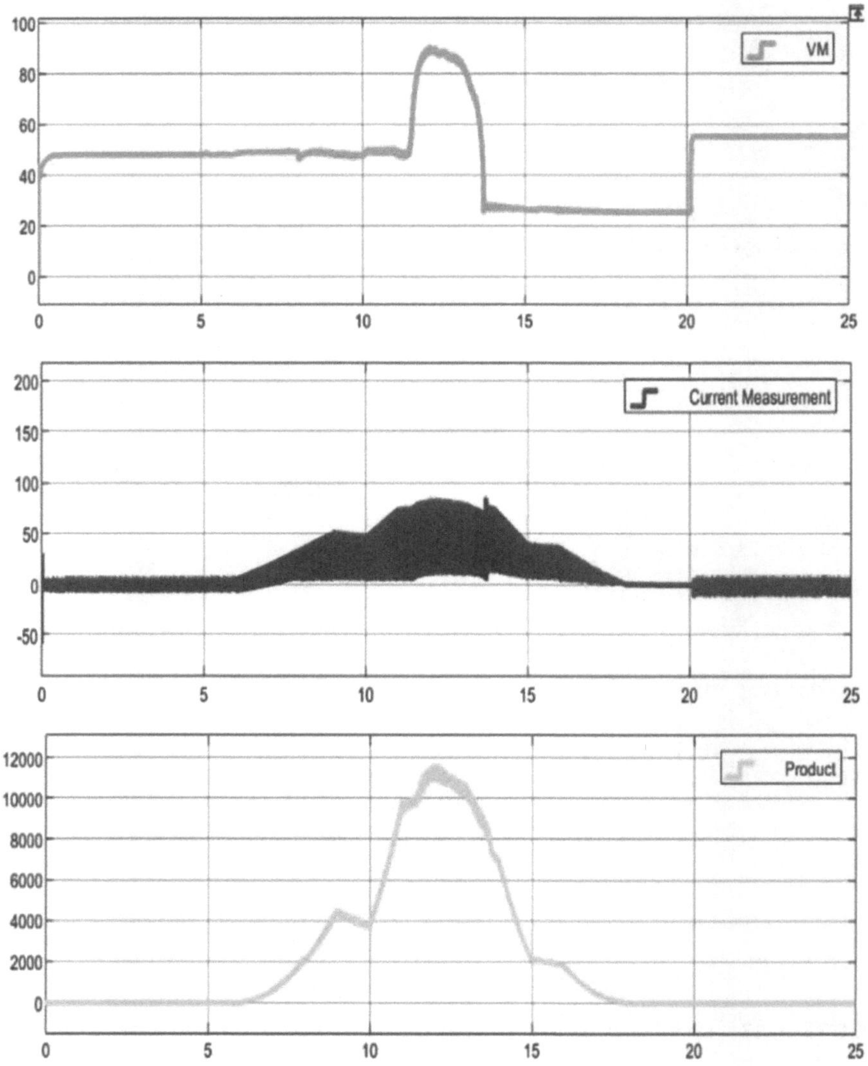

FIGURE 16.14 Boost converter output using FLC.

three controllers, MPPT, FLC, and PID, and all controllers are compared. As per the above discussion, we concluded that FLC gives better accuracy than other controllers. PID controller takes more time to achieve steady-state as compared to MPPT as well as FLC. By the use of FLC, we track the value of power at every instant of time change as compared to other controllers. Hence, we conclude that FLCs give better output value as compared to others.

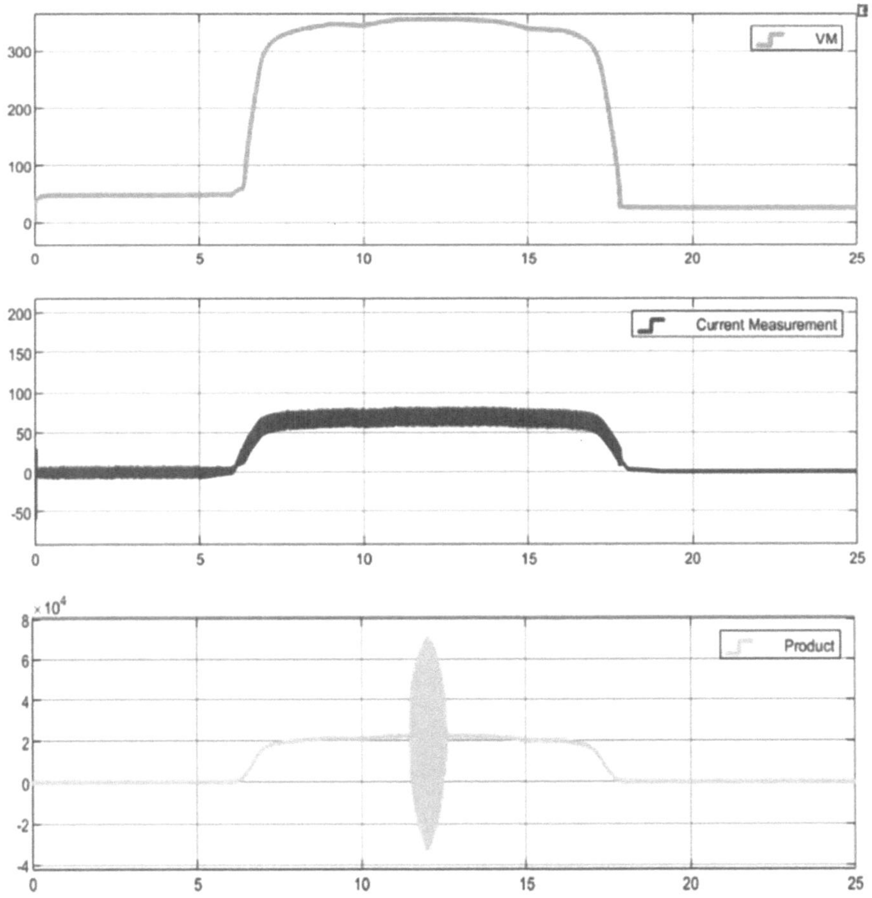

FIGURE 16.15 Boost converter (O/P) output using PID.

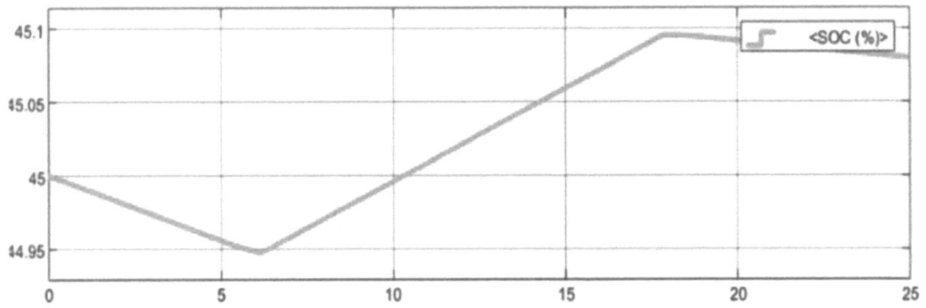

FIGURE 16.16 EV battery SOC using MPPT controller.

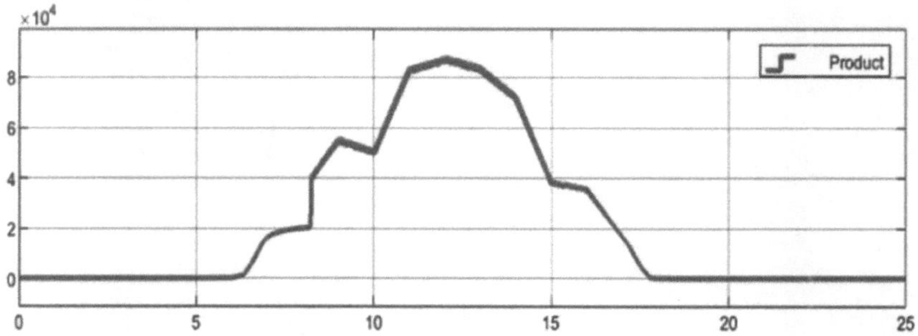

FIGURE 16.17 EV battery power using MPPT controller.

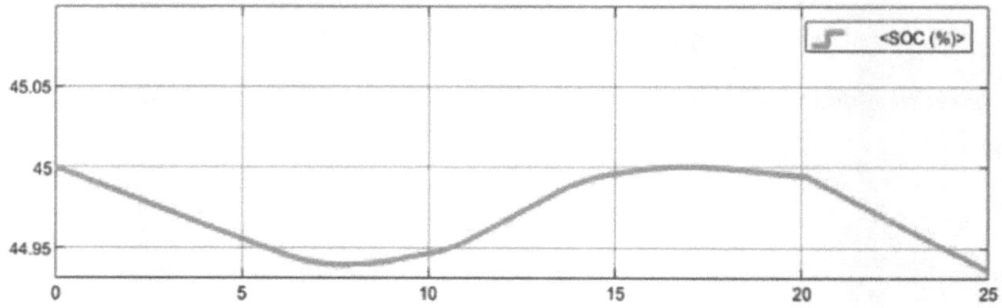

FIGURE 16.18 EV battery SOC using FLC controller.

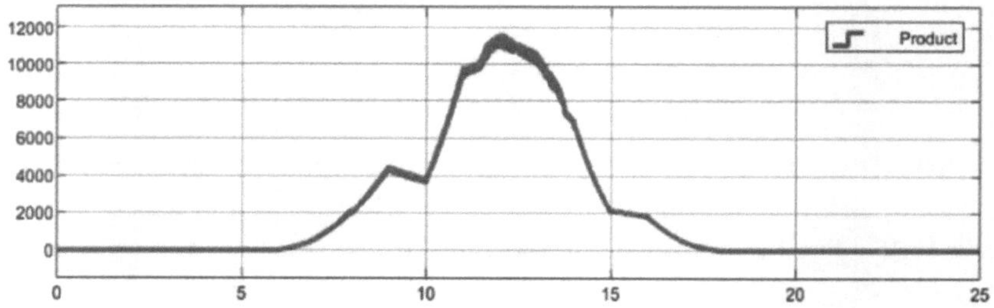

FIGURE 16.19 EV battery power using FLC controller.

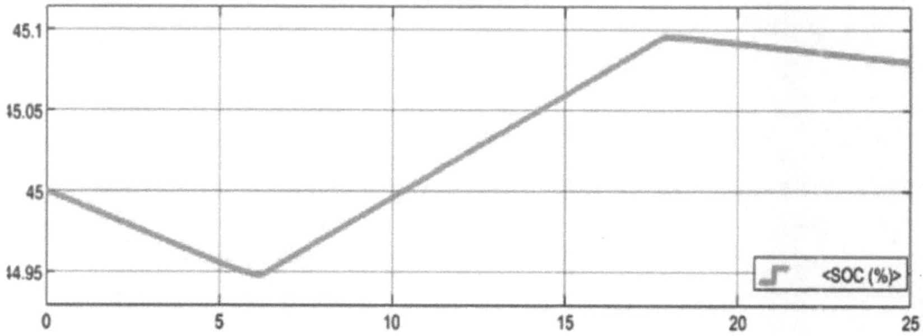

FIGURE 16.20 EV battery SOC using PID controller.

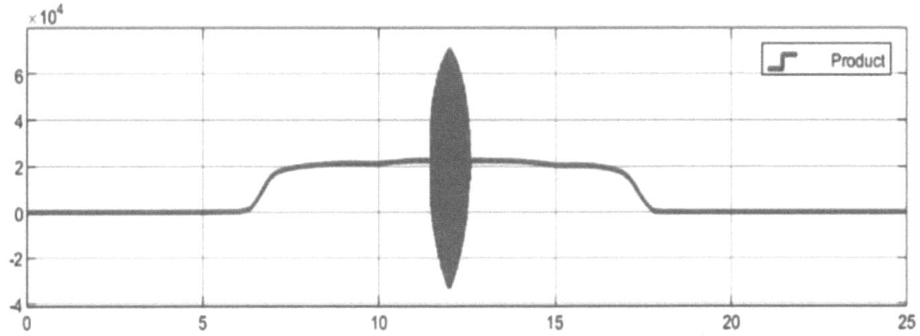

FIGURE 16.21 EV battery power using FLC controller.

REFERENCES

[1] A. Arancibia and K. Strunz, "Modeling of an Electrical Vehicle Charging Station for Fast DC Charging," in *Proceedings of IEEE International Electric Vehicle Conference*, pp. 1–6, 2012.

[2] A. Jain, A. Mani, A. S. Siddiqui, S. Sharma and H. P. Singh, "A Data Driven Approach for Scheduling the Charging of Electric Vehicles," in *2017 Recent Developments in Control, Automation & Power Engineering (RDCAPE)*, pp. 39–44, Noida. IEEE, 26–27 Oct., 2017. doi:10.1109/RDCAPE.2017.8358236

[3] R. K. Pachauri and Y. K. Chauhan, "Hybrid PV/FC Stand Alone Green Power Generation: A Perspective for Indian Rural Telecommunication Systems," in *Proc. IEEE Conference on Issues and Challenges in Intelligent Computing Techniques (ICICT)*, pp. 807–815, Ghaziabad. KIET, 7–8 Feb., 2014.

[4] S. Silvestre, A. Boronat and A. Chouder, "Study of Bypass Diodes Configuration on PV Modules," *Applied Energy*, vol. 86, no. 9, pp. 1632–1640, Sept. 2009.

[5] M. A. S. Masoum, H. Dehbonei and E. F. Fuchs, "Theoretical and Experimental Analyses of Photovoltaic Systems with Voltage and Current Based Maximum Power Point Tracking," *IEEE Transactions on Energy Conversion*, vol. 17, no. 4, pp. 514–522, Dec. 2002.

[6] T. Esram and P. L. Chapman, "Comparison of Photovoltaic Array Maximum Power Point Tracking Techniques," *IEEE Transactions on Energy Conversion*, vol. 22, no. 2, pp. 439–449, Jun. 2007.

[7] Y. C. Kuo, T. J. Liang and J. F. Cben, "Novel Maximum Power Point Tracking Controller for Photovoltaic Energy Conversion System," *IEEE Transactions on Industrial Electronics*, vol. 48, no. 3, pp. 594–601, Jun. 2001.

[8] ABB Conversations. http://www.abb-conversations.com/2013/10/before-buying-electric-vehicle-charging-stations-know-this/

[9] International Electrotechnical Commission. IEC 61851-1 Electric Vehicle Conductive Charging System–Part 1: General Requirements; IEC: Geneva, Switzerland, 2010; Volume 61851-1, p. 99.

[10] S.-K. Kim, J.-H. Jeon, C.-H. Cho, J.-B. Ahn and S.-H. Kwon, "Dynamic Modeling and Control of a Grid-Connected Hybrid Generation System with Versatile Power Transfer," *IEEE Transactions on Industrial Electronics*, vol. 55, no. 4, pp. 1677–1688, April 2008.

[11] R. R. N. Patel and S. K. Sinha, "Performance Analysis of PV System Integrated with Boost Converter for Low Power Applications," in M. Kumar, R. K. Pandey, and V. Kumar (Eds.), *Advances in Interdisciplinary Engineering*. Springer, Singapore, pp. 879–890, 2019.

[12] A. Gupta, P. Kumar, R. K. Pachauri and Y. K. Chauhan, "Performance Analysis of Neural Network and Fuzzy Logic Based MPPT Techniques for Solar PV Systems," in *6th IEEE Power India International Conference (PIICON)*, pp. 1–6, Delhi, India. IEEE, 5–7 Dec., 2014. doi: 10.1109/34084POWERI.2014.7117722

[13] O. Abdel-Rahim, M. Orabi, E. Abdelkarim, M. Ahmed and M. Z. Youssef, "Switched Inductor Boost Converter for PV Applications," in *2012 Twenty-Seventh Annual IEEE Applied Power Electronics Conference and Exposition (APEC)*, pp. 2100–2106, Orlando, FL, 2012. doi: 10.1109/APEC.2012.6166111

[14] F. M. Shakeel and O. P. Malik, "Vehicle-to-Grid Technology in a Micro-grid Using DC Fast Charging Architecture," in *IEEE Canadian Conference of Electrical and Computer Engineering (CCECE)*, pp. 1–4, Edmonton, AB, Canada. IEEE, 5–8 May, 2019. doi: 10.1109/CCECE.2019.8861592

[15] R. K. Pachauri and Y. K. Chauhan, "Fuzzy Logic Controlled MPPT Assisted PV-FC Power Generation for Water Pumping System," in *IEEE Students' Conference on Electrical, Electronics and Computer Science (SCEECS)*, pp. 1–6, Bhopal, India. MANIT, 2–3 March, 2014.

[16] MATLAB Simulink SimPowerSystems 7.6 (R2008a) "SimPowerSystems Library Documentation", 2008.

[17] Y. Jia Ying, V. K. Ramachandramurty, K. M. Tan and A. Atputharajah, "Modelling of Electric Vehicle Fast Charging Station and Impact on Network Voltage," in IEEE Conference on Clean Energy and Technology (CEAT), 2013.

[18] L. Magdalena, Fuzzy rule-based systems, in Kacprzyk J., Pedrycz W. (Eds), *Springer Handbook of Computational Intelligence*. Springer, Berlin, Heidelberg, pp. 203–218, 2015. https://link.springer.com/chapter/10.1007/978-3-662-43505-2_13

[19] Science Direct. https://www.sciencedirect.com/topics/engineering/fuzzification

[20] Massey. https://www.massey.ac.nz/~nhreyes/MASSEY/159741/Lectures/Lec2012-3-159741-FuzzyLogic-v.2.pdf

[21] Wikipedia. *Defuzzification*. https://en.m.wikipedia.org/wiki/Defuzzification

[22] Anjali Gupta and Anjali Jain, "Intelligent Control of Hybrid Power Systems for Load Balancing and Levelised Cost," in *IEEE 1st International Conference on Power Electronics, Intelligent Control and Energy Systems (ICPEICES)*, pp. 1–5, Delhi, India. IEEE, 4–6 July, 2016.

17 Joining of AA7075/SiC Composite Using Friction Stir Welding (FSW)

Sudhir Kumar, Vaibhav Gangwar, Ajay Kumar, and Dhairya Partap Singh

CONTENTS

DOI: 10.1201/9781003229018-17

17.1 INTRODUCTION OF COMPOSITE MATERIAL

A composite material is a mixture of more than two materials on a macroscopic level. Generally, it consists of matrix and reinforcements. The matrix holds the reinforcing particles to form required shape, size and improve the overall mechanical properties of composite materials [1–2].

The utilization of metal matrix composite (MMC) has increased day by day in many areas due to its effective mechanical properties. It attains higher strength, lighter in weight and has higher stiffness and anti-corrosion property than matrix aluminum alloy. Because of these outstanding properties, MMCs have potential to replace conventional materials in the field of automobile, aerospace and manufacturing organizations [3–4]. Commercially, matrix is available in the form of Al, Cu, Ag, Mg and Ti, etc. and reinforcements of SiC, B_4C and Al_2O_3. For fabrication of MMCs, some basic problems arise in proper mixing of the reinforcement like Al_2O_3 and SiC with liquid material due to poor wettability of reinforcement. So, proper selection of material matrix, reinforcement and fabrication technique is important in fabrication of composite [5–6].

In this present work, AA7075/10%SiC composite was developed using the method of friction stir casting. Experimentation was performed with four process variables at three levels, optimizing the process variables with respect to tensile strength and hardness by Taguchi method. SEM, EDAX, XRD and thermal analyses were carried out.

17.2 WELDING OF METAL MATRIX COMPOSITES

There are different types of welding used, which are as follows.

17.2.1 Conventional Welding

MMCs are new engineering material having higher hardness and toughness as compared to other alloys. Due to the frictional ability of MMC materials, these materials are difficult to weld by conventional welding methods such as Metal Inert Gas (MIG), Tungsten Inert Gas (TIG) and plasma arc welding.

17.2.2 Unconventional Welding

Unconventional welding like Electronic Beam welding, Laser Beam welding and Explosive welding are the major classifications of unconventional category. These processes are applicable to only those metals that have enough impact resistance and ductility. Operator requires more protection as compared to other conventional welding processes. FSW process is applicable where the geometries of work piece are simple and flat but cost is higher. It is a novel process which is used to join the composite materials.

17.3 PRINCIPLE OF FSW

In FSW, the tool is fed into a butt joint between two tightly clamped work pieces and a tool shoulder produces an axial force to work piece. The tool pin is slightly shorter (96–98% of work piece thickness) [7]. After a short dwell time, the probe of rotating

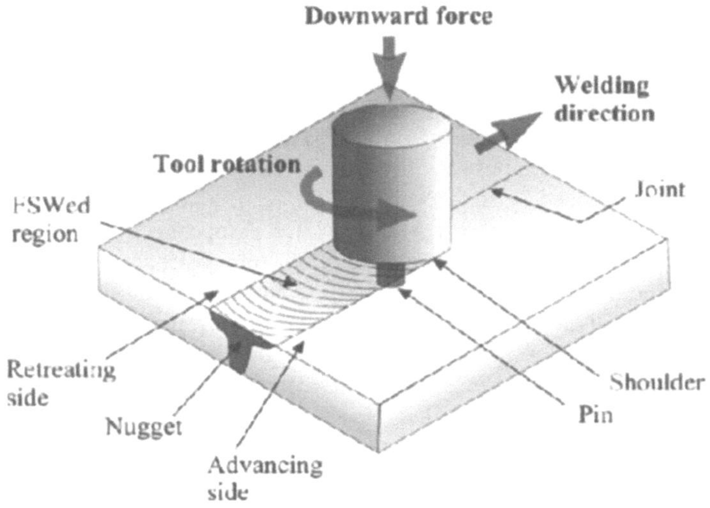

FIGURE 17.1 Operation of friction stir welding (adapted from Dhairya and Vikram 2019).

tool is inserted into the work piece, and the tool is moved forward along the joining line at the desired welding speed.

The basic principle and operation are shown in Figure 17.1. The tool advances in the forward direction and then reaches the retreating side. The next part of FSW process is where the solid material begins to change into a semi-solid shape and flows around the tool pin inserted into the material. The semi-solid material shrinks and cools in the retreating side. Therefore, in the friction stir welding process, the forward side has stronger properties than the moving side at any given time, so the advancing side produces more frictional stress (unbalanced frictional force) and ultimately more heat of plastic deformation compared to retreating side.

17.4 LITERATURE REVIEW

Hashim et al. [1999] studied a variety of composite materials using different matrix types, reinforcement sizes, shapes and quantities and processing technologies as per requirements and applications. To obtain the best performance of MMC materials, second phase in the matrix must be uniformly distributed and combination of these elements must be optimized [8].

Mishra and Ma [2005] fabricated aluminum MMC, which was oxidized or un-oxidized with SiC reinforcement with varying magnesium concentrations. The result revealed that magnesium segregated at the interface and prevents the formation of Al_4C_3 compound [9].

Pathak et al. [2006] explained the effect of dispersed silicon carbide particles in the grain boundary regions and synthesized silicon carbide particles and broken dendrites in aluminum-silicon alloy. Adding SiC particles to the matrix can improve the mechanical properties and wear resistance [10].

Bhushan et al. [2013] fabricated AA7075/10%SiC composite with particles size by mechanical stir casting technique. Experiments were performed at 500, 650 and 750 rpm rotational speed and stir time was 10 mins. The results showed the even distribution of SiC particles in the prepared composite material under the condition of stirring speed at 650 rpm and stirring time of 10 mins. On comparing the mechanical properties of fabricated composite with base alloy, results revealed that hardness and tensile strength increased by 10.48% and 12.74% respectively [11].

Bharath et al. [2014] developed Al6061/Al2O3 composite by liquid metallurgy, specifically combining the casting technology. The microstructural characterization of the composite material showed that the distribution of Al_2O_3 in the sample was fairly uniform and there was a certain amount of grain refinement. Further, tensile and hardness properties of composite were higher than Al6061 alloy [12].

Uzun [2007] fabricated AA2124/25%SiC composite through friction stir processing and characterized by the EDAX and microstructure analysis. The analyses confirmed the presence of fine as well as coarse silicon particulates reinforced with aluminum alloy matrix. A zone which is adjacent of the weld nugget, called thermally affected zone, was plastically deformed. Heat-affected zone formed in between thermally affected zone and base composite region. So, a similar microstructure was achieved on both advancing side and retreating side of the base composite [13].

Jain et al. [2018] explained the significance of FSW process variables, likes tool rotation, welding speed, tool geometry and diameter of tool shoulder, on weld quality of AA6082 and AA5083 alloys using Taguchi method and Grey relational analysis (GRA) approach. Analysis of variance (ANOVA) was used to study the effects of welding process variables such as elongation and tensile strength. Rotational speed of tool was found as having the most significant effect on the responses [14].

17.5 SELECTION OF MATERIAL

Aluminum alloy 7075 was selected as base material, which is the most commonly used and universally recognized in the production of lightweight bodies for low weight/strength ratio, high wear resistance, creep resistance and high corrosion resistance.

The composition of chemical elements of the AA7075 is shown in Table 17.1.

SiC, TiB, Al_2O_3, B_4C and TiC reinforcements are commonly used in commercial applications. Out of these reinforcements, SiC has the best mechanical properties (elastic modulus 350–450 GPa, hardness 2350–2850 HV, compressive strength 3850 MPa), including high thermal strength, chemical corrosion resistance and shock resistance [15–16]. It maintains high mechanical strength at a temperature as high as 1400°C. Hence, SiC has been selected as reinforcement with the size of 20–40 μm.

TABLE 17.1
Chemical Composition of AA7075 (data from Dhairya and Vikram 2019)

Element	Mg	Mn	Zn	Fe	Cu	Si	Cu	Al
Wt%	2.1	0.12	5.1	0.35	1.2	0.58	1.2	Bal

FIGURE 17.2 Experimental set-up of stir casting (adapted from Dhairya and Vikram 2019).

17.6 EXPERIMENTAL SET-UP OF MECHANICAL STIR CASTING

Mechanical stir casting process was designed to produce MMC. The experimental set-up of mechanical stir casting is shown in Figure 17.2. One kg of AA7075 was charged in graphite crucible. Temperature of furnace was increased to 700°C and AA7075 was allowed to convert into molten metal. After the melting of AA7075, the stirrer was placed in a crucible at the height of 10 mm from the base.

The stirrer rotated at 650 rpm and preheated SiC particles are mixed into the rotating molten metal. During this process, nitrogen gas (0.5 kg/cm^2) was added in the furnace. Stirring was done at 650 rpm for 10 mins. After 10 mins, heating was stopped and stirrer was taken out from the crucible. After solidification, ingots were taken out from crucible. The above procedure was adopted for preparing composites with 10% weight of SiC particulates. Fabricated composites are presented in Figure 17.3.

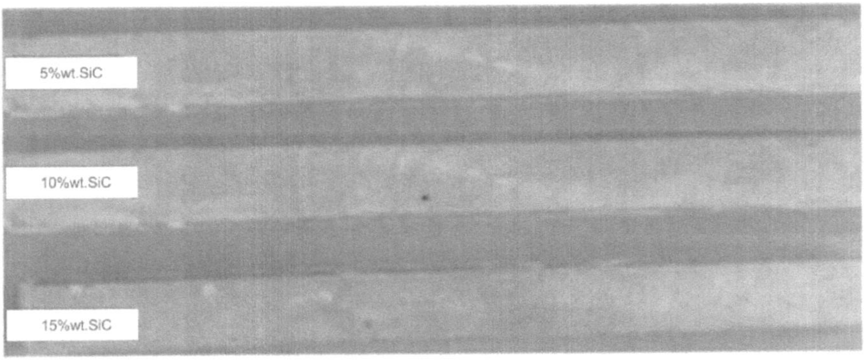

FIGURE 17.3 Fabricated composite with 10% SiC particulates (adapted from Dhairya and Vikram 2019).

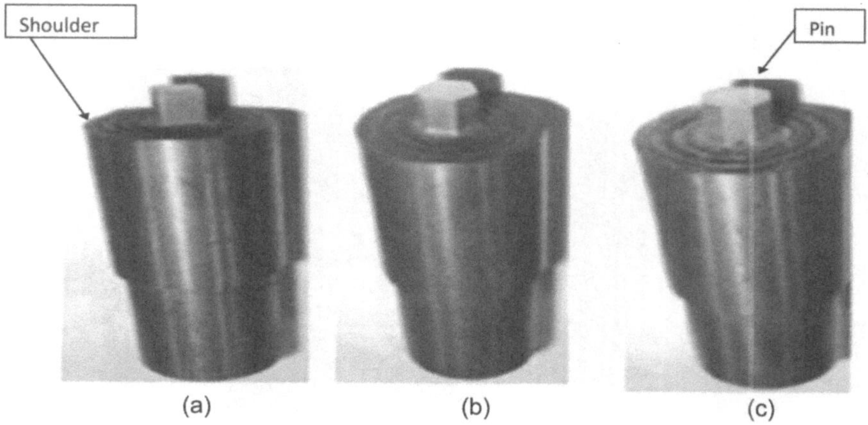

FIGURE 17.4 (a) Square pin profile, (b) Hexagonal pin profile, (c) Octagonal pin profile (adapted from Dhairya and Vikram 2019).

17.7 FABRICATION OF FRICTION STIR WELDING TOOLS

Non consumable high speed steel (HSS) tools with tool pin geometries like square, hexagonal and octagonal were fabricated for friction stir welding process. The schematic diagrams for the fabricated tools are shown in Figure 17.4. The total length of pin is 5.8 mm and pin diagonal is 6 mm. The length and diameter of shoulder are 60 mm and 18 mm respectively.

17.8 PROCEDURE OF FSW

Figure 17.5 shows that FSW process was performed on vertical milling machine. In this process, 5 HP of electric motor was fixed for rotating the spindle. Tool rotational speed may vary according to the rotation of motor. A rotating tool with a desired pin profile was inserted into a butt joint between two work materials until the shoulder touched the surface with sufficient thrust. The tool's pin was slightly shorter (97% of work piece thickness) than the work piece thickness to achieve excellent weld quality. After a short dwell time, the work table moves at a predetermined welding speed. When the inserted tool reaches the bottom of the part, the tool starts to retract. Repeat this process for other combinations of welding process variables, which are predefined.

17.9 ISHIKAWA DIAGRAM OF FSW

The significance of FSW process variables on the quality of FSW joint are presented by Ishikawa diagram as shown in Figure 17.6.

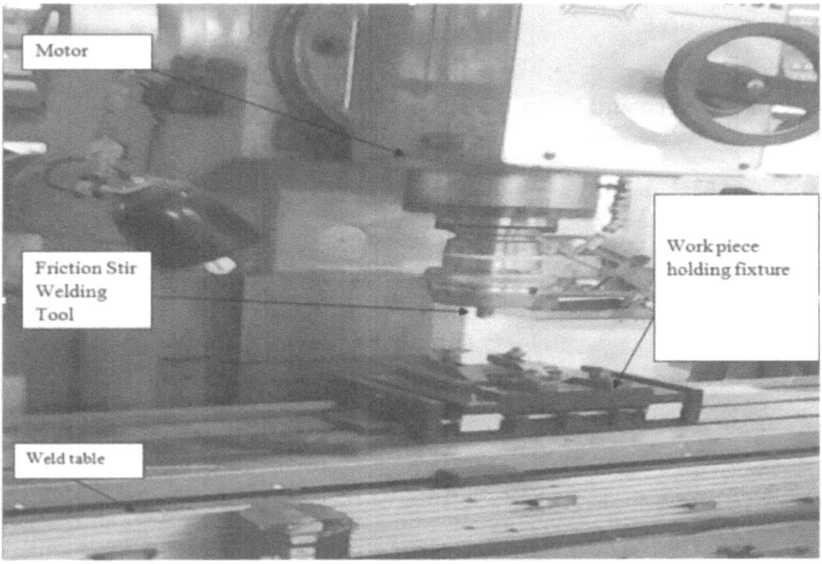

FIGURE 17.5 Friction stir welding set-up (adapted from Dhairya and Vikram 2019).

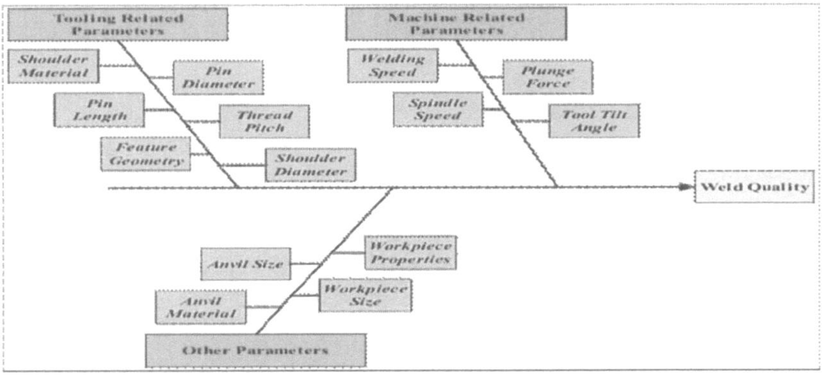

FIGURE 17.6 Ishikawa diagram of friction stir welding (adapted from Dhairya and Vikram 2019).

17.10 SELECTION OF PROCESS VARIABLES AND THEIR RANGES OF FRICTION STIR WELDING PROCESS

Pilot experiments are performed to identify the ranges of process variables for friction stir welding process. In this process, one process parameter varies from lower value to higher value and the other parameters remain constant at mid value and find the tensile strength and hardness [17–19]. These selected process variables and their ranges are mentioned in Table 17.2.

TABLE 17.2
Selected Process Parameters and Their Ranges Based on Pilot Experiments (data from Dhairya and Vikram 2019)

Symbol	Process Parameters	Level 1	Level 2	Level 3
A	Tool rotational speed (rpm)	1300	1500	1700
B	Welding speed (mm/sec.)	0.8	1.3	1.8
C	Axial force (KN)	5	7	9
D	Tool geometry	Square	Hexagonal	Octagonal

Twenty seven experiments were performed to obtain the weld joint and calculate the tensile strength and hardness, which are shown in Table 17.3.

17.11 RESULTS AND DISCUSSION

17.11.1 Characterization of AA7075/10%wt.SiC Composite

The standard specimens were prepared from AA7075/10%wt.SiC composite. Specimen was observed for microstructure through optical microscope analysis. Also, SEM, EDAX, XRD, DTA and SEM fractography analyses were carried out to analyze the specimen regarding elements, phases and thermal effects.

17.11.1.1 Microstructure

Optical microstructures were observed through microscope attached to a computer, at 100× shown in Figure 17.7 and at 400× shown in Figure 17.8. Figure 17.7 shows the microstructure of AA7075/10%wt.SiC composite material. The images clearly show the SiC is evenly distributed in AA7075.

Figure 17.8 shows that distribution of SiC particles is uniform and the aggregation of particles is limited. There is no porosity in the structures. Therefore, these composite materials have good mechanical properties and are suitable for welding experiments.

17.11.1.2 SEM and EDAX Analysis

Figures 17.9 and 17.10 show the SEM analysis and EDAX analysis of FSW AA7075/10%wt.SiC composite material respectively. According to both figures of AA7075/10%wt.SiC composite, SiC particulates are uniformly distributed throughout the matrix.

Result of EDAX revealed that the main constituents like Mg, Si, Zn and Cu are present in the majority. No new component and unfavorable reaction had been established in EDAX analysis of AA7075/10%wt./SiC composites.

17.11.1.3 X-ray Diffraction Analysis

XRD was done by using a Bruker ASX D-8X-ray diffractometer. Figure 17.11 shows that X-ray diffraction was done at scanning rate of 0.01° 2θ/sec. Peaks obtained in

TABLE 17.3
Measured Responses of Friction Stir Welding Process (data from Dhairya and Vikram 2019)

S. NO	Tool Rotational Speed	Welding Speed	Axial Force	Tool Geometry	Tensile Strength (Mpa)	Micro Hardness (VHN)	S/N Ratio for TS	S/N Ratio for Hardness
1	1300	0.8	5	S	221.11	90.55	46.89	39.1374
2	1300	0.8	7	H	249.14	100.90	47.93	40.0776
3	1300	0.8	9	O	235.00	95.17	47.42	39.5701
4	1300	1.3	5	H	251.32	101.78	48.00	40.1533
5	1300	1.3	7	O	269.10	108.98	48.60	40.7470
6	1300	1.3	9	S	243.41	98.58	47.73	39.8755
7	1300	1.8	5	O	216.00	90.25	46.69	39.1089
8	1300	1.8	7	S	237.62	96.23	47.52	39.6664
9	1300	1.8	9	H	229.16	93.81	47.20	39.4446
10	1500	0.8	5	H	234.14	94.82	47.39	39.5383
11	1500	0.8	7	O	267.00	108.13	48.53	40.6790
12	1500	0.8	9	S	247.14	100.09	47.86	40.0076
13	1500	1.3	5	O	288.00	116.64	49.19	41.3366
14	1500	1.3	7	S	311.00	125.95	49.86	42.0040
15	1500	1.3	9	H	297.10	120.32	49.46	41.6068
16	1500	1.8	5	S	283.15	114.67	49.04	41.1891
17	1500	1.8	7	H	292.24	118.35	49.31	41.4636
18	1500	1.8	9	O	285.16	115.49	49.10	41.2505
19	1700	0.8	5	O	241.11	100.65	47.64	40.0559
20	1700	0.8	7	S	279.75	113.29	48.94	41.0842
21	1700	0.8	9	H	249.40	101.00	47.94	40.0867
22	1700	1.3	5	S	268.13	108.59	48.57	40.7157
23	1700	1.3	7	H	284.11	115.06	49.07	41.2185
24	1700	1.3	9	O	261.25	105.80	48.34	40.4899
25	1700	1.8	5	H	243.00	100.85	47.71	40.0735
26	1700	1.8	7	O	260.00	105.30	48.30	40.4482
27	1700	1.8	9	S	249.43	101.02	47.94	40.0877

the diagram were analyzed. XRD peaks of the AA7075/10%wt.SiC are shown in Figure 17.11. The XRD peaks show the evidence of Al matrix and SiC particles in the developed composite material.

Figure 17.12 shows that the elements are correctly distributed in the AA7075/10%wt. SiC composite materials. In addition, no unfavorable reactions were found in the developed composite materials.

17.11.1.4 Thermal Analysis

The Thermo gravimetric (TG), Difference Thermo Gravimetric (DTG) and Differential Thermal Analysis (DTA) curves are shown in Figure 17.12. TG curve

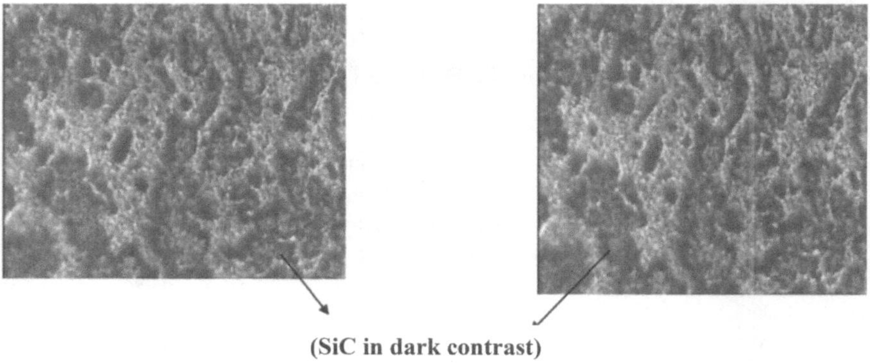

FIGURE 17.7 Optical microstructure of AA7075/10wt%SiC composite at 100× magnification (adapted from Dhairya and Vikram 2019).

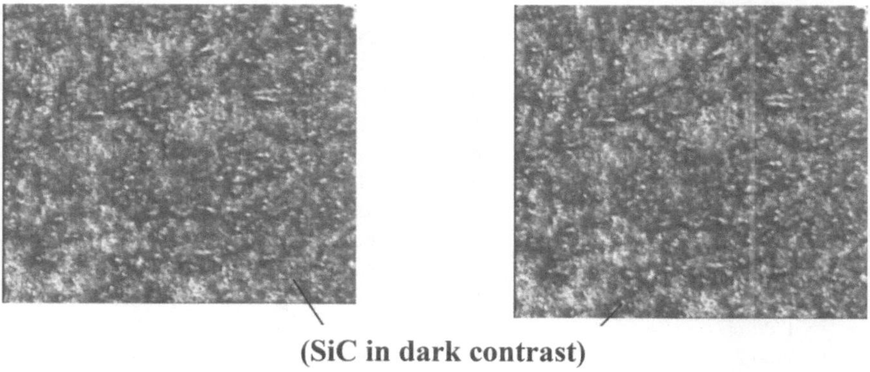

FIGURE 17.8 Optical microstructure of AA7075/10wt%SiC composite at 400× magnification (adapted from Dhairya and Vikram 2019).

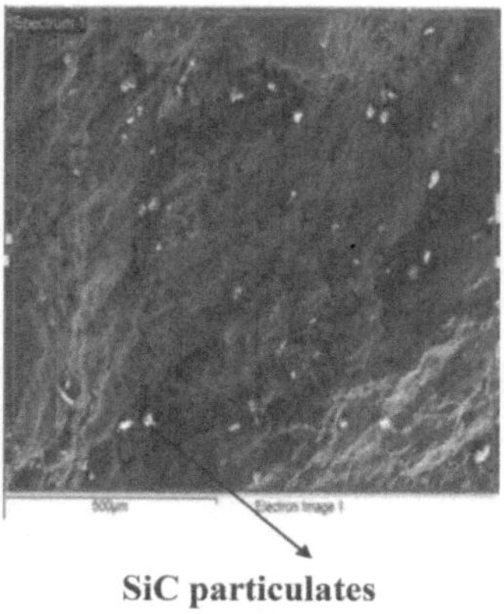

FIGURE 17.9 SEM of AA7075/10%wt.SiC (adapted from Dhairya and Vikram 2019).

TABLE 17.4
Weight % and atomic % for AA7075/SiC (adapted from Dhairya and Vikram 2019)

Element	Weight %	Atomic %
CK	9.77	11.95
OK	6.61	8.17
MgK	2.41	2.00
AlK	65.01	62.17
Sik	09.45	11.75
Fe k	0.71	0.50
Cuk	1.12	0.87
Zn k	4.92	1.49
Total	100	

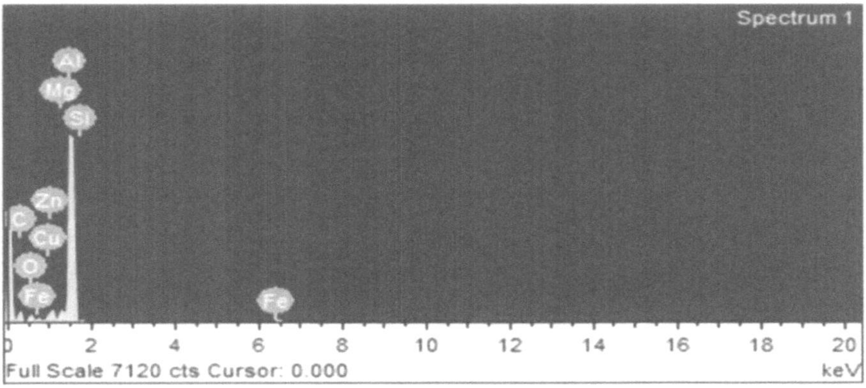

FIGURE 17.10 EDAX profile of AA7075/10%wt.SiC (adapted from Dhairya and Vikram 2019).

was carried out in the temperature of 22°C–850°C. The sample was examined through TG, differential thermo gravimetric and DTA to analyze their thermal degradation characteristics. A rate of heating at 10°C/min and air atmosphere is used to account for temperature differences and the degradation rate. The DTA curve shows a decline from the base. This shows that there is an endothermic reaction which happened at 642°C. The change in enthalpy on the DTA curve is 282 mJ/mg. In DTG curve, percentage of mass decreases from 99.2°C to 99.9°C. It observed around 0.7% due to the presence of moisture, as given in Table 17.4. After that, mass of the composite increases with increases in the temperature from 400°C to 826°C. It is around 2.3% because of zinc nitridation as constitute of SiC composite by nitrogen gas as details given in Table 17.4.

17.11.1.5 SEM Fractography

The SEM fractography analysis was performed on fractured tensile specimens of AA7075/10%wt.SiC welded joint. Figure 17.13 shows SEM fracture of

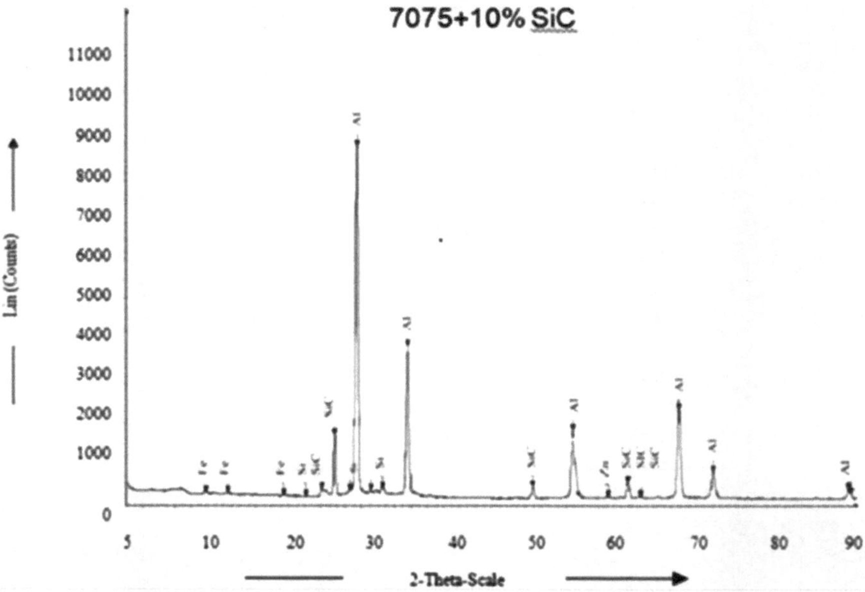

FIGURE 17.11 X-ray diffraction curves of AA7075/10%wt.SiC composite (adapted from Dhairya and Vikram 2019).

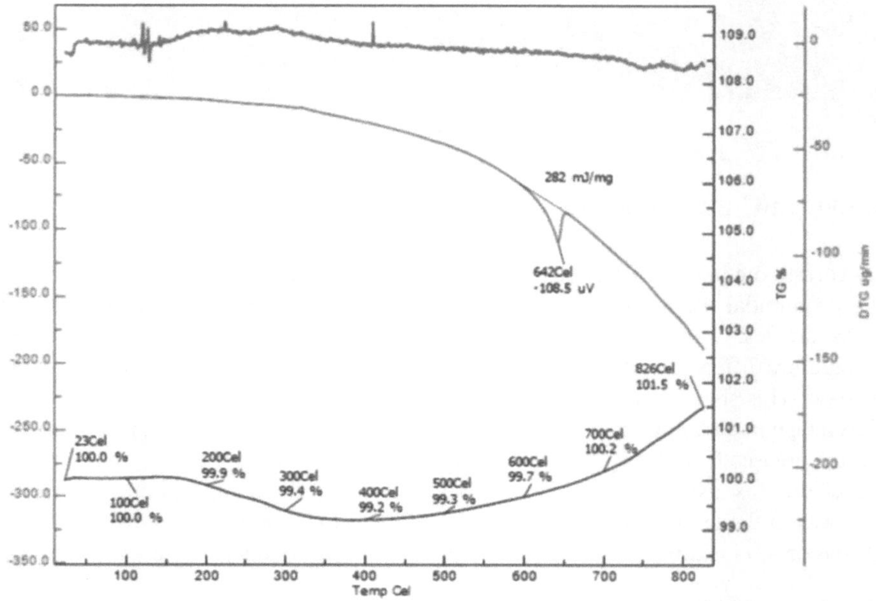

FIGURE 17.12 TG, DTG and DTA curve of AA7075/10%wt.SiC composite (adapted from Dhairya and Vikram 2019).

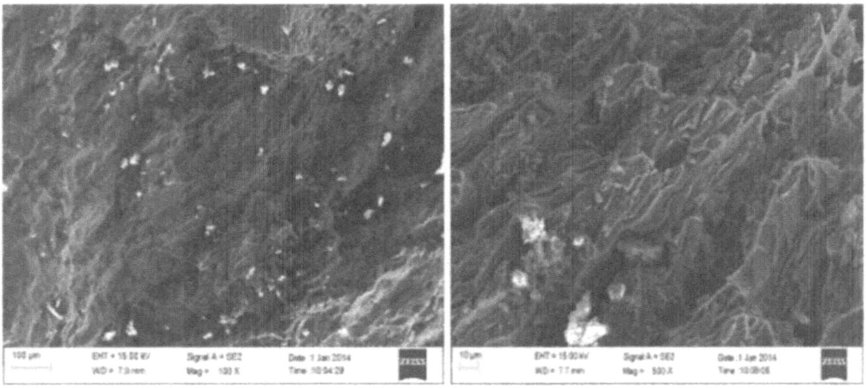

FIGURE 17.13 SEM fractography of AA7075/10%wt.SiC (adapted from Dhairya and Vikram 2019).

AA7075/10%wt.SiC composite. The fractured surface has a dimple which results in tensile fracture. Fracture analysis shows a strong alliance between reinforcing particle and matrix.

The stronger interfacial alliance between reinforcement and the matrix causes enhancement of the tensile strength. Reinforcing particle distribution is uniform in the matrix is necessary to achieve better mechanical properties [20].

17.12 ANALYSIS OF TENSILE STRENGTH AND HARDNESS

The data of tensile strength was analyzed to understand the role of FSW process variables. Table 17.3 shows the S/N ratio on the basis of experimental results. The mean analysis of each experiment will provide a better combination of process variables and confirm the higher tensile strength. At the levels 1, 2 and 3, the calculated S/N ratio of tensile strength of each process variable is shown in Table 17.5.

The hardness results were shown in Table 17.3. The value of S/N ratio of hardness for each variable at level 1, 2 and 3 were reported in Table 17.5. Response Table 17.5

TABLE 17.5
Mean Response Table for S/N Ratio (Data from Dhairya and Vikram 2019)

Process		Rotational Speed		Welding Speed		Axial Force		Tool Geometry	
Parameters	**Level**	**TS**	**Hardness**	**TS**	**Hardness**	**TS**	**Hardness**	**TS**	**Hardness**
Average value of tensile strength	L1	47.55	39.75	47.84	40.03	47.90	40.15	48.26	40.42
	L2	48.86	41.01	48.76	40.91	48.67	40.82	48.22	40.41
	L3	48.27	40.47	48.09	40.30	48.11	40.27	48.20	40.41
	Max-Min	1.31	1.25	0.92	0.88	0.77	0.68	0.06	0.01
	Rank	**1**	**1**	**2**	**2**	**3**	**3**	**4**	**4**

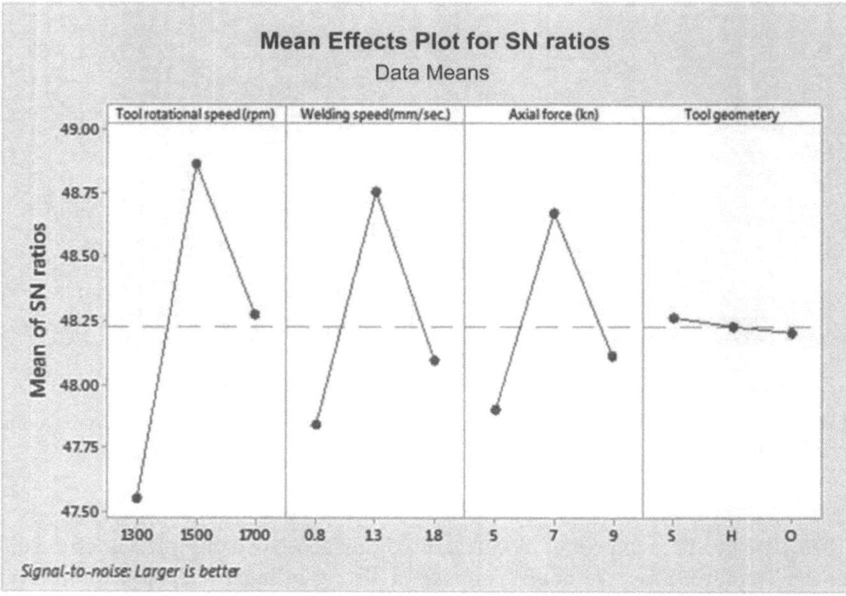

FIGURE 17.14 Mean effects plot for S/N (ratio) of tensile strength (adapted from Dhairya and Vikram 2019).

is graphically presented in Figures 17.14 and 17.15. These graphs were plotted through Minitab 17. The graphs indicate maximum tensile strength and hardness when rotational speed of tool, welding speed, axial force are at level 2, and tool geometry at level 1, i.e., rotational speed 1500 rpm, welding speed 1.3 mm/sec, axial force 7 KN and square tool geometry.

17.12.1 Results of Tool Rotational Speed on Tensile Strength and Hardness

Variation in tool rotational speed on tensile strength of AA7075/10%wt.SiC joints. The highest tensile strength was achieved at 1500 rpm tool rotational speed. The tensile strength of developed joint was very poor at 1300 rpm and 1700 rpm. With the increase in the tool rotational speed beyond 1300 rpm, the tensile strength also increased and achieved its highest value at rotational speed of 1500 rpm. Increase the tool rotational speed above 1500 rpm, the tensile strength of the joint will decrease. Rotational speed below 1300 rpm results in low heating conditions and poor movement of the pin and improper coupling of the work. So, lower tensile strength was achieved. Increment in the speed of rotation enhances the input heat along the joint length, thus improving the tensile strength. Increasing the rotational speed (more than 1500 rpm) can develop too much stirring on the top surface of the welded metal, generating microscopic pores in the stirring area. Rising temperature and poor cooling rate can also reduce tensile strength at higher tool rotational speeds.

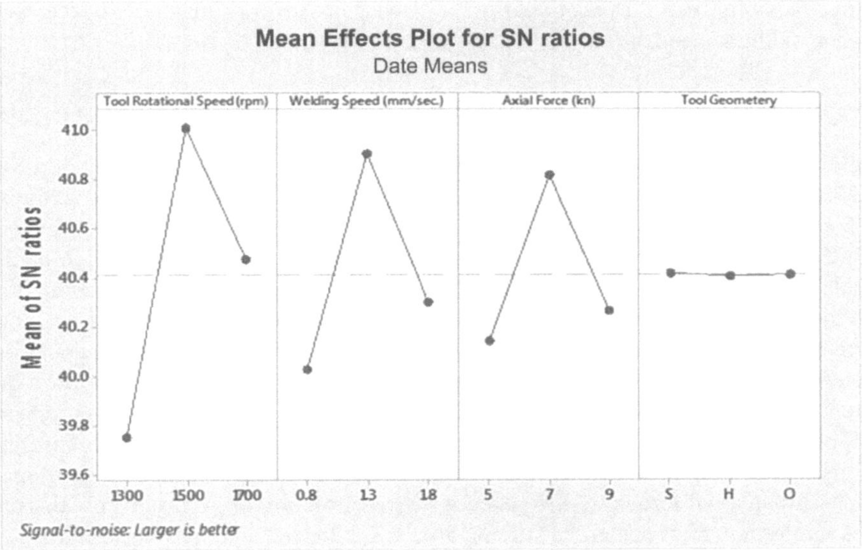

FIGURE 17.15 Mean effects plot for S/N (ratio) of hardness (adapted from Dhairya and Vikram 2019).

The effect of variation in tool rotational speed for hardness of AA7075/10%wt [23]. SiC composite welded joints is shown is Figure 17.17. A moderate quantity of heat is achieved at 1500 rpm, which is necessary to make a suitable welding joint and little change occurs in the mechanical properties of the joint, and defect-free weld joints were be produced. At this stage, re-crystallization of grain structure develops. The grains become fine and uniform. At 1300 rpm, sufficient heat is not produced, which causes improper mixing of plasticized metal. Due to this, flaws and gaps in the weld area are formed, which seem as defects [21]. Excessive heat is generated at 1700 rpm, which causes turbulence in the plasticized material, which tends to tunnel defects in the weld area.

17.12.2 Results of Welding Speed on Tensile Strength and Hardness

Variation in welding speed on tensile strength of AA7075/10%wt.SiC joints is shown in Figure 17.14. The tensile strength of fabricated joint is minimum at a low speed of welding (0.8 mm/s). Tensile strength increases because of increase in welding speed up to a value of 1.3 mm/s. A greater speed of welding will reduce heat input to the work piece, lead to low plastic flow of the metal and produce defects in the weld. This limits growth of grains and reduces the weld width. Thus, low tensile strength is achieved.

The variation of welding speed on hardness is shown in Figure 17.15. FSW joints have low hardness value at welding speed below 0.8 mm/s. The hardness value improves as the welding speed improves, until it reaches a maximum value of 1.3 mm/s. Further increasing the speed of welding will reduce the hardness of FSW joints. It is observed that frictional input of heat to the work material reduces at a higher welding speed.

17.12.3 Results of Axial Force on Tensile Strength and Hardness

Figure 17.14 shows the consequence of axial force on tensile strength of FSW AA7075/10%wt.SiC composite welded joints. Minimum tensile strength is achieved at 5 kN and 9 kN axial force. The maximum tensile strength of developed joint is achieved at 7 kN. If the axial load increases beyond 7 kN, tensile strength of the welded joint decreases. A huge quantity of input heat is produced by the rotation of the tool, which makes the metal plastic and very hot. The axial force has an influence on the insertion of tool pin. Welded joint of the material has dependency on the applied axial force of the extrusion process and the tool rotation. With low axial force such as 5 kN, minimum heat is generated, which is insufficient to produce a sufficient plastic condition. At a high axial force such as 9 kN, the tool is inserted deep into the work piece, which will greatly reduce the strength [22]. The developed joint with an axial force of 7 kN generates a fine structure of grains, evenly distributes reinforcing particles in the stirring area and achieves high tensile strength.

Figure 17.15 presents the effect for variation of axial force of FSW composite joints on hardness. At the axial force of 5 kN and 9 kN, lowest hardness values are achieved. The hardness of the joint increases at the axial force of 7 kN. The hardness of the joint decreases with increases in axial force. In the friction welding process, the tool rotation will generate a huge amount of heat, which makes the work material very hot and plastic.

17.12.4 Results of Tool Pin Geometry on Tensile Strength and Hardness

Figures 17.14 and 17.15 show the variation in tool pin profiles on the tensile strength and hardness. The square profile pin of tool provides maximum value of tensile strength and hardness due to proper material mixing during welding. In the square pin, the tool consists of four cutting edges. The cutting edges of each tool pin are treated like a separate cut, which will lead to maximum deformation of the material. Thus, welded joint achieved better tensile strength and hardness and no defect. The hexagonal and octagonal profiles of the pin will produce improper mixing because the tool pin may not deform the material correctly during rotation.

17.13 CONCLUSIONS

AA7075/10%wt.SiC (particulate size 20–40 μm) composite was successfully developed using stir casting process. SEM fractography and microstructure analysis of AA7075/10%wt.SiC reveal that distribution of reinforcing particles of SiC in matrix was uniform and homogeneous. The XRD of developed composite confirms the existence of aluminum and the other elements of matrix alloy. EDAX analysis confirms that the main components like Mg, Si, Zn and Cu are present in major quantity. Thermal analysis confirms no material loss in the fabricated composites. The developed AA7075/10%wt.SiC composite has an improvement of 44.23% and 51.37% in tensile strength and hardness, when compared to the Al7075 alloy.

REFERENCES

1. Kim, D., W. Lee, J. Kim, C. Kim, and K. Chung. "Formability evaluation of friction stir welded 6111-T4 sheet with respect to joining material direction." *International Journal Of Mechanical Sciences* 52, no. 4 (2010): 612–625.
2. Masounave, J., and F. G. Hamel. 1990. "Fabrication of Particulate Reinforced Metal Composites." *International, Montreal, Que, Canada*, 79–86. Proceedings of an International Conference Held September 17–29, 1990.
3. Azimzadegan, T., and S. Serajzadeh. "An investigation into microstructures and mechanical properties of AA7075-T6 during friction stir welding at relatively high rotational speeds." *Journal of Materials Engineering and Performance* 19, no. 9 (2010): 1256–1263.
4. Elangovan, K., V. Balasubramanian, and S. Babu. "Predicting tensile strength of friction stir welded AA6061 aluminum alloy joints by a mathematical model." *Materials & Design* 30, no. 1 (2009): 188–193.
5. Verma, S., and J. P. Misra. "Study on temperature distribution during friction stir welding of 6082 aluminum alloy." *Materials Today: Proceedings* 4, no. 2 (2017): 1350–1356.
6. Bozkurt, Y., and Z. Boumerzoug. "Tool material effect on the friction stir butt welding of AA2124-T4 Alloy Matrix MMC." *Journal of Materials Research and Technology* 7, no. 1 (2018): 29–38.
7. Sato, Y. S., H. Takauchi, S. H. C. Park, and H. Kokawa. "Characteristics of the kissing-bond in friction stir welded Al alloy 1050." *Materials Science and Engineering: A* 405, no. 1–2 (2005): 333–338.
8. Hashim, J., L. Looney, and M. S. J. Hashmi. "Metal matrix composites: Production by the stir casting method." *Journal of Materials Processing Technology* 92 (1999): 1–7.
9. Mishra, R. S., and Z. Y. Ma. "Friction stir welding and processing." *Materials Science and Engineering R: Reports* 50, no. 1–2 (2005): 1–78.
10. Pathak, J. P., J. K. Singh, and S. Mohan. "Synthesis and characterisation of aluminium-silicon-silicon carbide composite." *Indian Journal of Engineering and Materials Sciences* 13, no. 3 (2006): 238–246. Center of advanced study, department of metallurgical engineering, institute of technology Banaras Hindu university, Varanasi 221 005, India.
11. Bhushan, R. K., S. Kumar, and S. Das. "Fabrication and characterization of 7075 Al alloy reinforced with SiC particulates." *The International Journal of Advanced Manufacturing Technology* 65, no. 5–8 (2013): 611–624.
12. Bharath, V., M. Nagaral, V. Auradi, and S. A. Kori. "Preparation of 6061Al-Al2O3 MMC's by stir casting and evaluation of mechanical and wear properties." *Procedia Materials Science* 6 (2014): 1658–1667.
13. Uzun, H. "Friction stir welding of SiC particulate reinforced AA2124 aluminium alloy matrix composite." *Materials & Design* 28, no. 5 (2007): 1440–1446.
14. Jain, S., N. Sharma, and R. Gupta. "Dissimilar alloys (AA6082/AA5083) joining by FSW and parametric optimization using Taguchi, grey relational and weight method." *Engineering Solid Mechanics* 6, no. 1 (2018): 51–66.
15. Yunus, Mohammed, and Mohammad S. Alsoufi. "Mathematical modelling of a friction stir welding process to predict the joint strength of two dissimilar aluminium alloys using experimental data and genetic programming." *Modelling and Simulation in Engineering*, vol. 2018 (2018): article id 4183816.
16. Ramesh Babu, S., P. Karthik, S. Karthik, S. Arun Kumar, and J. Marris (editor name). "Optimization of Process Parameters during Friction Stir Welding of dissimilar Aluminium Alloys (AA 5083 & AA 6061) using Taguchi L9 Orthogonal Array." In *Applied Mechanics and Materials*, vol. 592, pp. 630–635. Trans Tech Publications Ltd, 2014.

17. Asif, M. M., K. A. Shrikrishna, and P. Sathiya. "Optimization of process parameters of friction welding of UNS S31803 duplex stainless steels joints." *Advances in Manufacturing* 4, no. 1 (2016): 55–65.
18. Raja, A. R., M. Vashista, and M. Z. Khan Yusufzai. "Estimation of material properties using hysteresis loop analysis in friction stir welded steel plate." *Journal of Alloys and Compounds* 814 (2020): 152265.
19. Veeresh Kumar, G. B., C. S. P. Rao, and N. Selvaraj. "Studies on mechanical and dry sliding wear of Al6061–SiC composites." *Composites Part B: Engineering* 43, no. 3 (2012): 1185–1191.
20. Huang, J.-h., Y.-l. Dong, Y. Wan, X.-k. Zhao, and H. Zhang. "Investigation on reactive diffusion bonding of SiCp/6063 MMC by using mixed powders as interlayers." *Journal of Materials Processing Technology* 190, no. 1–3 (2007): 312–316.
21. Barcellona, A., G. Buffa, L. Fratini, and D. Palmeri. "On microstructural phenomena occurring in friction stir welding of aluminium alloys." *Journal of Materials Processing Technology* 177, no. 1–3 (2006): 340–343.
22. McDanels, D. L. "Analysis of stress-strain, fracture, and ductility behavior of aluminum matrix composites containing discontinuous silicon carbide reinforcement." *Metallurgical Transactions A* 16, no. 6 (1985): 1105–1115.
23. Singh, Dhairya Pratap, Sudhir Kumar, and Vikram Singh. "Optimization of process parameters for friction stir welding AA7075/10% wt. SiC fabricated composite." *International Journal of Advances in Engineering Sciences* 8, no. 1-7 (2018).

18 Generation Scheduling of Solar-Wind-Hydro-Thermal Power System with Pumped Hydro Energy Storage using Squirrel Search Algorithm (SSA)

Chitralekha Jena, Pampa Sinha, Lipika Nanda, Sarita Samal, Babita Panda, and Arjyadhara Pradhan

CONTENTS

DOI: 10.1201/9781003229018-18

18.1 INTRODUCTION

Till now, the principal source of electric power production is the fossil-fuel-fired power plants. Sudden rise in electric power demand, diminution of fossil fuel and global warming have shown the path for utilizing green energy throughout the globe. The apprehension on climate change has gained acceptance for overcoming the demand for energy at reduced cost with no deleterious emissions by using clean and green energy. Incorporation of electric power sources like wind and solar which are climate-driven has resulted in higher uncertainties. The uncertain things that need to be taken into consideration are load variation, velocity of wind and solar irradiation. In generation scheduling problem, the irregularity and changeability of the resources confabulate a greater challenge which is to be trounced off. Entire grids may have a detrimental effect due to this blinking nature. Pumped hydro energy storage integration can overcome it, improving the supply variations as well as generation.

Optimal hydrothermal scheduling is a complicated problem that requires good-quality computational methods. Several classical deterministic techniques, for example, nonlinear programming with network flow, decomposition technique, dynamic programming, mixed integer programming, Lagrange relaxation [1–7], etc., have been used for solving this problem. Most of the classical deterministic techniques encounter difficulties in the fuel cost curves of thermal plants as well as hydro plants. Of these techniques, only dynamic programming is capable of handling non-linearity and non-convexity but it suffers from "curse of dimensional."

Evolutionary computation techniques are populace-based derivative-free search techniques applied to any optimization problem. Evolutionary computation methods, for example, evolutionary programming technique, simulated annealing technique, differential evolution, particle swarm optimization (PSO) [8–13], modified chaotic differential evolution [14] are used for scheduling hydrothermal units optimally evading all limitations.

Optimal generation scheduling in a sovereign system having renewable energy sources is discussed in [15]. Although renewable energy sources are pollution as well as dirt free, power generation capacity is less. Usage of amalgam sources of energy, for example, thermal with wind power [16], solar plant with thermal [17], integration of hydrothermal with wind power [18] and solar power with hydrothermal [19], has come up.

The pumped-storage-hydraulic (PSH) unit achieves large attention due to its energy storage features [20]. A PSH unit's key function [21] is to store the extra energy available during off-peak load time at cheaper cost. Thus, typically a PSH unit operates every day or monthly. The average cost of fuel can be minimized in a power plant by the PSH unit.

For finding the optimal scheduling hydrothermal plants with PSH unit considering constraints, Lagrangian multiplier along with gradient search techniques [22] are used. Khandualo et al. [23] utilized evolutionary programming methodology to resolve the problem of hydrothermal system output/pumping preparation for pumped storage facilities. Mohan et al. [24] addressed the importance of PHU unit. Ma et al. [25] has studied a solar energy penetration hydro-pumped storage unit and in particular small self-sufficient systems in remote regions. Surender Reddy [26] addresses optimal scheduling of thermal-wind-solar power system with storage. Lakshminarasimman and S. Subramanian [27] discuss short-term scheduling of hydrothermal power system problem with cascaded reservoirs by using modified differential evolution. Now a nature-inspired optimization technology, the Squirrel search algorithm (SSA), has arisen from the complex scavenging behavior, which Jain, Singh and Rani have demonstrated by drawing on southern flying squirrels as well as their successful locomotive methods [28]. Thomas and Weigl [29] address the dynamic foraging behavior in the southern flying squirrel.

The searching system starts when flying squirrels start foraging. When the weather is warm, squirrels find sources of food and they change their sites as well as move around the regions of the forest. As the weather is warm, they gather acorns quickly which are available in abundance, and eat the acorns instantly. They wander for the optimal food supply (hickory nuts) for winter after meeting their everyday energy requirements. When the nuts are processed, they maintain their energy demand in poor weather, thus increasing the possibility of existence. Finally, flying squirrels become animated at the end of winter. This procedure is cyclic and goes on until the squirrel's natural life, and the foundation of SSA is built on this process.

In this situation, SSA is utilized for solving the scheduling of a solar-wind-hydrothermal system having a pumped hydroelectric storage scheme with multiple hydropower storage systems and thermal units taking into consideration the valve-point consequence. Thermal generator's spectrum limits are also considered, like volatility of solar and wind supplies and transmission losses. Here there are two evaluation devices. The outcomes of simulation were compared with PSO as well as grey wolf optimization (GWO). The proposed SSA represents an advantageous approach.

18.2 FORMULATION OF PROBLEM

Because of the insignificant cost of hydro-power the solar-wind-hydro-thermal including pumped hydro-energy storage is considered to decrease the fuel cost of thermal plant, cost of wind generating unit and Solar Photovoltaic plant, The following objective function as well as shortcomings are included in the problem formulation definition:

$$F_C = \sum_{t=1}^{T} \left[\sum_{i=1}^{N_t} \{ f_{sit}(\mathrm{P}_{sit}) \} + \sum_{k=1}^{N_W} \{ \mathrm{K}_{wk} \times \mathrm{P}_{wkt} + O_{wkt}(\mathrm{P}_{wkt}) + U_{wkt}(\mathrm{P}_{wkt}) \} \right.$$
$$\left. + \sum_{m=1}^{N_{PV}} \{ \mathrm{K}_{sm} \times \mathrm{P}_{PVmt} + O_{PVmt}(\mathrm{P}_{PVmt}) + U_{PVmt}(\mathrm{P}_{PVmt}) \} \right] \tag{18.1}$$

A repeated, rectified sinusoidal contribution complements the quadratic function to model. The thermal generator's fuel cost feature is suggested as taking the valve-point effect into account.

$$\begin{aligned} f_{sit}(\mathrm{P}_{sit}) &= a_{si} + b_{si}\mathrm{P}_{sit} + c_{si}\mathrm{P}_{sit}^2 \\ &+ \left| d_{si} \times \sin\left\{ e_{si} \times \left(\mathrm{P}_{si}^{\min} - \mathrm{P}_{sit} \right) \right\} \right| \end{aligned} \tag{18.2}$$

The overestimation reserve costs and the underestimation penalty cost for wind power dispatch are modeled in Equations (18.3)–(18.4),

$$O_{wkt}(\mathrm{P}_{wkt}) = o_{wk} \times \int_{\mathrm{P}_{wkt}^{\min}}^{\mathrm{P}_{wkt}} (\mathrm{P}_{wkt} - y) \times f_w(y)\, dy \tag{18.3}$$

$$U_{wkt}(\mathrm{P}_{wkt}) = u_{wk} \times \int_{\mathrm{P}_{wkt}}^{\mathrm{P}_{wkt}^{\max}} (y - \mathrm{P}_{wkt}) \times f_w(y)\, dy \tag{18.4}$$

The overestimation reserve costs and the underestimation penalty cost for solar power dispatch are modeled in Equations (18.5)–(18.6).

$$O_{PVmt}(\mathrm{P}_{PVmt}) = o_{PVm} \times \int_{\mathrm{P}_{PVmt}^{\min}}^{\mathrm{P}_{PVmtt}} (\mathrm{P}_{PVmt} - x) \times f_{PV}(x)\, dx \tag{18.5}$$

$$U_{PVmt}(\mathrm{P}_{PVmt}) = u_{PVm} \times \int_{\mathrm{P}_{PVmt}}^{\mathrm{P}_{PVmt}^{\max}} (x - \mathrm{P}_{PVmt}) \times f_{PV}(x)\, dx \tag{18.6}$$

subject to different constraints.

18.2.1 Different Constraints

18.2.1.1 Constraints of Power Balance

$$\sum_{i=1}^{\mathrm{N}_t} \mathrm{P}_{sit} + \sum_{j=1}^{\mathrm{N}_h} \mathrm{P}_{hjt} + \sum_{k=1}^{\mathrm{N}_w} \mathrm{P}_{wkt} + \sum_{m=1}^{\mathrm{N}_{PV}} \mathrm{P}_{PVmt} + \mathrm{P}_{ght} - \mathrm{P}_{Dt} - \mathrm{P}_{Lt} = 0, t \in \mathrm{T}_{gen} \tag{18.7}$$

$$\sum_{i=1}^{\mathrm{N}_t} \mathrm{P}_{sit} + \sum_{j=1}^{\mathrm{N}_h} \mathrm{P}_{hjt} + \sum_{k=1}^{\mathrm{N}_w} \mathrm{P}_{wkt} + \sum_{m=1}^{\mathrm{N}_{PV}} \mathrm{P}_{PVmt} - \mathrm{P}_{pht} - \mathrm{P}_{Dt} - \mathrm{P}_{Lt} = 0, t \in \mathrm{T}_{pump} \tag{18.8}$$

Hydroelectric power generation is based on the discharge water rate and the head of water reservoir, which as a result is a storage function.

$$P_{hjt} = C_{1j}V_{hjt}^2 + C_{2j}Q_{hjt}^2 + C_{3j}V_{hjt}Q_{hjt} + C_{4j}V_{hjt} + C_{5j}Q_{hjt} + C_{6j}, j \in N_h\ t \in T \quad (18.9)$$

B-coefficient can be used to measure overall transmission loss P_{Lt}. Here, "total number of plants $N_T = N_t + N_h + N_w + N_{PV}$ and P_{im} is the respective thermal, hydro, wind power, solar PV generation."

$$P_{Lt} = \sum_{i=1}^{N_T}\sum_{j=1}^{N_T} P_{it}B_{ij}P_{jt} + \sum_{i=1}^{N_T} B_{0i}P_{it} + B_{00} \quad (18.10)$$

18.2.1.2 Wind Power Model

At time t, the o/p power [30] of k-th wind generating power unit corresponding to a certain speed of wind is defined by

$$\begin{aligned} P_{wkt} &= 0, && \text{for } v_{wt} < v_{in} \text{ and } v_{wt} > v_{out} \\ P_{wkt} &= P_{wrk}\left(\frac{v_{wt} - v_{in}}{v_r - v_{in}}\right), && \text{for } v_i \le v_{wt} \le v_r \\ P_{wkt} &= P_{wrk}, && \text{for } v_r \le v_{wt} \le v_{out} \end{aligned} \quad (18.11)$$

18.2.1.3 Solar Power Model

Power output [31] from PV cell is shown by

$$\begin{aligned} P_{PVmt} &= P_{sr}\left(\frac{G^2}{G_{std}R_c}\right), && \text{for } 0 < G < R_c \\ P_{PVmt} &= P_{sr}\left(\frac{G}{G_{std}}\right), && \text{for } G > R_c \end{aligned} \quad (18.12)$$

18.2.1.4 Pumped-Storage Constraints

PSH unit is totally dependent on water pumped from lower reservoir to upper reservoir.

$$V_{res(t+1)} = V_{res,t} + Q_{pht}\left(P_{pht}\right), t \in T_{pump} \quad (18.13)$$

$$V_{res(t+1)} = V_{res,t} - Q_{ght}\left(P_{ght}\right), t \in T_{gen} \quad (18.14)$$

$$P_{gh}^{\min} \le P_{ght} \le P_{gh}^{\max} \quad t \in T_{gen} \quad (18.15)$$

$$P_{ph}^{\min} \le P_{pht} \le P_{ph}^{\max} \quad t \in T_{pump} \quad (18.16)$$

$$V_{res}^{\min} \leq V_{res,t} \leq V_{res}^{\max}, t \in \mathrm{T} \tag{18.17}$$

The initial as well as final water volume of the upper unit of PSH reservoir should be equal to zero in this issue. Thus, the overall net water volume of the PSH unit should be equivalent to zero.

$$V_{res,0} = V_{res,\mathrm{T}} = V_{res}^{start} = V_{res}^{end} \tag{18.18}$$

18.2.1.5 Generation Limits

$$\mathrm{P}_{hj}^{\min} \leq \mathrm{P}_{hjt} \leq \mathrm{P}_{hj}^{\max} \quad j \in \mathrm{N}_h \quad t \in \mathrm{T} \tag{18.19}$$

$$\mathrm{P}_{si}^{\min} \leq \mathrm{P}_{sit} \leq \mathrm{P}_{si}^{\max} \quad i \in \mathrm{N}_t, t \in \mathrm{T} \tag{18.20}$$

18.2.1.6 Limits of Ramp Rate

$$\mathrm{P}_{sit} - \mathrm{P}_{si(t-1)} \leq UR_i, \quad i \in \mathrm{N}_t, t \in \mathrm{T}$$

$$\mathrm{P}_{si(t-1)} - \mathrm{P}_{sit} \leq DR_i, \quad i \in \mathrm{N}_t, t \in \mathrm{T} \tag{18.21}$$

18.2.1.7 Constraints of Hydraulic Network

The hydraulic outfitted constraints comprise the water balance equations for each hydro unit in addition to the bounds on reservoir storage and release targets. These bounds are decided by the physical reservoir and plant limitations as well as the multipurpose necessity of the hydro system. The hydraulic machinery limitations contain water balance equations of every hydro unit, taken with limits to storage reservoir and release priorities. The physical reservoir along with plant limits and also the multifunctional need for the hydro system assess the limitations. These drawbacks contain:

a. Physical drawbacks on reservoir discharge rates and storage volumes,

$$V_{hj}^{\min} \leq V_{hjt} \leq V_{hj}^{\max} \quad j \in \mathrm{N}_h, t \in \mathrm{T} \tag{18.22}$$

$$Q_{hj}^{\min} \leq Q_{hjt} \leq Q_{hj}^{\max} \quad j \in \mathrm{N}_h, t \in \mathrm{T} \tag{18.23}$$

b. Continuity equation for the hydro reservoir network

$$V_{hj(t+1)} = V_{hjt} + I_{hjt} - Q_{hjt} - S_{hjt} + \sum_{l=1}^{R_{uj}} (Q_{hl(t-\tau_{lj})} + S_{hl(t-\tau_{lj})}), j \in \mathrm{N}_h, t \in \mathrm{T} \tag{18.24}$$

18.3 SQUIRREL SEARCH ALGORITHM

SSA has recently been developed by Jain, Singh and Rani as a populace-based swarm intelligence algorithm [28]. It draws on south-flying squirrel's complex scavenging behavior and proficient patterns of flight called gliding. It begins with the initial flight squirrel random location (SQs). The place of an SQ in *dm* dimensional search space is stated as a vector. The SQs will then slide into one space, two, three or more dimensions and can change their vectors in their place. The number of SQs is 100 in this situation.

18.3.1 Initialization

Taking *N* as the number of SQs, the *a*-th position of SQ may be defined in the vector form. As shown in Equation (18.25), the matrix gives information about each and every SQ position.

$$\begin{bmatrix} SQ_{1,1} & SQ_{1,2} & \cdots & \cdots & SQ_{1,dm} \\ SQ_{2,1} & SQ_{2,2} & \cdots & \cdots & SQ_{2,dm} \\ \vdots & \vdots & \vdots & \vdots & \vdots \\ \vdots & \vdots & \vdots & \vdots & \vdots \\ SQ_{N,1} & SQ_{N,2} & \cdots & \cdots & SQ_{N,dm} \end{bmatrix} \quad (18.25)$$

where $SQ_{a,b}$ shows "b-th dimension of a-th SQ. The initial location assignment for SQs takes place in the same way as defined in Equation (18.26).

$$SQ_a = SQ_{al} + U(0,1) \times (SQ_{au} - SQ_{al}) \quad (18.26)$$

where SQ_{au} and SQ_{al} indicate limits of maxim and minim "imposed on the a-th SQ in the b-th dimension respectively. U(0,1) defines the uniformly distributed random number in [0,1]."

18.3.2 Calculation of Value of Fitness

Equivalent to every position of Squirrel location, the fitness value computation is gained by substituting the decision variables' values such as "solution vectors in a user-specified fitness function." In the array mentioned in Equation (18.27), relevant values get stored.

$$\begin{bmatrix} f_1\left(SQ_{1,1} \quad SQ_{1,2} \quad \cdots \quad \cdots \quad SQ_{1,dm} \right) \\ f_2\left(SQ_{2,1} \quad SQ_{2,2} \quad \cdots \quad \cdots \quad SQ_{2,dm} \right) \\ \vdots \\ \vdots \\ f_N\left(SQ_{N,1} \quad SQ_{N,2} \quad \cdots \quad \cdots \quad SQ_{N,dm} \right) \end{bmatrix} \quad (18.27)$$

Fitness value in the location of each of the SQ indicates the characteristic of the food source that it is hunting, respectively, its optimum source of food (hickory tree), normal source of food (acorn tree) as well as its survival probability.

18.3.3 Announcement, Sorting, and Selection

The first step is to find the fitness value conforming to every SQ location, and then order the list from low value to high value. SQ is represented to be on hickory nut tree with lowest fitness rating. SQ. The next nine SQs are therefore supposed to be present in the nut trees of the acorn, while these nine SQs are still supposed to be going in the path of hickory nut tree. Remnant SQs on common trees are known to be present. In view of random selection, certain SQs are expected to switch to a hickory nut tree on the premise that their everyday energy needs have come to fruition. The remaining SQs are going to move in the same path as that of acorn nut trees. SQs are still affected by living predators. The incorporation of location updates with probability of deprecatory habitation models this natural reality (PD_{prb}).

18.3.4 Creation of New Positions

Considering the above facts, three circumstances can be possible during the SQs' dynamic scavenging. In every condition, during the absence of the predator, it is assumed that SQs glide and explore efficiently inside the whole forest for finding their eatable(s). Throughout "the predator existence, SQs are forced to shift in small walking steps arbitrarily finding a nearby hiding location. Dynamic foraging mathematical modeling incorporating expressions is given.

Case 1. SQs on acorn nut trees (SQ_{acn}) walk in the direction of hickory nut tree. As a result, immediate position of these SQs is shown as

$$SQ_{acn}^{ni+1} = \begin{cases} SQ_{acn}^{ni} + d_{gl} \times G_{cn} \times \left(SQ_{hck}^{ni} - SQ_{acn}^{ni} \right) & RN_1 \geq PD_{prb} \\ Arbitrary \quad position & Otherwise \end{cases} \tag{18.28}$$

where gliding distance is defined by di_{gl}, arbitrary number is defined by RN_1 having [0, 1], SQ position is represented by SQ_{hck} who came to the hickory nut tree, and on-going iteration is represented by ni. GL_{cn} helps in attaining the balance between exploitation as well as exploration. The efficiency of the proposed algorithm has a considerable impact. Now, the GL_{cn} value is considered as 1.9.

Case 2. (SQ_{nrm}), squirrels on normal trees, walk to the acorn nut trees for fulfilling the need of daily energy. Hence, the current position of SQs is given as:

$$SQ_{nrm}^{ni+1} = \begin{cases} SQ_{nrm}^{ni} + d_{gl} \times G_{cn} \times \left(SQ_{acn}^{ni} - SQ_{nrm}^{ni} \right) & RN_2 \geq PD_{prb} \\ Arbitrary \quad position & Otherwise \end{cases} \tag{18.29}$$

where RN_2 indicates any number between $[0,1]$.

Case 3. A few SQs present on normal and acorn nuts can take and advance to the Hickory Nut Tree which may be helpful when sources are unavailable. The current location of the SQs can be represented in this scenario

$$SQ_{nrm}^{ni+1} = \begin{cases} SQ_{nrm}^{ni} + d_{gl} \times G_{cn} \times \left(SQ_{hck}^{ni} - SQ_{nrm}^{ni}\right) & RN_3 \geq PD_{prb} \\ Arbitrary \quad position & Otherwise \end{cases} \tag{18.30}$$

where RN_3 indicates any number within $[0,1]$.

18.3.5 Aerodynamics of Gliding

SQs' gliding technique is described as EG "equilibrium glide." In EG, an (L_f) addition of lift with (D_r) drag force gives a (R_f) resultant force. R_f value is same as well as opposite in direction to the SQ weight (W_{SQ}). Hence, a linear gliding path is acquired by the SQs having CV "constant velocity." SQ, which is gliding at constant "speed, always go down at θ angle to horizontal. L_f to D_r proportion is given as per Equation (18.31).

$$\frac{L_f}{D_r} = \frac{1}{\tan\theta} \tag{18.31}$$

By generating a smaller glide angle (θ), the SQs make an increment in glide-path length. L_f to D_r ratio is also enhanced by this increment.

L_f which is obtained from air's downward deflection of air, whereas path over the wings are identified as [29]

$$L_f = 1/2\rho C_{L_f} (CV)^2 S_A \tag{18.32}$$

where ρ gives the density of the air (1.204 kg/m^3), lift coefficient is defined by C_{L_f}, CV is speed (5.25 m/s), and S_A gives the surface area of body. The frictional D_r can be stated as per (18.33).

$$D_r = 1/2\rho (CV)^2 S_A C_{FD_r} \tag{18.33}$$

where C_{FD_r} shows coefficient of frictional drag. This drag component remains dominant at low speed. Component dominance is not high during high speed. θ at the steady state is given below from equation (18.29),

$$\theta = \arctan\left(\frac{D_r}{L_f}\right) \tag{18.34}$$

The di_{gl} estimated roughly is computed as per Equation (18.35).

$$d_{gl} = \left(h_{gl} / \tan\theta\right) \tag{18.35}$$

where h_{gl} gives the height loss after the gliding.

SQ has the glide length's changing ability or d_{gl} by creating modifications in the L_f to D_r ratio taking into consideration the position meant for landing. To attain satisfactory algorithmic outcome, d_{gl} value is scaled down. "Scaling down means division by a non-zero numeral called as scaling factor (sc_f). C_{L_f} variations between 1.5 and 0.675 remain convenient, while C_{D_r} is kept fixed at 0.60."

18.3.6 Condition of Seasonal Monitoring

Foraging SQs' activity is significantly influenced by seasonal variations [29]. SQs' movements are also driven by various weather conditions. Taking such actions into account is a practical optimization strategy. There is also no consequence of local Optima being stuck in the inclusion of a seasonal oversight condition(s). This is addressed by the calculations below.

A. Seasonal constant calculation (Se_c)

$$Se_c^{ni} = \sqrt{\sum_{m=1}^{e}\left(SQ_{acn,m}^{ni} - SQ_{hck,m}^{ni}\right)^2} \text{ for } \forall tm = 1,2,... \tag{18.36}$$

B. Checking seasonal monitoring condition, ensuring that

$$Se_c^{tm} \prec Se_{\min} \tag{18.37}$$

where $Se_{\min}$ gives the seasonal constant's least possible value, which is calculated as:

$$Se_{\min} = \frac{10E^{-6}}{(365)^{ni/(ni_{\max}/2.5)}} \tag{18.38}$$

where $ni_{\max}$ and ni give the maximum iteration and present value respectively.

18.3.7 Relocating Arbitrarily Subsequently Ending of the Winter Season

When the winter season is over, the SQs who cannot find food in the winter season are randomly dispersed. The foraging SQs' activity is significantly influenced by seasonal variations [29]. This is addressed by the calculations below.

$$SQ_{nrm}^{new} = SQ_l + Lévy(t) \times (SQ_u - SQ_l) \tag{18.39}$$

According to Equation (18.39), SQs are relocated where Lévy delivery facilitates increased and effective research space exploration. This distribution is defined in Equation (18.40).

$$L(s) \sim |s|^{-1-\alpha} \tag{18.40}$$

where $0 \prec \alpha \leq 2$ signifies an index. Mathematical expression is given as per Equation (18.41)

$$L(s,\beta,\mu) = \begin{cases} \sqrt{\frac{\beta}{2\pi}} \exp\left(\frac{-\beta}{2(s-\mu)}\right)\frac{1}{(s-\mu)^{1.5}} & 0 \prec \mu \prec s \prec \infty \\ 0 & \textit{otherwise} \end{cases} \tag{18.41}$$

where $\beta, \mu \succ 0$ with μ and β gives shift and scale "parameter respectively. The Lévy flight" calculation is done as per Equation (18.42).

$$Lévy(x) = 0.01 \times \frac{RN_a \times \upsilon}{|RN_b|^{\frac{1}{\gamma}}} \tag{18.42}$$

RN_a and RN_b demonstrate "two normally distributed arbitrary numbers in [0, 1]. γ is a constant, and υ can found as per Equation (18.43) with $\Gamma(g) = \lfloor g-1$ (with 'g' as any particular number)."

$$\upsilon = \left(\frac{\Gamma(1+\gamma) \times \sin\left(\frac{\pi\gamma}{2}\right)}{\Gamma\left(\frac{1+\gamma}{2}\right) \times \gamma \times 2^{\left(\frac{\gamma-1}{2}\right)}} \right)^{\frac{1}{\gamma}} \tag{18.43}$$

18.4 NUMERICAL RESULTS

SSA was implemented to solve two various test systems. Results of simulation are matched up for finding the efficiency of the proposed SSA with GWO and PSO. The developed SSA, PSO and GWO were compared through operating MATLAB 7.0 on a PC (3.0 GHz, 1 TB, Pentium-IV). The overall working flow is described in Figure 18.1

18.4.1 First Test System

It consists of a cascaded chain of four major reservoir hydro plants, one solar PV plant, one equivalent generating unit of wind power, one pumped storage plant as well as three thermal power plants. Scheduling period is taken as 1 day, which is distributed into 24 intervals. Valve point ramp rate and loading effect limits of thermal generator and transmission loss are considered here. Hydrothermal systems details are taken from [27]. P_{wr} = 150 MW is the wind power generator rating. v_r = 15 m/s, v_o = 25 m/s and v_{in} = 4 m/sec are the wind speed rate, cut out, and cut in respectively. K_w (direct cost coefficient) is taken as 7 for wind power generator whereas u_{wk} (penalty cost) along with o_{wk} (reserve cost) are taken as 1 and 2 respectively.

P_{PVr} = 175MW is the solar PV generator rating. For the solar PV generator, K_s (direct cost coefficient) is taken as 6 whereas u_{PVm} (penalty cost) and O_{PVm} (reserve cost) are taken as 1 and 2 respectively. In R_c (definite radiation point) as well as G_{std} (standard environment), solar radiation were taken as 120 W/m^2 and 1000 W/m^2 respectively.

Figures A.1 and A.2 respectively indicate the higher as well as lower forecast limit for solar irradiation along with wind velocity. The overall time demand for load is shown in Table A.1. The characteristics of the pumped hydro storage facility are as follows:

Generating mode

Q_{ght} is positive while generating, P_{ght} is positive and $0 \le P_{ght} \le 100$ MW, $Q_{ght}(P_{ght}) = 70 + 2P_{ght}$ acre-ft/hr

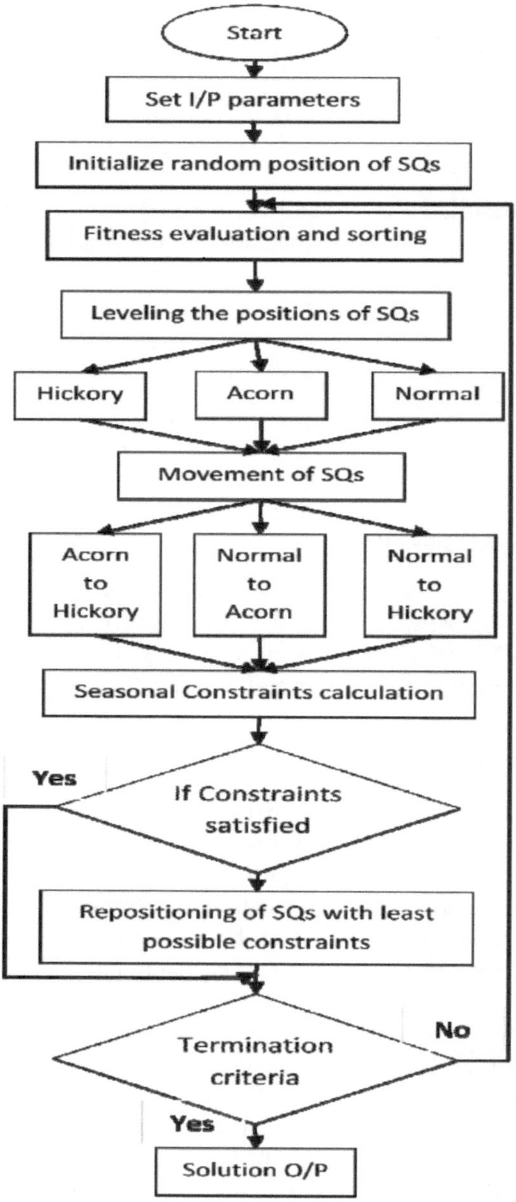

FIGURE 18.1 Flowchart of Squirrel Search Algorithm. How the Algorithm works, showing step by step procedure, starting from how the algorithm starts and ending at getting output.

Pumping mode

Q_{pht} is negative while pumping, P_{pht} is negative and $-100MW < P_{pht} \leq 0MW$, $Q_{pht}(P_{pht}) = -200$ acre-ft/hr with $P_{pht} = -100$ MW

Operating constraints: PHP is given permission to operate at –100 MW when pumping. Reservoir starts at 3000 acre-ft and at the end of the 24 hours, it must be at 3000 acre-ft. Water in flow rate is neglected and spillage is not considered.

SSA, GWO and PSO are used to solve the problem. Squirrel number is chosen as 100 in case of SSA. 10 is taken as nutritious food sources with 1 hickory and 9 acorn nut trees. 90 is the number of trees with no source of food. (G_c), the gliding constant, is chosen to be 1.9 with predator presence probability (P_{dp}) as 0.1. Wolves number is taken as 100 in GWO. In case of PSO, the parameters are chosen to be $N_P = 100$, $w_{max} = 0.25$, $w_{min} = 0.05$, $c_1 = 0.5$. For all these techniques, maximum iteration number is taken as 200.

Table 18.1 shows the solar-wind-hydro-thermal-pumped storage generations relating to best cost obtained with SSA. The best, average and worst costs from 100 runs of solution obtained from SSA, GWO and PSO are presented in Table 18.2. In Figures 18.2 and 18.3, the optimal hourly discharges and storage volume of reservoir are respectively shown for four hydro plants obtained from SSA. The characteristics of cost convergence acquired from recommended SSA, GWO and PSO are shown in Figure 18.4.

18.4.2 Second Test System

It consists of a multi-chain cascade of reservoir of four hydro plants, one equivalent wind power generating unit, one equivalent solar PV plant, ten thermal power plants and one PSH plant. Period of scheduling is 1 day and divided into 24 intervals. Valve point loading effect as well as limits of ramp rate of thermal generator are considered. Thermal power plant data is shown in Table A.2 in appendix. Wind power generating unit, solar PV plant and PSH plant are similar to the first test system. Table A.3 shows the total hourly load demand.

SSA, GWO and PSO are used for solving the problem. Number of squirrels is taken as 100 in SSA. 10 is assumed as the number of nutritious food sources with 9 acorn and 1 hickory nut trees. The number of trees having no food source is 90. Gliding constant (G_c) and probability of predator presence (P_{dp}) are chosen as 1.9 and 0.1 respectively. Number of wolves taken in GWO are 100, and in PSO, the parameters taken are $N_P = 100$, $w_{max} = 0.25$, $w_{min} = 0.05$ and $c_1 = 0.5$. Maximum iteration number is 300 for all methods.

Table 18.3 shows the wind-hydro-solar-pumped storage generations related to best cost using SSA. The thermal generations corresponding to best cost obtained from recommended SSA are shown in Table 18.4. The best, average and worst costs with 100 runs of solution obtained from recommended SSA, GWO and PSO are shown in Table 18.5. The optimal hourly discharges and reservoir storage volumes of four hydro plants obtained from SSA are depicted in Figures 18.5 and 18.6 respectively. Cost convergence characteristics obtained from recommended SSA, GWO and PSO are depicted in Figure 18.7.

TABLE 18.1
Hydro-Thermal-Wind-Solar-Pumped Storage Generation (MW) Found from SSA of First Test System 1

Hour	P_{h1}	P_{h2}	P_{h3}	P_{h4}	P_{s1}	P_{s2}	P_{s3}	P_w	P_{PV}	P_{gh}	Ploss
1	58.8871	52.8636	20.9686	190.8077	210.7345	137.4034	23.1230	66.2561	0	−100.0000	11.0440
2	69.3857	51.6773	38.8239	226.5466	97.6447	65.0050	136.1445	100.4250	0	−100.0000	5.6527
3	77.3630	58.6358	44.0454	157.2826	133.2536	55.6841	28.9158	149.5846	0	−100.0000	4.7647
4	68.0793	57.4981	10.7044	154.3289	52.5919	156.2845	104.3909	150.0000	0	−100.0000	3.8780
5	75.4946	85.1369	32.2220	177.6826	40.0000	104.0819	98.7470	150.0000	9.9237	−100.0000	3.2889
6	66.7618	77.4784	46.5515	200.1240	60.6572	50.0000	117.6610	150.0000	34.7003	−100.0000	3.9343
7	54.8073	63.9171	48.9600	149.9620	90.9610	124.2020	109.0556	150.0000	62.8673	−100.0000	4.7323
8	93.9223	67.2854	31.5869	133.6970	147.5195	131.3733	140.9392	120.5825	91.7177	−100.0000	8.6239
9	86.2015	52.2351	43.8352	191.8059	40.0000	127.8760	161.8570	93.4504	117.1527	81.0657	5.4794
10	66.9339	59.5096	50.2765	181.8217	117.3832	251.0373	38.1274	24.2110	136.7314	60.8115	6.8435
11	89.3628	84.4456	52.6847	221.6820	111.4213	120.7339	26.7409	67.8264	166.4047	63.6589	4.9612
12	59.6033	48.2627	53.3568	186.7368	135.6634	222.8045	110.4272	0	156.7280	84.8529	8.4355
13	93.6584	42.9272	47.8001	163.4343	40.0000	270.8849	105.3462	13.4219	137.8475	100.0000	5.3207
14	61.9443	68.4874	55.8829	269.6054	155.2933	127.5492	129.2399	54.9799	108.4748	7.6217	9.0787
15	79.8103	52.5316	59.4153	279.2370	148.3405	132.8681	121.0009	40.5019	86.0162	18.8878	8.6097
16	76.7667	69.0080	60.5713	264.9535	40.0000	216.0143	80.5481	150.0000	39.2713	67.4962	4.6294
17	78.8820	55.9769	6.4535	195.5020	148.4544	350.7767	110.2622	35.4435	34.6814	46.1439	12.5765
18	96.7255	56.4154	59.6259	269.2742	67.9056	268.5988	91.2844	0	29.6097	87.0090	6.4485
19	68.7080	74.9735	54.4985	137.5686	40.0000	330.0581	91.7171	92.6814	3.0427	82.7643	6.0121
20	71.0346	53.4297	55.5513	266.2019	66.2451	238.5000	118.3693	7.3047	0	79.7116	6.3483
21	106.3291	63.6843	59.8101	189.6262	159.1822	202.5057	104.9861	33.4955	0	−100.0000	9.6191
22	89.5787	49.8742	48.5647	277.1426	150.8483	141.0448	27.9032	82.5631	0	−100.0000	7.5196
23	86.9642	60.8005	15.0723	220.9401	40.0000	229.5930	172.0069	31.8614	0	−100.0000	7.2383
24	84.3076	50.8268	5.8540	281.0085	124.4152	161.5644	81.4485	17.5501	0	−100.0000	6.9752

TABLE 18.2
Comparison of First Test System Performance

Techniques	Best Cost ($)	Average Cost ($)	Worst Cost ($)
SSA	57561.34	57562.67	57563.95
GWO	58147.69	58149.51	58152.37
PSO	58361.21	58364.85	58368.05

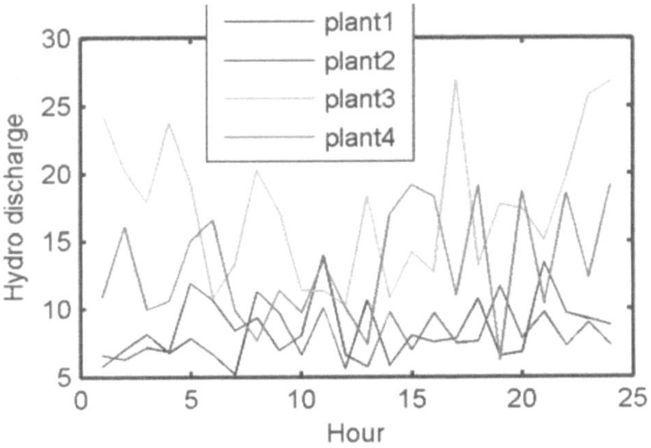

FIGURE 18.2 Hydro plant discharge ($\times 10^4 m^3$) of first test system using SSA. The optimal hourly discharges are shown for four hydro plants obtained from Squirrel search algorithm in a first test system

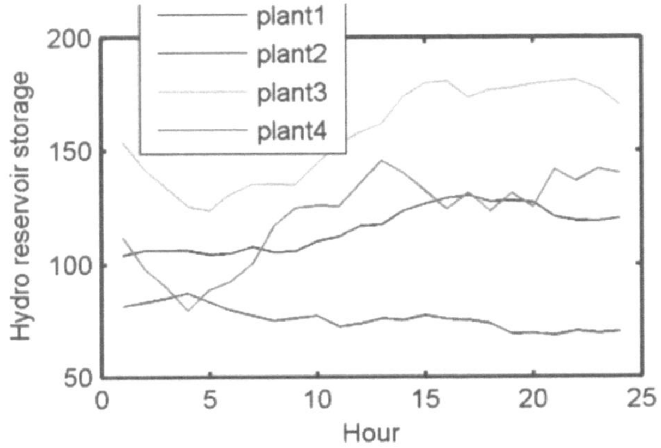

FIGURE 18.3 Hydro reservoir storage volumes ($\times 10^4 m^3$) of first test system using SSA. Storage volume of reservoir for four hydro plants obtained from Squirrel search algorithm for first test system

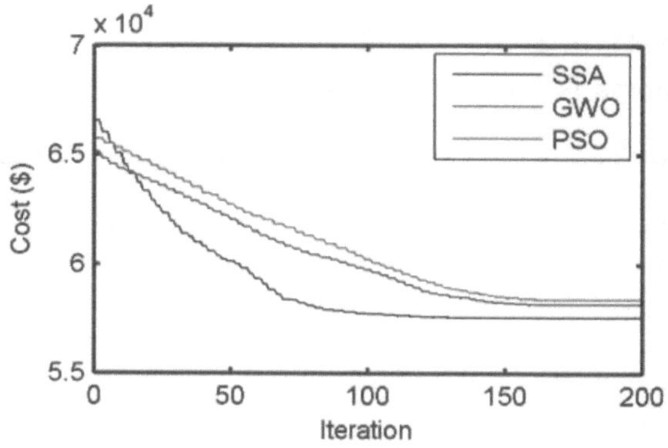

FIGURE 18.4 Cost characteristic convergence for first test system. Comparison between the cost characteristics obtained using particle swarm optimization and grey wolf optimization is shown for first test system

TABLE 18.3
Hydro-Wind-Solar-Pumped Storage Generation (MW) Taken from SSA of Second Test System

Hour	P_{h1}	P_{h2}	P_{h3}	P_{h4}	P_w	P_{PV}	P_{gh}
1	84.6950	41.9587	49.2559	164.0958	143.6643	0	−100.0000
2	72.4929	40.4222	56.7245	225.9519	150.0000	0	−100.0000
3	89.0215	51.2461	8.7917	183.7700	150.0000	0	−100.0000
4	88.8141	95.5803	52.6637	177.5252	150.0000	0	−100.0000
5	60.3908	78.9855	53.3645	201.4997	150.0000	15.0227	−100.0000
6	65.2594	59.9532	54.7760	101.5834	150.0000	23.9074	−100.0000
7	62.0009	24.0213	36.9264	210.5691	150.0000	71.5851	−100.0000
8	62.7769	56.5265	42.7316	84.6744	148.0307	97.8216	−100.0000
9	56.4907	76.4252	29.0099	124.8829	71.5226	112.4449	8.2707
10	100.2130	29.8691	41.3510	190.4814	82.8157	131.1476	30.7648
11	95.9578	65.5424	51.9928	165.2194	67.0295	162.4021	24.2347
12	52.5859	64.3362	50.7663	234.6590	0	155.9797	34.3270
13	68.2113	86.2360	45.9666	163.6574	5.7393	117.1615	98.7835
14	91.2097	62.5206	10.3548	238.7643	90.7793	110.1801	86.7332
15	63.6070	91.2369	35.0301	279.2274	88.4801	94.9215	100.0000
16	54.8662	67.1502	18.5737	228.0539	150.0000	36.0136	45.2298
17	86.7106	43.3969	48.4563	257.7199	42.9604	39.2153	70.5348
18	58.4334	44.3095	50.3907	243.5786	66.9974	26.0105	100.0000
19	75.6332	90.4680	24.5181	217.3584	123.7226	5.3817	100.0000
20	102.7052	33.3319	54.2433	171.3730	0	0	81.5114
21	76.6746	49.3894	53.7549	274.3151	67.3837	0	−100.0000
22	83.6501	37.8787	54.4145	315.0735	92.6036	0	−100.0000
23	54.7268	83.2322	58.7571	235.6456	92.9894	0	−100.0000
24	87.0620	72.5490	56.4922	275.1494	12.3544	0	−100.0000

TABLE 18.4
Thermal Generation (MW) Using SSA of Second Test System

Hour	P_{s1}	P_{s2}	P_{s3}	P_{s4}	P_{s5}	P_{s6}	P_{s7}	P_{s8}	P_{s9}	P_{s10}
1	42.1968	97.3341	62.7051	88.7458	97.0000	68.0000	134.7882	170.2469	254.5188	130.7946
2	36.0000	70.4343	100.4318	113.9452	90.5886	74.3441	110.0000	186.2908	211.3531	131.0206
3	66.1736	108.6381	96.0217	100.6728	83.1101	68.0000	124.1969	147.3712	217.5187	210.4675
4	102.0375	89.2116	119.2589	82.0316	85.9728	102.8241	118.1220	135.0000	161.7213	149.2370
5	101.7277	79.2105	93.3063	96.9235	97.0000	129.8053	110.0000	205.0950	135.0000	172.6684
6	98.4822	114.0000	104.7284	147.4696	86.0905	129.7128	183.6925	200.8557	143.6110	175.8780
7	80.5035	97.1793	92.6458	118.1188	70.4561	130.1370	180.0821	259.7365	172.5731	143.4649
8	108.2835	94.4866	60.0000	120.6841	94.2255	124.3114	211.7936	266.3784	215.2828	171.9923
9	108.1447	103.8290	65.6360	80.0000	68.6694	101.5940	256.8770	270.4077	182.4101	183.3852
10	68.2332	67.0831	94.7695	82.8016	97.0000	140.0000	201.6588	286.5462	205.3993	139.8657
11	85.7792	62.9550	63.5631	130.2677	95.2994	133.7353	226.6924	287.1809	135.5609	196.5873
12	109.1606	80.0398	63.6334	151.1979	85.3647	121.3954	252.5505	274.0851	198.7671	211.1514
13	114.0000	105.3427	67.1154	123.4743	97.0000	78.0476	234.5042	300.0000	266.1978	228.5625
14	76.0685	97.6126	60.0000	133.2751	84.9879	74.6198	186.2746	265.7036	287.1598	193.7559
15	114.0000	81.1853	87.4687	109.3338	97.0000	108.4211	181.7586	300.0000	276.0677	142.2617
16	109.3793	74.9294	60.6678	115.7396	85.6689	119.7440	145.2188	288.5563	299.5628	200.6457
17	103.6319	109.9035	60.0000	104.6422	97.0000	140.0000	183.5439	255.8880	276.7623	159.6339
18	110.2459	71.8661	64.0352	91.7006	83.4358	96.7974	187.1393	288.5625	198.7548	137.7423
19	97.2171	111.7658	74.7105	80.0000	97.0000	68.2123	174.1555	300.0000	240.4984	139.3584
20	94.9078	106.7000	69.9256	113.9506	85.9796	109.2586	204.3318	254.4965	295.1550	202.1297
21	114.0000	78.1183	102.9903	119.6205	85.9926	98.5733	147.4450	300.0000	260.9767	200.7654
22	103.2270	68.1963	94.5543	136.8389	73.6485	101.6642	179.1842	228.4695	234.7309	135.8659
23	111.6810	82.1636	60.0000	125.1376	47.0000	68.0000	257.4024	214.4484	236.0882	132.7278
24	76.4413	80.7129	81.1067	80.0000	47.9040	113.2923	179.8007	263.2953	159.7142	204.1255

TABLE 18.5
Performance Comparison of Second Test System

Techniques	Best Cost ($)	Average Cost ($)	Worst Cost ($)
SSA	410363	410365	410369
GWO	411105	411109	411115
PSO	411689	411694	411698

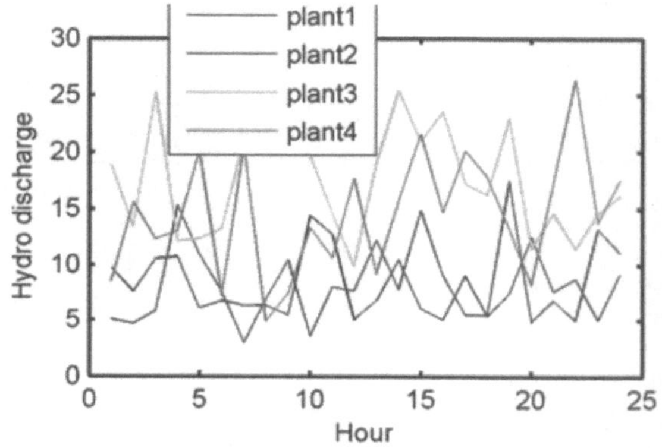

FIGURE 18.5 Hydro plant discharges ($\times 10^4 m^3$) of second test system using SSA. The optimal hourly discharges are shown for four hydro plants obtained from Squirrel search algorithm in a second test system

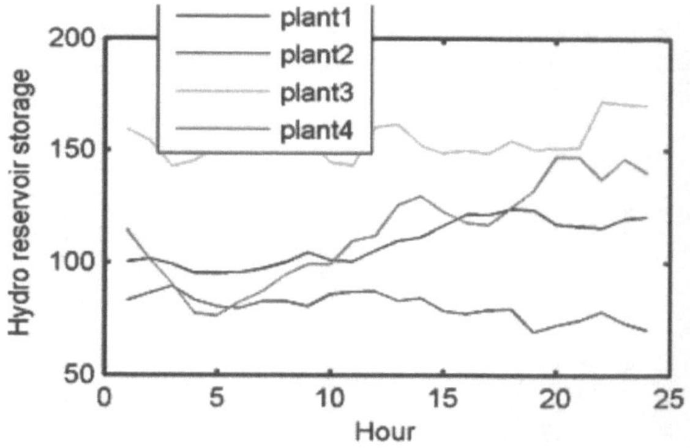

FIGURE 18.6 Hydro reservoir storage volumes ($\times 10^4 m^3$) of second test system using SSA. Storage volume of reservoir for four hydro plants obtained from Squirrel search algorithm in a second test system

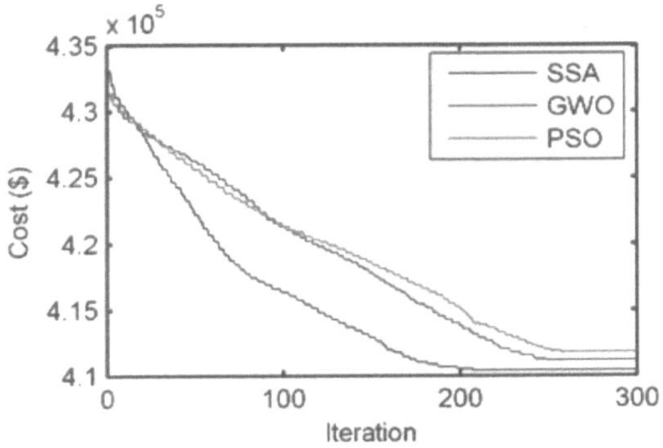

FIGURE 18.7 Cost convergence characteristics of second test system. Comparison between the cost characteristics obtained using particle swarm optimization and grey wolf optimization is shown for second test system

18.5 CONCLUSION

SSA is emphasized for optimal scheduling of generation of a solar-wind-hydro-thermal power system considering PSH unit and uncertainty of solar and wind energy sources. Using SSA, GWO and PSO techniques, two systems are solved. It is observed from simulation results after comparisons that SSA gives results superior to GWO and PSO.

ABBREVIATIONS

F_c	Cost function
a_{si}, b_{si}, c_{si}, d_{si}, e_{si}	i^{th} thermal generator cost coefficients
C_{1j}, C_{2j}, C_{3j}, C_{4j}, C_{5j}, C_{6j}	Coefficients of jth hydro unit power generation
I_{hjt}	At time t, j^{th} reservoir's inflow rate
Q_{hjt}	At time t, j^{th} reservoir's rate of water discharge
Q_{hj}^{min}, Q_{hj}^{max}	Minimum and maximum j^{th} reservoir's discharge rate of water
R_{uj}	No. of upstream units above j^{th} hydro-plant
S_{hjt}	At time t, j^{th} reservoir spillage
τ_{lj}	Water transport delay from i to j reservoir
V_{hjt}	At time t, j^{th} reservoir's volume of storage
V_{hj}^{min}, V_{hj}^{max}	Minimum as well as maximum j^{th} reservoir's volume storage
V_{hj0}	Initial volume of jth reservoir's storage V_{hjM}: final volume of jth reservoir's storage
P_{sit}	At time t, i^{th} thermal unit's o/p power
P_{si}^{min}, P_{si}^{max}	Minimum as well as maximum generation limit for i^{th} thermal unit
DR_i, UR_i	i^{th} thermal unit's ramp-down and ramp-up limits
P_{hjt}	At time t, j^{th} hydro unit's o/p power
P_{hj}^{max}, P_{hj}^{min}	Maximum and minimum generation limit for j^{th} hydro unit

P_{wkt}	At time t, wind power available of kth wind power generating unit
P_{wk}^{max}, P_{wk}^{min}	Maximum and minimum generation limit for k^{th} wind power generating unit
P_{wrk}	k^{th} generating unit's wind power rate
K_{wk}	Coefficient of direct cost for k^{th} wind power generator
o_{wk}, u_{wk}	Reserve cost and penalty cost for the kth wind power generator
v_r, v_{out}, v_{in}	Wind speed rate, cut-out, cut-in
v_{wt}	Forecast wind speed at time t
P_{PVmt}	Power o/p from m^{th} solar PV plant at time t
P_{sr}	PV generator's equivalent rated power output
G	Forecast solar irradiation
G_{std}	Solar irradiation in the standard environment
R_c	Positive irradiation point
K_{sm}	Direct cost coefficient for the mth solar PV plant
O_{PVm}, u_{PVm}	Reserve cost as well as penalty cost for the mth solar PV plant
P_{Dt}	Load demand at time t
P_{Lt}	Total transmission line losses at time t
P_{ght}	PSP power generation at time t
P_{pht}	PSP pumping power at time t
P_{gh}^{min}, P_{gh}^{max}	Minimum, maximum power generation limits of PSP
P_{ph}^{min}, P_{ph}^{max},	Minimum, maximum pumping power limits of PSP
Q_{ght} (P_{ght})	Rate of discharge of PSP at time t
Q_{pht} (P_{pht})	Rate of discharge of PSP at time t
$V_{res,t}$	Volume of water in upper reservoir of pumped storage plant at time t
V_{res}^{min}, V_{res}^{max}	Minimum, maximum upper reservoir storage limits of PSP
V_{res}^{start}, V_{res}^{end}	Specified initial and final stored water volumes in upper reservoir of PSP
T,t	Scheduling period and time index
T_{gen}	Set containing all time intervals where PSP operated is in generation mode
T_{pump}	Set containing all time intervals where PSP operated is in pumping mode
N_t, N_h: N_w, N_{PV}	Number of thermal, hydro, solar and wind PV plant generating units

REFERENCES

[1] O. Nilsson and D. Sjelvgren, "Mixed-integer programming applied to short-term planning of a hydro-thermal system," IEEE Trans. Power Syst., vol. 11, no. 1, pp. 281–286, Feb. 1996.

[2] R. W. Ferrero, J. F. Rivera, and S. M. Shahidehpour, "A dynamic programming two-stage algorithm for long-term hydrothermal scheduling of multi-reservoir systems," IEEE Trans. Power Syst., vol. 13, no. 4, pp. 1534–1540, Nov. 1998.

[3] J. S. Yang and N. M. Chen, "Short term hydrothermal coordination using multi-pass dynamic programming," IEEE Trans. Power Syst., vol. 4, no. 3, pp. 1050–1056, Aug. 1989.
[4] Q. Xia, N. Xiang, S. Wang, B. Zhang, and M. Huang, "Optimal daily scheduling of cascaded plants using a new algorithm of nonlinear minimum cost network flow," IEEE Trans. Power Syst., vol. 3, no. 3, pp. 929–935, Aug. 1988.
[5] M. V. F. Pereira and L. M. V. G. Pinto, "A decomposition approach to the economic dispatch of the hydrothermal system," IEEE Trans. Power App. Syst., vol. PAS-101, no. 10, pp. 3851–3860, Oct. 1982.
[6] S. Al-Agtash and S. Renjeng, "Augmented Lagrangian approach to hydro-thermal scheduling," IEEE Trans. Power Syst., vol. 13, no. 4, pp. 1392–1400, Nov. 1998.
[7] S. Ruzic and R. Rajakovic, "Optimal distance method for Lagrangian multipliers updating in short-term hydro-thermal coordination," IEEE Trans. Power Syst., vol. 13, no. 4, pp. 1439–1444, Nov. 1998.
[8] K. P. Wong and Y. W Wong, "Short-term hydrothermal scheduling part 1: Simulated annealing approach," IEE Proceedings – Generation, Transmission and Distribution, vol. 141, no. 5, pp. 497–501, 1994.
[9] S. O. Orero and M. R. Irving, "A genetic algorithm modeling framework and solution technique for short term optimal hydrothermal scheduling," IEEE Trans. PWRS, vol. 13, no. 2, pp. 501–518, May 1998.
[10] N. Sinha, R. Chakrabarti, and P. K. Chattopadhyay, "Fast evolutionary programming techniques for short-term hydrothermal scheduling," IEEE Trans. PWRS, vol. 18, no. 1, pp. 214–220, Feb. 2003.
[11] L. Lakshminarasimman and S. Subramanian, "Short-term scheduling of hydrothermal power system with cascaded reservoirs by using modified differential evolution," IEEE Proceedings – Generation, Transmission and Distribution, vol. 153, no. 6, pp. 693–700, Nov. 2006.
[12] R. K. Swain, A. K. Barisal, P. K. Hota, and R. Chakrabarti, "Short-term hydrothermal scheduling using clonal selection algorithm," Int. J. Electr. Power Energy Syst., vol. 33, pp. 647–656, 2011.
[13] J. Zhang, J. Wang, and C. Yue, "Small population-based particle swarm optimization for short-term hydrothermal scheduling," IEEE Trans. PWRS, vol. 27, no. 1, pp. 142–152, Feb. 2012.
[14] J. Zhang, S. Lin, and W. Qiu, "A modified chaotic differential evolution algorithm for short-term optimal hydrothermal scheduling," Int. J. Electr. Power Energy Syst., vol. 65, 159–168, 2015.
[15] A. G. Bakirtzis and E. S. Gavanidou, "Optimum operation of a small autonomous system with unconventional energy sources," Elect. Power Syst. Res., vol. 23, no. 1, pp. 93–102, 1992.
[16] S. Mondal, A. Bhattacharya, and S. Nee Dey Halder, "Multi-objective economic emission load dispatch solution using gravitational search algorithm and considering wind power penetration," Int. J. Electr. Power Energy Syst., vol. 44, no. 1, pp. 282–292, 2013.
[17] N. A. Khan, A. B. Awan, A. Mahmood, Razzaq Sohail, Zafar Adnan, and G. A. S. Sidhu, "A combined emission economic dispatch of power system including solar photo voltaic generation," Energy Convers. Manag., vol. 92, no. 1, pp. 82–91, 2015.
[18] H. M. Dubey, M. Pandit, and B. K. Panigrahi, "Ant lion optimization for short-term wind integrated hydrothermal power generation scheduling," Electr. Power Energy Syst., vol. 83, no. 1, pp. 158–174, 2016.
[19] R. Singh Patwal, N. Narang, and H. Garg, "A novel TVAC-PSO based mutation strategies algorithm for generation scheduling of pumped storage hydrothermal system incorporating solar units," Energy, vol. 142, pp. 822–837, 2018.

[20] J. I. Perez-Diaz and J. Jim, "Contribution of a pumped-storage hydropower plant to reduce the scheduling costs of an isolated power system with high wind power penetration," Energy, vol. 109, no. 1, pp. 92–104, 2016.
[21] S. Fadil and B. Urazel, "Solution to security constrained non-convex pumped-storage hydraulic unit scheduling problem by modified sub-gradient algorithm based on feasible values and pseudo water price," Electr. Power Compon. Syst., vol. 41, pp. 111–135, 2013.
[22] A. J. Wood and B. F. Wollenberg, Power Generation, Operation and Control, 2nd ed, New York: Wiley, 1984, ch. 7, pp. 230–239.
[23] S. K. Khandualo, A. K. Barisal, and P. K. Hota, "Scheduling of pumped storage hydrothermal system with evolutionary programming," J. Clean Energy Technol., vol. 1, no. 4, pp. 308–312, 2013.
[24] M. R. Mohan, S. R. Paranjothi, and S. Prince Israel, "Use of pumped hydro peak-load management plant in optimal scheduling of power systems," Electr. Mach. Power Syst., vol. 25, no. 10, pp. 1047–1061, 2007.
[25] T. Ma, H. Yang, L. Lu, and J. Peng, "Pumped storage-based standalone photovoltaic power generation system: Modeling and techno-economic optimization," Appl. Energy, vol. 137, no. 1, pp. 649–659, 2015.
[26] S. Surender Reddy, "Optimal scheduling of thermal-wind-solar power system with storage," Renew. Energy, vol. 101, pp. 1357–1368, 2017.
[27] L. Lakshminarasimman and S. Subramanian, "Short-term scheduling of hydrothermal power system with cascaded reservoirs by using modified differential evolution," IEEE Proceedings – Generation, Transmission and Distribution, vol. 153, no. 6, pp. 693–700, Nov. 2006.
[28] M. Jain, V. Singh, and A. Rani, "A novel nature-inspired algorithm for optimization: Squirrel search algorithm," Swarm Evol. Comput., vol. 44, pp. 148–175, 2019.
[29] R. B. Thomas and P. D. Weigl, "Dynamic foraging behavior in the southern flying squirrel (Glaucomys volans): Test of a model," Am. Midl. Nat., vol. 140, no. 2, pp. 264–270, 1998.
[30] J. Hetzer, D. C. Yu, and K. Bhattarai, "An economic dispatch model incorporating wind power," IEEE Trans. Energy Convers., vol. 23, no. 2, pp. 603–611, June 2008.
[31] R.-H. Liang and J.-H. Liao, "A fuzzy-optimization approach for generation scheduling with wind and solar energy systems," IEEE Trans. PWRS, vol. 22, no. 4, pp. 1665–1674, Nov. 2007.

APPENDIX

TABLE A.1
Hourly Load Demand of First Test System

Hour	P_D	Hour	P_D	Hour	P_D
1	650	9	990	17	1050
2	680	10	980	18	1020
3	600	11	1000	19	970
4	650	12	1050	20	950
5	670	13	1010	21	810
6	700	14	1030	22	760
7	750	15	1010	23	750
8	850	16	1060	24	700

TABLE A.2
Thermal Generator Characteristics of the Second Test System

	P_s^{min}	P_s^{max}	a_s	b_s	c_s	d_s	e_s	UR	DR
Unit	MW	MW	$/h	$/MWh	$/MW²h	$/h	rad/MW	MW/h	MW/h
1	36	114	94.705	6.73	0.00690	100	0.084	40	40
2	36	114	94.705	6.73	0.00690	100	0.084	40	40
3	60	120	309.540	7.07	0.02028	100	0.084	40	40
4	80	190	369.030	8.18	0.00942	150	0.063	60	60
5	47	97	148.890	5.35	0.01140	120	0.077	30	30
6	68	140	222.330	8.05	0.01142	100	0.084	50	50
7	110	300	287.710	8.03	0.00357	200	0.042	80	80
8	135	300	391.98	6.99	0.00492	200	0.042	80	80
9	135	300	455.76	6.60	0.00573	200	0.042	80	80
10	130	300	722.82	12.9	0.00605	200	0.042	80	80

TABLE A.3
Hourly Load Demand of Second Test System

Hour	P_D	Hour	P_D	Hour	P_D
1	1530	9	1900	17	2080
2	1570	10	1990	18	1920
3	1605	11	2050	19	2020
4	1610	12	2140	20	1980
5	1680	13	2200	21	1930
6	1740	14	2150	22	1840
7	1800	15	2250	23	1760
8	1860	16	2100	24	1690

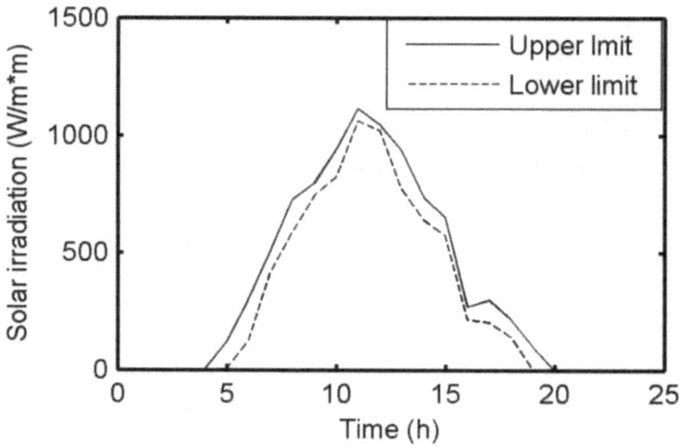

FIGURE A.1 The upper and lower forecast limits of solar irradiation. It shows the curve between lower and upper limits in hour under X-axis with solar irradiation in W/m-square in Y-axis.

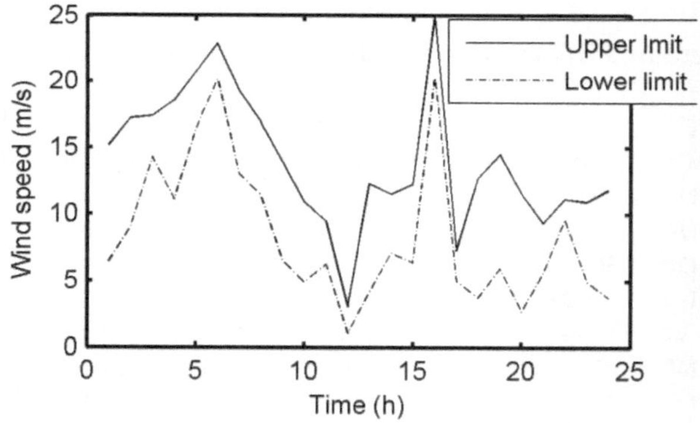

FIGURE A.2 The upper and lower forecast limits of wind speed. It shows the curve between lower and upper limits in hour under X-axis with wind speed in m/sec in Y-axis.

Index

C

D

E

F

G

H

N

O

P

T